Advanced Structured Materials

Volume 248

Series Editors

Common engineering materials are reaching their limits in many applications, and new developments are required to meet the increasing demands on engineering materials. The performance of materials can be improved by combining different materials to achieve better properties than with a single constituent, or by shaping the material or constituents into a specific structure. The interaction between material and structure can occur at different length scales, such as the micro, meso, or macro scale, and offers potential applications in very different fields.

This book series addresses the fundamental relationships between materials and their structure on overall properties (e.g., mechanical, thermal, chemical, electrical, or magnetic properties, etc.). Experimental data and procedures are presented, as well as methods for modeling structures and materials using numerical and analytical approaches. In addition, the series shows how these materials engineering and design processes are implemented and how new technologies can be used to optimize materials and processes.

Advanced Structured Materials is indexed in Google Scholar and Scopus.

Shiv Kumari Panda · Visakh P. M.
Editors

Nanocomposites for Defense, Aeronautical and Aerospace Applications

Springer

Editors
Shiv Kumari Panda
Department of Chemistry
Udayanath Autonomous College of Science
and Technology
Adaspur, Cuttack, Odisha, India

Visakh P. M.
Natural Bioactive Materials Laboratory
Ege University
Bornova, Izmir, Türkiye

ISSN 1869-8433 ISSN 1869-8441 (electronic)
Advanced Structured Materials
ISBN 978-3-032-11567-6 ISBN 978-3-032-11568-3 (eBook)
https://doi.org/10.1007/978-3-032-11568-3

This Springer imprint is published by the registered company Springer Nature Switzerland AG
The registered company address is: Gewerbestrasse 11, 6330 Cham, Switzerland

Contents

About the Editors

Dr. Shiv Kumari Panda earned her Ph.D. from KIIT University in 2018, following her M.Sc. from Sambalpur University, Odisha, in 2011. She has been serving as Assistant Professor at Udayanath Autonomous College of Science and Technology, Odisha, since 2018. Dr. Shiv Kumari Panda is actively engaged in research, with a primary focus on the synthesis and transformation of waste materials into high-performance biomaterials with novel applications. In recognition of her contributions, she was achieved a four-year research grant from the Odisha State Higher Education Council under the Odisha University Research and Innovation Incentivisation Plan (OURIIP). She has published many research articles in various Scopus-indexed journals. She remains highly active in the research community, contributing as Peer Reviewer for leading scientific journals, including *Bioresource Technology Reports* (Elsevier), *Fibers and Polymers* (Springer Nature), and *ACS Sustainable Chemistry and Engineering* (American Chemical Society). Additionally, she has served as Reviewer for the International Conference on Materials Engineering and Manufacturing Systems, published in IOP Science: Materials Science and Engineering. She has published more than 15 research articles and more than 10 book chapters. She is also working with more than 10 books with Dr. Visakh P. M. (3 books already in press).

Dr. Visakh P. M. (M.Sc., M.Phil., Ph.D.) is Prolific Editor with more than 53 books already published. Currently he is working in Natural Bioactive Materials Laboratory Department of Bioengineering, Ege University, Bornova/Izmir, Turkey. He was working as Associate Professor in TUSUR University, Tomsk, Russia, from 2017 till 2023.He did his Post doc. Research in Tomsk Polytechnic University, Tomsk, Russia (2014–2017). He obtained his Ph.D., M.Phil., and M.Sc. degrees from School of Chemical Sciences, Mahatma Gandhi University, Kottayam, Kerala, India. He edited **50 books** from Scrivener (Wiley), Springer, Royal Society of Chemistry, and Elsevier and more than **25 books** in press, (from Wiley, Springer, Royal Society of Chemistry, and Elsevier). He has been invited as Visiting Researcher in Russia (2014–2023), Portugal (2013, 2014), Czech Republic (2012, 2013), Italy (2009, 2012), Argentina (2010), Sweden (2010–2012), Switzerland (2010), Spain (2011, 2012), Slovenia (2011), France (2011), Belgium (2012), and Austria (2012) for his research work. He visited 12 countries; he visited 15 universities in Europe. He published 20 publications, 4 reviews, and more than **53** book chapters. He has attended and presented more than 28 conferences, he has **3808** citations, and his **h-index is 29**. He acts as Guest Editor for 4 International Journals. Listed in the "**World Ranking of Top** 2% **Scientists-2023–2025**" Source-**Stanford University, USA**. His interest of research: polymer sciences, polymer nanocomposites, material sciences, bio-nanocomposites, rubber-based nanocomposites, fire retardant polymers and liquid crystalline polymers, and silicon sensors.

Chapter 1
Nanocomposites for Defense, Aeronautical, and Aerospace Applications: Scope and Challenges

Shiv Kumari Panda and Visakh P. M.

Abstract In this chapter providing and short version of all chapters, here we are writing about the different chapter topics such as high-performance polymer materials for aeronautical engineering applications, mechanical behavior for hybrid metal nanoparticles-based thermoplastic laminates for aeronautical applications, development and characterization of novel DOPO-based phosphorus tetraglycidyl epoxy nanocomposites for aerospace applications, nanocomposites using in the parts of aeronautical engineering in radio frequency identification stage, nanocomposites for defense, aeronautical, and aerospace applications: challenges self-healing aeronautical nanocomposites, future and challenging attributes of aeronautical nanocomposites, green composites and nanocomposites for aerospace applications, multifunctional hybrid sol–gel coatings in aeronautical and aerospace applications, nanocomposites for defense, aeronautical, and aerospace applications: challenges.

Keywords High-performance polymer · Materials aeronautical engineering · Mechanical behavior · Hybrid Metal · Nanoparticles · Thermoplastic laminates · Phosphorus tetraglycidyl · Nanocomposites · Radio Frequency · Green composites · Hybrid Sol–gel

S. K. Panda (✉)
Udayanath Autonomous College of Science and Technology, Adaspur, Cuttack, Odisha, India
e-mail: shivkumaripanda@gmail.com

Visakh P. M.
Natural Bioactive Materials Laboratory, Department of Bioengineering, Ege University, Bornova/Izmir, Turkey

S. K. Panda (ed.), *Nanocomposites for Defense, Aeronautical and Aerospace Applications*, Advanced Structured Materials 248,
https://doi.org/10.1007/978-3-032-11568-3_1

1.1 High-performance Polymer Materials for Aeronautical Engineering Applications

High-performance polymers (HPPs) materials are designed to withstand extreme conditions that include high temperature, aggressive chemicals, and mechanical stresses. High-performance polymers are most widely used in several industries such as oil and gas, aerospace, automobile, and so on. Among these fields, one of the significant areas of advancement for high-performance polymers is aerospace engineering applications. In addition, the HPPs material has dimension stability and flexibility that can easily designed, molded, and fabricated into different complex geometries to satisfy the demand of the industrial applications [1–3]. One of the major drawbacks of aromatic content on the HPPs material is that the polymer becomes very hard and stiff. In order to overcome this problem, the heterogeneous structure such as nitrogen, oxygen, and sulfur elements are incorporated into the aromatic structure. As a result, the processing of the polymer material becomes easier with decrease in the thermal and chemical resistance due to the bond strength. The factors that contribute to enhance the bonding strength are molecular weight of the polymer, crosslink density, crystallinity, and the incorporation of the reinforcements. Polyetherimide is a type of high-performance amorphous thermoplastic that has excellent heat resistance, strength, and chemical resistance making it a suitable to use in the field of aerospace structural components and interior parts. PEI possess significant mechanical and chemical properties that includes high strength, creep resistance dimension stability, and thermal resistance. Because of this the PEI polymer material is most frequently used in the structural components. PEI is most resistance chemicals such as alcohols, halogenated solvents, and hydrocarbon agents. It also has the ability to retard flame and produce very less smoke. Shangqin Yuan et al. [4] investigated on the MWCNT/polymer composites fabricated through selective laser sintering on the electrical and thermal properties of the composites. The CNT were added in different ratios of 0.1, 0.5, & 1 wt.% to PA12 and PU powders. It is found that the 1wt.% of CNT showed significant results in the electrical properties. However, the thermal properties were not improved significantly. In addition, the microstructure and its corresponding mechanism related to thermal and electrical properties were elucidated. The different combination of the polymer material such as high-performance carbon fibers/polyamide12/epoxy were fabricated by selective laser sintering process.

Joe et al. [5] studied on the fabrication of 2wt.% of graphene/polyurethane resin using digital light processing technique. The results showed that there is 41% of increase in the tensile strength of the polymer composite than that of pure polymer matrix. In addition, the electric/magnetic fields applied on the DLP process in order to improve the properties and the material to be more bionic which can result in the application of 3D printed composites. Sofiane Guessasma et al. [6] reported on the 3D printing of the acrylate-modified PLA/lignin blends fabricated through digital light procession examined on the microstructure and mechanical properties of the composites. Commercial fabrics composed of ultra-high molecular weight polyethylene

(UHMW-PE) fibers are now used in sports equipment and personal body armor due to their exceptional strength, durability, and abrasion resistance. UHMW-PE fibers exhibit exceptionally high axial thermal conductivity and outstanding mechanical properties due to a high degree of polymer chain orientation. However, these materials' potential for usage in flexible and wearable temperature control systems has not been fully explored [7]. Selecting the appropriate materials has proven to be one of the most important aspects of developing equipment for chemical processes. Particularly in the oil and gas industry, non-metals (such advanced and high-performance polymeric materials) have replaced metals in the majority of structural elements, coatings, equipment parts, and the like. Therefore, it is crucial to comprehend the advantages and disadvantages of these polymer materials before they can be employed successfully for a given purpose [8].

The coated surfaces were ready for characterization and checked for surface flaws. The development and testing of composites' mechanical properties. 86 shore-D is the hardness, 16.74 MPa is the lap shear strength, and 98 MPa is the flexural strength of the pure epoxy composite coating material. Epoxy's higher hardness and poorer flexural and shear strength values result on its high brittle molecular structure [9]. Semiconductor carbon nanotubes (s-CNTs) have emerged as a promising alternative to traditional silicon in ultrascaled field-effect transistors (FETs) due to their exceptional properties. Hongji Zhang et al. [10] developed a continuous carbon-fiber-reinforced magnesium matrix composite (C_f/Mg) by using epoxy resin as a matrix polymer phase, which can focus on the high-impact resistance in aircraft applications such as landing gear doors. According to this research, it reveals that the epoxy-based bonding increases the peak impact loading by 14.5% under repeated impacts. In high-heat zones, polyimide (PI) and PMR-15-based composites have been utilized due to their superior thermal stability and oxidative resistance, sustaining performance at temperatures exceeding 300 °C. As noted by Ghosh and Mittal [11], these materials are ideal for use in fairings located near engine exhaust zones or heat-sensitive landing mechanisms, although they may require careful processing due to issues like brittleness and toxicity in older formulations.

1.2 Mechanical Behavior for Hybrid Metal Nanoparticles-Based Thermoplastic Laminates for Aeronautical Applications

Nanoparticles are essential materials, despite their nanoscale size. Their spectrum of nanoscale sizes allows the relationships between different constituents to be identified. The size of the nanoparticle determines the categories into which the nanocomposite materials are separated. Because of the material's dimensional properties, the link between volume and surface area is especially important. Nanomaterials play an important role in the field of contemporary nanomaterials and are becoming increasingly popular among researchers due to their degradability, biocompatibility, and economic benefits [12]. Polymer-oriented nanoparticles are regarded biodegradable and biocompatible, and their fascinating features make them viable choices for a variety of applications. Nanocomposites made with metallic, nonmetallic, and polymer-oriented materials offer amazing advantages, like surface deformities and other material properties [13]. These composites are unique due to their increased surface area, significant mechanical properties, resistance to wear and scratches, and improved optical properties. Dimensional quantities, quality, resemblance to the microstructural lattice, and the dispersion of nanomaterials are the characteristics that determine the selection of nanocomposites. Previous studies have discovered that evenly distributed oxide components in nanoparticles enhanced the mechanical characteristics, thermal stability, and fire resistance of the composites over micro-composites [14].

In certain applications, nanoparticles can also be designed to increase thermal conductivity. High thermal conductivity nanoparticles, for example, can be utilized in thermal management systems to effectively dissipate heat and prevent the overheating of delicate components [15]. Nanoparticles help materials have better fatigue resistance, which is important for materials that are subjected to cyclic loads. Because operating circumstances vary often, aerospace components are subjected to repetitive stress. Nanoparticles assist to distribute these stresses more evenly throughout the material, lowering the possibility of fatigue-related failure. Another property of nanoparticles is their ability to increase a material's resistance to gradual deformation over time under continuous tension. This feature is especially crucial for high-temperature applications where materials must withstand extended loads. Nanoparticles contribute to the long-term preservation of the material's mechanical characteristics and shape by strengthening the matrix [15]. In aerospace applications, most engineering composite materials consist of continuous carbon or glass fiber reinforcing a polymeric matrix. Because of its excellent strength-to-weight ratio and ability to improve mechanical and thermal properties by changing its form, orientation, and composition, carbon fiber is frequently utilized to create composites for high-end applications [16]. Carbon fiber is distinguished by its exceptional strength, high modulus, low thermal expansion, superior creep resistance, and resistance to corrosion. Currently, carbon fiber-reinforced plastics are increasingly replacing metallic components in high-temperature and corrosion-resistant applications. Among the composites used in aerospace and structural applications, carbon fiber-reinforced epoxies are preferred for their properties such as high failure strength, low creep, high modulus, and relatively low cost [17].

Natural fibers are classified based on their sources, which can be plants, animals, or minerals. The characteristics of hybrid composites depend on their individual sustainable components and the fiber/matrix interfacial bonding. Natural cellulosic plant fibers in these composites range in size from micro to centimeters [18]. Synthetic fibers possess superior mechanical properties and damping resistance, whereas natural fibers do not. Hybrid composites (HCs) are multifunctional materials used in advanced structural applications where multiple benefits from constituent materials are needed. HCs are favorable and well-designed materials that incorporate one type of filler or reinforcement in a blend of two similar or dissimilar polymer matrices more than one filler in a single polymer resin, or a combination of these methods. The properties of HCs are influenced by the weight, geometry, and specific applications in various structural engineering components [19]. The potent antibacterial qualities of silver nanoparticles are widely recognized; they work well against a variety of bacteria, viruses, and fungi. This makes them very useful in the aerospace sector, where it is crucial to keep everything clean, notably in passenger cabins, crewed spacecraft, and medical transport aircraft. For instance, surfaces of medical equipment and airplane cabins are coated with coatings based on silver nanoparticles to prevent the growth and spread of dangerous bacteria. Moreover, air filtration systems employ silver nanoparticles to improve their capacity to eradicate or neutralize airborne infections, guaranteeing a safer atmosphere for travelers and workers. Surface plasmon resonance (SPR), one of the special optical characteristics of silver nanoparticles, is used in the creation of sophisticated sensors for structural health monitoring in airplanes [20].

1.3 Development and Characterization of Novel DOPO-Based Phosphorus Tetraglycidyl Epoxy Composites for Aerospace Applications

Epoxy resins are widely recognized for their exceptional mechanical properties, chemical resistance, and ease of processing, making them a preferred choice in the aerospace industry [21, 22]. These thermosetting polymers are commonly used in the fabrication of composite materials, coatings, adhesives, and structural components. The versatility of epoxy resins allows them to be tailored for specific applications, providing the necessary strength and durability required in aerospace engineering [23, 24]. Their ability to form strong adhesive bonds and maintain structural integrity under extreme conditions makes them indispensable in the construction of aircraft, spacecraft, and related components. Conventional fire retardant materials exhibit several significant drawbacks that emphasize the need for better alternatives. A major concern is their high toxicity, particularly in halogenated compounds, which release harmful gases like hydrogen chloride (HCl) and dioxins during combustion. These toxic byproducts present serious health risks and environmental hazards. Moreover, many of these fire retardants persist in the environment, leading to bioaccumulation and ecological damage, with some contributing to ozone layer depletion.

In addition to toxicity, conventional fire retardants often require high loading levels—typically 20–30 wt%—to achieve effective flame retardancy. This high filler content negatively affects the mechanical properties of materials, resulting in reduced strength, flexibility, and durability. The excessive incorporation of fire retardants can also lead to deterioration in mechanical properties, such as increased brittleness and lower tensile strength, which is particularly problematic in applications requiring both fire resistance and structural integrity, like aerospace and construction. Furthermore, many conventional fire retardants exhibit poor thermal stability at high temperatures, compromising their effectiveness in critical conditions. Processing challenges also arise due to the high filler content, which can cause uneven dispersion and weak spots in the final material. Phosphorus-based additives, such as DOPO, play a vital role in enhancing the fire resistance of epoxy resins without compromising their mechanical properties. DOPO, in particular, is effective due to its unique chemical structure, which allows it to act both as a reactive and additive flame retardant [25, 26]. When incorporated into epoxy resins, DOPO improves fire resistance by promoting the formation of a protective char layer during combustion. This char layer acts as a barrier, reducing the release of flammable gases and heat, thereby slowing down the spread of fire. Moreover, DOPO enhances the mechanical properties of epoxy nanocomposites. It contributes to better dispersion of nanoparticles within the resin matrix, resulting in a more uniform and robust composite material. This improved dispersion leads to enhanced tensile strength, flexibility, and toughness, making the material suitable for aerospace structural applications [27]. Additionally, DOPO's environmental benefits, such as reduced toxicity and compliance with regulatory standards, make it a preferable choice over traditional halogenated flame retardants [28]. In summary, the incorporation of DOPO-based phosphorus additives in epoxy resins addresses the critical need for materials with both enhanced fire resistance and superior mechanical properties in aerospace applications. This chapter explores the development and characterization of novel DOPO-based phosphorus tetraglycidyl epoxy nanocomposites, highlighting their potential to meet the demanding requirements of modern aerospace engineering.

The DOPO-based phosphorus tetraglycidyl epoxy nanocomposites exhibit excellent compatibility with various aerospace structural materials, making them highly suitable for integration into advanced aerospace applications [29]. Their enhanced thermal stability and mechanical properties complement the high-strength, lightweight materials commonly used in aerospace, such as carbon fiber and glass fiber composites. The improved adhesion and uniform dispersion of the nanocomposites within these structural materials ensure that they maintain structural integrity under extreme conditions, including high mechanical stresses and elevated temperatures. This compatibility facilitates the use of DOPO-based nanocomposites as matrix resins or coatings, where they can effectively bind and reinforce other materials, enhancing overall performance and reliability. The enhanced fire resistance and mechanical properties of DOPO-based phosphorus tetraglycidyl epoxy nanocomposites open up several potential applications within aircraft interiors and structural components. In aircraft interiors, these nanocomposites can be used for manufacturing lightweight, fire-resistant panels and components, such as seat frames, cabin partitions, and overhead storage bins [30, 31].

1.4 Nanocomposites Used in the Parts of Aeronautical Engineering in Radio Frequency Identification (RFID) Tag

Application-focused studies with RFID in supply chains [32–34], health care [35], food safety [36], and construction [37] have demonstrated their potential to improve efficiency. These studies frequently trace the benefits of RFID systems, including real-time object identification without line-of-sight, enhanced tracking accuracy, and operational efficiency, while also noting implementation challenges such as interference, cost barriers, and privacy risks [38–40]. Recent contributions emphasize the development of emerging extensions to RFID systems, such as sensor-integrated tags and chipless technologies [41]. Together, these studies form a well-rounded corpus, introducing the architecture, functionality, and practical relevance of RFID technology. They lay the foundation to understand operational principles of RFID systems, the contexts in which they are deployed, and the opportunities and constraints that shape their evolution. Polymers are preferred as primary substrate materials for sensor tags due to their flexibility, processability, and cost-efficiency [42]. Commonly used polymer substrates include polyethene terephthalate, polyethene naphthalate, polyethersulfone, polyimide, polycarbonate, and polydimethylsiloxane [43]. The polyethene terephthalate-based RFID sensor demonstrated excellent mechanical stability under bending conditions and maintained structural and functional integrity when exposed to various environmental stresses [44]. The dielectric constant of the substrate plays a critical role in determining the sensor's sensitivity, as it directly influences the electromagnetic response mechanism of the device under varying environmental conditions such as temperature, humidity, and mechanical stress, which can alter the dielectric properties and thereby affect signal strength and accuracy [45]. Additionally, the dielectric properties of the substrate significantly influence critical performance parameters of an RFID system, including overall readout efficiency, impedance matching, and antenna bandwidth, thereby affecting the system's communication reliability and sensing accuracy [46]. Hence, careful selection of polymers is essential during design and fabrication.

Carbon-based materials are gaining fast attention due to their low density with high specific strength, offering a combination of desirable properties, including lightweight nature, superior mechanical strength, excellent corrosion resistance, and outstanding thermal stability. These characteristics make carbon-based materials highly suitable for the production of components in sensor systems, namely sensor antennas and sensing interfaces, where durability, sensitivity, and environmental stability are critical [47]. The antenna is a critical component of an RFID tag, responsible for sending and receiving signals. Incorporating nanomaterials such as graphene and carbon nanotubes into antenna structures improves electrical conductivity, resulting in longer read ranges and higher signal sensitivity. For instance, graphene-based conductive inks enable inkjet-printed antennas that are flexible, lightweight, and more efficient than traditional metal-based antennas [48]. Nanomaterials, namely graphene and silver nanowires, substantially reduce electrical resistance, thereby improving antenna efficiency and enabling thinner, lightweight designs [49]. High-gain antennas enhance power conversion efficiency and enable increased power acquisition, making them the preferred choice in many applications. The performance of an antenna can be characterized by parameters such as gain, operating frequency band, radiation pattern, polarization, physical dimensions, and intended field of application [50]. Nanocomposite-based substrates exhibit superior flexibility, durability, and resistance to mechanical stress, moisture, and temperature fluctuations, making them suitable for harsh environments such as aerospace and biomedical applications [51].

1.5 Nanocomposites for Defense, Aeronautical, and Aerospace Applications: Challenges

The relentless pursuit of enhanced performance, durability, and efficiency across defense, aeronautical, and aerospace sectors has propelled materials science to the forefront of innovation [52]. Among the most promising avenues of exploration lies the study of nanocomposites—materials engineered at the nanoscale to exhibit superior properties compared to their conventional counterparts. The integration of these advanced materials holds the potential to revolutionize critical applications, from next-generation military armaments to lightweight, fuel-efficient aircraft, and robust spacecraft [53]. However, the path to widespread implementation is fraught with challenges that demand meticulous research and innovative solutions. Nanocomposites, by definition, are multiphase solid materials where at least one dimension of one of the phases is in the nanometer scale (1–100 nm). This nanoscale manipulation imparts unique characteristics, such as increased strength, enhanced thermal stability, improved electrical conductivity, and superior wear resistance. In defense, this translates to lighter yet more resilient armor, advanced sensor systems, and high-performance structural components. In aeronautics, nanocomposites offer the prospect of reduced aircraft weight, leading to significant fuel savings and decreased emissions, alongside improved structural integrity and resistance to environmental degradation. In aeronautics, the drive for fuel efficiency and reduced emissions has led to a growing interest in lightweight materials. Nanocomposites offer the potential to create aircraft structures that are lighter and stronger than traditional materials, leading to significant fuel savings and reduced environmental impact. The use of carbon nanotube-reinforced polymer composites in aircraft wings and fuselages is one example of this trend. Furthermore, nanocomposites are being explored for use in aircraft engines, where their enhanced thermal stability can enable higher operating temperatures and improved efficiency. To enhance the mechanical performance and extend the operational lifespan of military hardware, nanocomposites are increasingly used in structural components such as aircraft fuselages, unmanned aerial vehicles (UAVs), missiles, and armored vehicle panels. One prominent example is the use of carbon fiber-reinforced polymers (CFRPs) enhanced with nanoclay or graphene nanoplatelets, which significantly improve fatigue resistance, stiffness, and damage tolerance, all while keeping the structure lightweight [54]. These enhancements are crucial for enduring the high mechanical stress and dynamic loads encountered in military operations. Another widely used material is the aluminum matrix nanocomposite, particularly aluminum reinforced with silicon carbide (SiC) or aluminum oxide (Al_2O_3) nanoparticles [55]. Nanocomposites are at the forefront of innovation in the aerospace industry, providing transformative advantages in terms of weight reduction, fuel efficiency, mechanical strength, and thermal resistance. In the aeronautical sector, where every gram matters and structural integrity is critical under extreme operational conditions, nanocomposites enable the design of next-generation aircraft that are both lighter and more resilient. Their ability to combine multiple functionalities in a single material system makes them particularly attractive for use in airframes, propulsion systems, avionics, and environmental control systems [56].

Modern aircraft increasingly rely on embedded electronics and sensors for performance monitoring, navigation, and communication. Nanocomposite materials allow for the development of flexible, lightweight, and durable sensors that can be seamlessly integrated into the aircraft structure. Conductive polymer nanocomposites using silver nanoparticles or carbon-based fillers enable the creation of stretchable circuits and antennas that conform to complex geometries. These smart materials can monitor strain, temperature, and damage in real time, supporting predictive maintenance and enhancing operational safety. Nanocomposite coatings are used to line fuel tanks and pipelines, offering enhanced barrier properties against fuel permeation and corrosion. For example, epoxy resins embedded with layered silicates or graphene oxide create dense molecular structures that minimize fuel vapor leakage and resist chemical degradation. These systems ensure long-term reliability and safety in aviation fuel systems while complying with stringent environmental regulations. Conductive nanocomposites, such as those containing carbon nanotubes, graphene, or metallic nanoparticles (e.g., silver or copper), are used as EMI-shielding layers within avionics enclosures, cable coatings, and fuselage linings. These materials not only reduce electromagnetic noise but also provide weight savings compared to traditional metallic shields. Furthermore, aerospace vehicles are at risk of lightning strikes, especially at high altitudes. Nanocomposites with enhanced electrical conductivity are used in outer fuselage layers to safely dissipate electrical discharges, protecting the underlying structure and systems.

1.6 Self-healing Aeronautical Nanocomposites

The aerospace industry continually faces challenges related to material durability and maintenance, which are critical for ensuring safety and performance due their exposure to extreme heat and mechanical stress conditions [57, 58]. Traditional aerospace materials, while strong and lightweight, are susceptible to damage from mechanical stress, thermal cycling, and environmental exposure [59–62]. Such damage can lead to costly repairs, increased downtime, and in extreme cases, catastrophic failures. To address these issues, there is a growing demand for the development of advanced functional materials that possess self-healing capabilities. Polymer-based self-healing materials are among the most extensively studied for aerospace applications due to their versatility and ease of processing. Polymers can be engineered to exhibit self-healing properties through various mechanisms, such as thermally activated healing, photoresponsive healing, and chemical-based healing systems. These mechanisms enable polymers to autonomously repair damage, making them highly desirable in industries where material integrity is crucial. The self-healing process in polymer materials can be achieved through several mechanisms, including reversible covalent bonding, hydrogen bonding, or supramolecular interactions. A common approach involves embedding microcapsules filled with a healing agent within the polymer matrix. When damage occurs, these capsules rupture, releasing the healing agent into the crack. External stimuli, such as heat or light or/ and a catalyst, are often required to activate the healing agent, which then forms new bonds to repair the material. This versatile approach allows polymers to heal in response to different types of damage and environmental conditions.

Polymer composites are widely used in aerospace due to their lightweight nature and high strength-to-weight ratio. The integration of self-healing capabilities into these materials makes them even more appealing, particularly for components that are frequently subjected to stress and potential damage, such as fuselage skins, interior panels, and fuel tanks. The ability of these materials to autonomously heal reduces the need for frequent inspections and repairs, improving operational efficiency and enhancing safety in aerospace applications [63–70]. However, there are challenges in developing self-healing polymer composites. Ensuring the longevity and reliability of the healing process is a key concern, as is maintaining the mechanical properties of the material after repeated healing cycles. Advances in polymer chemistry and material engineering are expected to address these challenges, paving the way for the next generation of high-performance, self-healing polymer composites in aerospace, with enhanced durability and operational lifespan.

1.7 Future and Challenging Attributes of Aeronautical Nanocomposites

From a state-of-the-art perspective, research in aeronautical nanocomposites has progressed significantly in recent years. Advancements in nanofabrication techniques have enabled the production of high-quality nanoparticles with controlled dimensions and surface chemistries. This has facilitated the development of nanocomposites with tailored properties for specific applications. For instance, CNT-reinforced polymer composites have shown remarkable improvements in mechanical strength and electrical conductivity, making them attractive for use in aircraft structural components and electromagnetic shielding [71]. Similarly, graphene-based nanocomposites have demonstrated exceptional thermal conductivity and barrier properties, opening up possibilities for advanced thermal management systems and corrosion protection.

Despite these advancements, several challenging attributes remain that hinder the widespread implementation of aeronautical nanocomposites. One of the primary challenges is achieving uniform dispersion of nanoparticles within the matrix material. Agglomeration of nanoparticles can lead to stress concentration points, compromising the overall mechanical properties of the composite. Researchers are exploring various strategies to address this issue, including surface functionalization of nanoparticles, solution mixing, and melt blending. Furthermore, the interfacial bonding between nanoparticles and the matrix material is crucial for effective load transfer and overall performance [72, 73]. Achieving strong interfacial bonding often requires the use of coupling agents or surface treatments, which add complexity to the manufacturing process. At the core of nanocomposite development lies the pursuit of superior mechanical performance without the penalty of increased mass. Future aeronautical nanocomposites will incorporate advanced nanofillers such as carbon nanotubes (CNTs), graphene nanoplatelets, and nanoclay reinforcements, enhancing tensile strength, stiffness, and fracture toughness [74, 75].

These nanostructures enable efficient load transfer and crack-bridging mechanisms, reducing the propagation of microdefects under cyclic loading. The result is a new class of ultra-lightweight materials that meet or exceed the structural integrity requirements of modern aerospace components, including fuselages, wings, and turbine blades [76]. In aerospace applications, managing thermal loads and ensuring electromagnetic compatibility are crucial. Future nanocomposites will offer enhanced thermal conductivity by integrating carbon-based nanomaterials and metallic nanoparticles. These materials facilitate efficient heat dissipation from critical zones such as engines and avionics, mitigating thermal fatigue. Additionally, nanocomposites will be tailored to achieve controlled electrical conductivity. Through percolation networks of conductive nanoparticles, the material can be engineered for lightning strike protection, electrostatic discharge mitigation, and even electromagnetic interference (EMI) shielding. These attributes are particularly valuable for composite-based airframes that otherwise lack intrinsic conductivity 77, 78. One of the most persistent problems in nanocomposite fabrication is the uniform dispersion of nanoparticles within the host matrix. Nanoparticles such as carbon nanotubes (CNTs), graphene nanoplatelets (GNPs), or metal oxides exhibit a strong tendency to agglomerate due to their high surface energy and strong van der Waals forces. Despite the use of advanced dispersion techniques—including ultrasonication, mechanical shear mixing, solution casting, and in situ polymerization—perfect homogeneity remains difficult to achieve [79, 80].

For instance, in aerospace-grade epoxy-CNT nanocomposites, insufficient dispersion has been shown to result in a reduction of tensile strength by up to 30% compared to well-dispersed counterparts. Agglomerates act as stress concentrators that initiate cracks and delamination under cyclic loading. To counter this, researchers have experimented with surface functionalization methods (e.g., oxidation or polymer grafting) to enhance compatibility with the matrix. However, such modifications may alter the intrinsic electrical and mechanical properties of the fillers, creating a trade-off scenario [81]. Concurrently, polymeric nanocomposites are making remarkable strides, presenting viable, lightweight, corrosion-resistant, and radiation-tolerant alternatives to conventional metallic structures. Sophisticated synthesis and fabrication methodologies, coupled with meticulous characterization techniques such as AFM, TEM, and XRD, guarantee the uniform dispersion of reinforcing agents, thereby maximizing the inherent potential of these materials. Notable examples, such as Ag/cellulose-polydopamine for electromagnetic interference shielding, SiO_2-CNTs/PVDF-HFP for robust, superhydrophobic corrosion-resistant coatings, and PZT/PDMS for advanced structural health monitoring, underscore the diverse applications and significant performance enhancements achievable with polymeric nanocomposites.

1.8 Green Composites and Nanocomposites for Aerospace Applications

The aerospace industry has consistently driven innovation in advanced materials, primarily to meet demanding criteria such as high strength-to-weight ratio, thermal stability, and resistance to fatigue and corrosion. Traditionally, materials like aluminum alloys, titanium, and carbon fiber-reinforced polymers have dominated this field due to their proven performance in extreme environments [82]. Green composites have opened a new horizon for developing innovative solutions to conventional issues. These materials are composed of natural fibers and bio-based or biodegradable polymers, as viable alternatives for certain aerospace applications. These materials also offer several advantages, including reduced environmental footprint, renewability, and lower cost, while still meeting structural and functional demands in non-primary load-bearing components [83]. Their relevance has grown particularly in the context of interior panels, seat structures, and insulation systems where sustainability can be prioritized without significant trade-offs in mechanical performance. Natural fiber-reinforced composites are becoming increasingly attractive due to their renewable nature, ease of recycling, low cost, and lightweight properties [84].

Fibers like jute, hemp, kenaf, bamboo, banana, and sisal have historically been used as reinforcement materials. More recently, other natural fibers such as flax, wheat straw, sugarcane, coconut, silk, cotton, and oil palm have drawn interest for use in sophisticated engineering applications [85, 86]. Some of these fibers, especially when combined with biopolymers like polylactic acid (PLA), have shown promising results in thermoplastic and thermoset matrices. Their growing annual production reflects the increasing environmental consciousness driving their adoption. In addition to addressing ecological concerns, biocomposites show potential in reducing packaging waste and can be engineered for use in automotive structures, particularly in electric vehicles, where reducing overall weight is crucial for battery efficiency. These materials are also finding applications in biomedical fields, especially in creating biodegradable products. Cellulose, in its natural or chemically modified state, is among the most widely used polysaccharides due to its outstanding benefits, including high specific strength, biocompatibility, non-toxicity, abundant availability, and low cost. Structurally, it is a syndiotactic homopolymer composed of D-glucopyranose units connected by β-1,4-glycosidic bonds [87].

The cellulose chains feature a high density of hydroxyl groups, which engage in extensive intra- and intermolecular hydrogen bonding. This hydrogen bonding imparts key properties such as structural stability, hydrophilicity, and a high number of reactive sites for functionalization. From an application standpoint, esterification and etherification are the most important chemical modification techniques used to enhance cellulose performance [20]. Another study developed high-performance composites by incorporating nanocellulose extracted from water hyacinth via acid hydrolysis and sonication. Fullerenes are composed of rings that form a geodesic spheroidal structure made up of sp^2-hybridized carbon atoms with significant isomeric diversity, often assembling into multi-layered formations. While carbon nanotubes (CNTs) are structurally related to fullerenes and exhibit quasi-one-dimensional characteristics, fullerenes have garnered considerable interest due to their distinct geometry. Their ability to undergo endohedral, substitutional, and exohedral doping makes them valuable for applications in nanotechnology, electronics, and biotechnology, offering unique structural, magnetic, superconducting, electrochemical, and physicochemical properties [88].

Metal and metal oxide nanoparticles have gained significant attention for their exceptional characteristics, particularly in fields like optics, catalysis, electronics, and biomedicine [39]. Typically derived from noble metals such as silver, gold, and platinum, these nanoparticles are synthesized via two main strategies: top-down and bottom-up approaches. Top-down methods reduce bulk materials into nanoscale forms using techniques like milling, laser ablation, or sputtering, whereas bottom-up approaches build nanoparticles from atoms or molecules through methods like vapor deposition, chemical reduction, or electrochemical processes [89].

In line with sustainable practices, green synthesis methods—avoiding hazardous chemicals—are increasingly favored. One such eco-friendly technique involves using plant extracts as reducing agents to convert metal ions into nanoparticles [90]. This biogenic method not only eliminates the need for toxic reagents but also proves suitable for biomedical applications. Additionally, the type of plant extract used can influence the nanoparticles' morphology, size distribution, and antibacterial effectiveness [91].

1.9 Multifunctional Hybrid Sol–gel Coatings in Aeronautical and Aerospace Applications

The aeronautical and aerospace sectors function in among the most severe and challenging environments, where parts are constantly subjected to varying temperatures, ultraviolet (UV) radiation, corrosive environments, mechanical stress, and erosive particulate matter. For maintaining structural integrity, mission safety, and durability of these high-performance systems, the use of strong surface protection measures is essential [92, 93]. Protective coatings act as the initial line of defense against such environmental stressors, adding a service life to metallic substrates and composite materials and keeping them in their optimal operational performance. Traditional coatings such as chromate-based primers, epoxy resins, and thermal spray coatings have been used extensively for corrosion protection, thermal insulation, wear resistance, and surface functionalization on aerospace structures. Although these coatings have shown high levels of effectiveness, they are subject to many limitations that limit their versatility and longevity. Chromate-based coatings, while very effective corrosion inhibitors, are carcinogenic and ecotoxic, leading to strict regulatory control under directives like REACH and RoHS [94, 95].

Sol–gel coatings can be tailored to provide a broad variety of protective properties depending on the precursors and modifiers used, such as corrosion inhibition, thermal barrier effects, hydrophobicity, UV shielding, and mechanical reinforcement [96, 97]. One of the most appealing characteristics of sol–gel coatings is that they can be designed at the molecular level, enabling nanoparticles (e.g., ceria, silica, zirconia, and graphene) to be included, corrosion inhibitors, self-healing agents, and organic functionalities in a single coating system. This molecular-scale design freedom provides for the production of multifunctional surfaces that react smartly to environmental stimuli, withstand degradation due to cyclic loading, and demonstrate enhanced adhesion to aerospace-grade materials like aluminum alloys, titanium, and carbon fiber composites. Sol–gel is a general and common chemical route to produce inorganic and hybrid organic–inorganic materials as powders, thin films, or monoliths. The two building-block reactions are the center of this process: hydrolysis and condensation, by which molecular precursors are converted into a crosslinked network creating a gel [98].

In aerospace applications, sol–gel-derived coatings are valued for their ability to act as multifunctional barrier layers on light metals such as aluminum, magnesium, and titanium alloys, which are prone to corrosion and oxidation under aggressive environmental conditions. By adjusting the sol–gel chemistry, it is possible to achieve dense, crack-free, and adherent coatings with strong interfacial bonding to metallic substrates. Moreover, the low-temperature processing characteristic of sol–gel coatings makes them suitable for integration with sensitive structural components and composite materials used in modern aircraft and space vehicles. The sol–gel process represents a powerful and flexible route for developing high-performance coatings with controlled composition, structure, and functionality. Its compatibility with a wide range of materials and ease of processing make it an attractive technique for next-generation protective coatings in aeronautical and aerospace systems [99, 100]. The deposition process used is crucial in determining the uniformity, thickness, adhesion, and functional properties of films deposited through sol–gel coatings. Aerospace applications have large surface areas, complex geometries, and high-performance demands being common approaches to coating requiring versatility, scalability, and precision. Four of the most widely used deposition techniques for sol–gel coatings are dip-coating, spin-coating, spray-coating, and electrophoretic deposition, each with its relative advantages and limitations [101, 102].

Spin-coating is particularly ideally suited for producing thin and uniform coatings on planar substrates and is widely employed in laboratory-scale production and sensor fabrication. In spin-coating, a small amount of sol is placed on the center of the substrate and rapidly spun. Centrifugal force drives the sol radially out onto the surface to form a thin film. The thickness of the final coating is a function of the spin rate, viscosity of the sol, and duration. While spin-coating yields coatings with excellent uniformity and reproducibility, it is less suitable for intricate geometries and big parts, limiting its use in large-scale aerospace manufacturing. Still, it remains extremely handy for functional coating testing in small device applications such as aerospace sensors, dielectric films, or EMI-shielding coatings [103, 104].

Surface densification techniques are also critical to promote functional performance of the coating. These include additional thermal annealing, laser sintering, and microwave-assisted curing, which reduce porosity, drive away trapped gases, and create smoother and denser surfaces. Laser sintering, for instance, is advantageous because it has high-precision localized heating, which is useful in spot treatment or patterning. Microwave cure is increasingly being recognized for its capability to achieve volumetric curing, reducing cure time and enabling consistency of crosslinking across coating thickness [105, 106]. Hybrid sol–gel coatings present powerful physical barriers to suppress the entry of corrosive species such as water, oxygen, and chloride ions. The impervious inorganic–organic network developed by hydrolysis and condensation reactions imparts low porosity and high substrate adhesion, which minimizes under-film corrosion. Adding silanes like TEOS and GPTMS causes strong siloxane bonds, which strongly affix the coating to the metallic surface, so enhancing barrier qualities [107].

In advanced aerospace engine components, parts are often subjected to severe thermal, oxidative, and mechanical loading. Such unfavorable operating conditions, frequently involving temperatures higher than 1200 °C and accelerated thermal cycling, require tough surface protection schemes. Thermal barrier coatings (TBCs) made of ceramic materials have been extensively utilized in turbine blades, exhaust nozzles, and combustion chambers to thermally insulate structural substrates from harmful heat flux. Yet, traditional plasma-sprayed or electron-beam physical vapor deposition (EB-PVD) coatings, though effective, are afflicted with high processing costs, low conformity to complex geometries, and delamination susceptibility due to thermal expansion mismatches. Sol–gel-derived ceramic hybrids provide an alternative solution in this situation. Sol–gel TBCs use organo-silicate or metal alkoxide precursors (e.g., tetraethyl orthosilicate or zirconium propoxide) to deposit homogeneous ceramic networks by controlled hydrolysis and condensation reactions.

References

1. Hergenrother PM (2003) The use, design, synthesis, and properties of high performance/high temperature polymers: an overview, High Perform Polym 15:3–45. https://doi.org/10.1177/095400830301500101
2. Caldona EB, De Leon ACC, Pajarito BB, Advincula RC (2017) A review on rubber-enhanced polymeric materials. Polym Rev 57:311. https://doi.org/10.1080/15583724.2016.1247102
3. Tsai C-L, Yen H-J, Liou G-S (2016) Highly transparent polyimide hybrids for optoelectronic applications. React Funct Polym 108:2–30. https://doi.org/10.1016/j.reactfunctpolym.2016.04.021
4. S Yuan, Y Zheng, CK Chua, Q Yan, K Zhou (2018) Electrical and thermal conductivities of MWCNT/polymer composites fabricated by selective laser sintering, Composites: Part A 105:203–213. https://doi.org/10.1016/j.compositesa.2017.11.007
5. Joo H, Cho S (2020) Comparative studies on polyurethane composites filled with polyaniline and graphene for DLP-type 3D printing. Polymers 12. https://doi.org/10.3390/polym12010067
6. Guessasma S, Stephant N, Durand S, Belhabib S (2024) Digital light processing route for 3D printing of acrylate-modified PLA/Lignin blends: microstructure and mechanical performance. Polymers 16:1342. https://doi.org/10.3390/polym16101342
7. Candadai AA, Nadler EJ, Burke JS, Weibel JA, Marconnet AM (2021) Thermal and mechanical characterization of high-performance polymer fabrics for applications in wearable devices. Sci Rep 11. https://doi.org/10.1038/s41598-021-87957-7
8. De Leon ACC, Da Silva ÍG, Pangilinan KD, Chen Q, Caldona EB, Advincula RC (2021) High performance polymers for oil and gas applications. React Funct Polym 162(2021):104878. https://doi.org/10.1016/j.reactfunctpolym.2021.104878
9. Thavasilingam K, Senthil Kumar A, Sakthimurugan D, Meenatchisundaram P (2023) Synthesis and characterization of fumed silica dispersed bees/carnauba wax-epoxy biocomposite coating: a comparative study. Biomass Convers Biorefinery, 13:7375–7383. https://doi.org/10.1007/s13399-023-04112-0
10. Zhang H, Zhao P, Li J, Cao J (2024) Impact response characteristics and damage mechanism of continuous carbon-fiber-reinforced magnesium matrix composite materials. J Reinf Plast Compos. https://doi.org/10.1177/07316844241256157
11. Ghosh M (2018) Polyimides: fundamentals and applications. CRC Press 912. https://doi.org/10.1201/9780203742945
12. Hassan T, Salam A, Khan A, Khan SU, Khanzada H, Wasim M, Kim IS et al (2021) Functional nanocomposites and their potential applications: a review. J Polym Res 28(2):36
13. Kashfipour MA, Mehra N, Zhu J (2018) A review on the role of interface in mechanical, thermal, and electrical properties of polymer composites. Adv Compos Hybrid Mater 1:415–439
14. Zhang H, Zhu H, Xu C, Li Y, Liu Q, Wang S, Yan S (2022) Effect of nanoparticle size on the mechanical properties of polymer nanocomposites. Polymer 252:124944
15. Zhou K, Xie F, Wu X, Wang S (2019) Fabrication of high temperature oxidation resistance nanocomposite coatings on PEO treated TC21 alloy. Materials 13(1):11
16. Hassanzadeh-Aghdam MK, Ansari R, Mahmoodi MJ, Darvizeh A (2018) Effect of nanoparticle aggregation on the creep behavior of polymer nanocomposites. Compos Sci Technol 162:93–100
17. Mrazova M (2013) Advanced composite materials of the future in aerospace industry. Incas bulletin 5(3):139
18. Ahmad F, Awadh MA, Abas M, Noor S, Hameed A (2022) Optimization of carbon fiber reinforced plastic curing parameters for aerospace application. Appl Sci 12(9):4307
19. Gholampour A, Ozbakkaloglu T (2020) A review of natural fiber composites: properties, modification and processing techniques, characterization, applications. J Mater Sci 55:829–892

20. Mohd Bakhori SN, Hassan MZ, Mohd Bakhori N, Jamaludin KR, Ramlie F, Md Daud MY, Abdul Aziz SA (2022) Physical, mechanical and perforation resistance of natural-synthetic fiber interply laminate hybrid composites. Polymers, 14(7):1322
21. Rycenga M, Cobley CM, Zeng J, Li W, Moran CH, Zhang Q, Xia Y et al (2011) Controlling the synthesis and assembly of silver nanostructures for plasmonic applications. Chem Rev 111(6):3669–3712
22. Wazalwar R, Sahu M, Raichur AM (2021) Mechanical properties of aerospace epoxy composites reinforced with 2D nano-fillers: current status and road to industrialization. Nanoscale Adv 3(10):2741–2776
23. Hamerton I, Kratz J (2018) The use of thermosets in modern aerospace applications. In: thermosets. Elsevier, pp 303–340
24. Balakrishnan P, John MJ, Pothen L, Sreekala MS, Thomas S (2016) Natural fibre and polymer matrix composites and their applications in aerospace engineering. In: Advanced composite materials for aerospace engineering. Woodhead Publishing, pp 365–383
25. Bello SA, Agunsoye JO, Hassan SEA, Kana MZ (2015) Epoxy resin based composites, mechanical and tribological properties: a review. Tribology in Industry 37(4):500
26. Liu Q, Wang D, Li Z, Li Z, Peng X, Liu C, Zhang Y, Zheng P (2020) Recent developments in the flame-retardant system of epoxy resin. Materials 13(9):2145
27. Liu S, Fang Z, Yan H, Wang H (2016) Superior flame retardancy of epoxy resin by the combined addition of graphene nanosheets and DOPO. RSC Adv 6(7):5288–5295
28. Joshi M, Chatterjee U (2016) Polymer nanocomposite: an advanced material for aerospace applications. In: Advanced composite materials for aerospace engineering, pp 241–264. Woodhead Publishing
29. Yang Y, Wang DY, Jian RK, Liu Z, Huang G (2023) Chemical structure construction of DOPO-containing compounds for flame retardancy of epoxy resin: a review. Prog Org Coat 1(175):107316
30. Huo S, Song P, Yu B, Ran S, Chevali VS, Liu L, Fang Z, Wang H (2021) Phosphorus-containing flame retardant epoxy thermosets: Recent advances and future perspectives. Prog Polym Sci 1(114):101366
31. Uddin F (2016) Flame-retardant fibrous materials in an aircraft. J Ind Text 45(5):1128–1169
32. Pinni M, Scatteia L Advanced materials in space architecture: new concepts for space tourism in the coming decades. In: 57th international astronautical congress, pp E3–4
33. Weinstein R (2005) RFID: a technical overview and its application to the enterprise. IT Professional 7(3):27–33. https://doi.org/10.1109/MITP.2005.69
34. Li S, Visich JK (2006) Radio frequency identification: supply chain impact and implementation challenges. Int J Integrated Supply Manage 2(4). https://doi.org/10.1504/IJISM.2006.009643
35. Lee YM, Cheng F, Leung (2004) Exploring the impact of RFID on supply chain dynamics. In; Proceedings of the 2004 winter simulation conference. https://doi.org/10.1109/WSC.2004.1371441
36. Ahsan K, Shah H, Kingston P (2010) RFID applications: an introductory and exploratory study. Int J Comput Sci Iss 7(1). https://doi.org/10.48550/arXiv.1002.1179
37. Kumar P, Reinitz HW, Simunovic J, Sandeep KP, Franzon PD (2009) Overview of RFID technology and its applications in the food industry. J Food Sci 74:R101–R106. https://doi.org/10.1111/j.1750-3841.2009.01323.x
38. Sun C, Jiang F, Jiang S (2013) Research on RFID applications in construction industry. J Networks 8:1221–1228. https://doi.org/10.4304/jnw.8.5.1221-1228
39. Prakashe VA, Rathor KO (2014) RFID technology: a performance differentiator, Library Herald, 52. https://doi.org/10.5958/j.0976-2469.52.1.006
40. Kaur M,. Sandhu MK, Mohan N, Sandhu P (2011) RFID technology principles, advantages, limitations and its applications. Int J Comput Electr Eng 3(1):151–157 https://doi.org/10.7763/IJCEE.2011.V3.306
41. Xiao Y, Yu S, Wu K, Ni Q, Janecek C, Nordstad J (2007) Radio frequency identification: technologies, applications, and research issues. Wirel Commun Mob Comput 7:457–472. https://doi.org/10.1002/wcm.365

42. Ezzeddine H, Merakeb Y, Huillery J, Breard A, Duroc Y (2023) Pulsing the passive RF identification technology: the pulsed wave mode. IEEE Microwave Mag 24(10):40–50. https://doi.org/10.1109/MMM.2023.3293613
43. Yakimets I, MacKerron D, Giesen P, Kilmartin KJ, Goorhuis M, Meinders E, MacDonald WA (2010) Polymer substrates for flexible electronics: achievements and challenges. Adv Mater Res 93–94:5–8. https://doi.org/10.4028/www.scientific.net/AMR.93-94.5
44. Zhang L, Tan Q, Kou H, Wu D, Zhang W, Xiong J (2019) Highly sensitive NH3 wireless sensor based on Ag-RGO composite operated at room temperature. Sci Rep 9:9942. https://doi.org/10.1038/s41598-019-46374-5
45. Sharma AK, Singh AK, Akhtar MJ (2014) Dielectric characterisation of substrate materials for RFID applications and their effect on tag performance. IEEE Trans Antennas Propag 62(1):230–238. https://doi.org/10.1109/TAP.2013.2287285
46. Zaid J, Farahani M, Denidni TA (2017) Magneto-dielectric substrate-based microstrip antenna for RFID applications. IET Microw Antennas Propag 11:1389–1392. https://doi.org/10.1049/iet-map.2016.0799
47. Yoon H, Jang J (2009) Conducting-polymer nanomaterials for high-performance sensor applications: issues and challenges. Adv Funct Mater 19:1567–1576. https://doi.org/10.1002/adfm.200801324
48. Finkenzeller K (2010) RFID handbook: fundamentals and applications in contactless smart cards and identification, 3rd ed. Wiley
49. Mousavi SM et al (2021) Graphene-based materials for advanced RFID applications: a review. Carbon 182:837–857
50. Cui L, Zhang Z, Gao N, Meng Z, Li Z (2019) Radio frequency identification and sensing techniques and their applications—a review of the state-of-the-art. Sensors 19(18):4012. https://doi.org/10.3390/s19184012
51. Hoa S (2022)Advanced carbon-fibre composite materials for RFID tag antenna applications. Appl Comput Electromagnet Soc J
52. Brinker CJ et al (2012) Nanocomposite dielectric materials for miniaturised RFID tags. Adv Funct Mater 22(13):2769–2776. https://doi.org/10.1002/adfm.201102566
53. Loukil R, Gazehi W, Besbes M (2024) Improving the classification of a nanocomposite using nanoparticles based on a meta-analysis study, recurrent neural network and recurrent neural network Monte-Carlo algorithms. Nanocomposites 10(1):309–337
54. Gazehi W, Loukil R, Besbes M (2025) Classification of a nanocomposite using a combination between recurrent neural network based on transformer and Bayesian network for testing the conductivity property. Expert Syst Appl 270:126518
55. Pang W, Li Y, DeLuca LT, Liang D, Qin Z, Liu X, Fan X et al (2021) Effect of metal nanopowders on the performance of solid rocket propellants: a review. Nanomaterials 11(10):2749
56. Dong J, Wang W, Wang X, Zhou Q, Miao R, Du M, Cao F et al (2021) Effects of nano aluminum powder on the mechanical sensitivity of RDX-based explosives. Nanomaterials 11(9):2182
57. Zhu M, Zhang P, Gao F, Bai Y, Zhang H, Zu M, Zhang Z et al (2024) Advanced lightweight lightning strike protection composites based on super-aligned carbon nanotube films and thermal-resistant zirconia fibers. Nanosc Adv 6(19):4858–4864
58. Williams JC, Starke EA Jr (2003) Progress in structural materials for aerospace systems. Acta Mater 51(19):5775–5799
59. Zhang X, Chen Y, Hu J (2018) Recent advances in the development of aerospace materials. Prog Aerosp Sci 97:22–34
60. Boyer RR, Cotton JD, Mohaghegh M, Schafrik RE (2015) Materials considerations for aerospace applications. MRS Bull 40(12):1055–1066
61. Singh J, Srivastawa K, Jana S, Dixit C, Ravichandran S (2024) Advancements in lightweight materials for aerospace structures: a comprehensive review. Acceleron Aerospace J 2(3):173–183

62. Nagaraju SB, Priya HC, Girijappa YGT, Puttegowda M (2023) Lightweight and sustainable materials for aerospace applications. In: Lightweight and sustainable composite materials. Woodhead Publishing, pp 157–178
63. Blanco D, Rubio EM, Lorente-Pedreille RM, Sáenz-Nuño MA (2021) Lightweight structural materials in open access: latest trends. Materials 14(21):6577
64. Lee MW, An S, Yoon SS, Yarin AL (2018) Advances in self-healing materials based on vascular networks with mechanical self-repair characteristics. Adv Coll Interface Sci 252:21–37
65. Shields Y, De Belie N, Jefferson A, Van Tittelboom K (2021) A review of vascular networks for self-healing applications. Smart Mater Struct 30(6):063001
66. An S, Lee MW, Yarin AL, Yoon SS (2018) A review on corrosion-protective extrinsic self-healing: comparison of microcapsule-based systems and those based on core-shell vascular networks. Chem Eng J 344:206–220
67. Zanjani JSM, Okan BS, Menceloglu Y, Yildiz M (2018) Self-healing thermosetting composites: concepts, chemistry, and future advances. Hybrid Organic-Inorganic Interfaces: Towards Adv Funct Mater 121–150
68. Toohey KS, Sottos NR, Lewis JA, Moore JS, White SR (2007) Self-healing materials with microvascular networks. Nat Mater 6(8):581–585
69. Sridhar LM, Oster MO, Herr DE, Gregg JB, Wilson JA, Slark AT (2020) Re-usable thermally reversible crosslinked adhesives from robust polyester and poly (ester urethane) Diels-Alder networks. Green Chem 22(24):8669–8679
70. Samui AB (2022) Self-healing polymers. In: Smart polymers. CRC Press, pp 115–142
71. de Souza FM, Sulaiman MR, Gupta RK (2021) Materials and chemistry of polyurethanes. In: Materials and chemistry of flame-retardant polyurethanes volume 1: A fundamental approach. Am Chem Soc 1–36
72. Liu Y, Rajadas A, Chattopadhyay A (2013) Self-healing nanocomposite using shape memory polymer and carbon nanotubes. In: Sensors and smart structures technologies for civil, mechanical, and aerospace systems. SPIE, pp 34–42
73. Bogue R (2011) Nanocomposites: a review of technology and applications. Assem Autom 31(2):106–112
74. Kausar A (2022) Future and challenging attributes of aeronautical nanocomposites. Polymeric nanocomposites with carbonaceous Nanofillers for aerospace applications, 317
75. Guney Yilmaz S et al (2024) High-performance thermoplastic nanocomposites for aerospace applications. J Reinforced Plastics and Compos
76. Yilmaz G, Sukran F, Erdem B, Berat S et al (2024) High-performance thermoplastic nanocomposites for aerospace applications: a review of synthesis, production, and analysis. J Reinforced Plast Compos 07316844241272035
77. Ramachandran K, Boopalan V, Bear JC, Subramani R (2022) Multi-walled carbon nanotubes (MWCNTs)-reinforced ceramic nanocomposites for aerospace applications: a review. J Mater Sci 57(6):3923–3953
78. Kausar A, Ahmad I, Zhao T (2023) Corrosion-resisting nanocarbon nanocomposites for aerospace application: an up-to-date account. Applied Nano 4(2):138–158
79. Nazarychev VM (2024) Enhanced thermal conductivity of thermoplastic polyimide nanocomposites: effect of using hexagonal nanoparticles. Polymers, 16(23):3231
80. Fenta EW, Mebratie BA (2024) Advancements in carbon nanotube-polymer composites: Enhancing properties and applications through advanced manufacturing techniques. Heliyon 10(16)
81. Kannan S, Siva V, Murugan A, Shameem AS, Thangamani A, Thangarasu S, Raja A (2023) Nanotechnology: nanocomposites processing. In: Innovative development in micromanufacturing processes. CRC Press, pp 372–392
82. Zhao J et al (2017) Polymer nanocomposites: computer simulation understanding of properties. Polymer 133:272–287
83. Campbell FC (2010) Structural composite materials. ASM International

84. Satyanarayana KG, Arizaga GGC, Wypych F (2009) Biodegradable composites based on lignocellulosic fibers—an overview. Prog Polym Sci 34(9):982–1021. https://doi.org/10.1016/J.PROGPOLYMSCI.2008.12.002
85. Puttegowda M (2025) Eco-friendly composites: exploring the potential of natural fiber reinforcement. Discover Appl Sci 7(5):1–24. https://doi.org/10.1007/S42452-025-06981-8/FIGURES/10
86. Peças P, Carvalho H, Salman H, Leite M (2018) Natural fibre composites and their applications: a review. J Compos Sci 2(4). https://doi.org/10.3390/JCS2040066
87. Kurien RA, Maria Anil M, Sharan Mohan SL, Anna Thomas J (2023) Natural fiber composites as sustainable resources for emerging applications- a review. Mater Today Proc https://doi.org/10.1016/J.MATPR.2023.04.363
88. Dahman Y (2017) Smart nanomaterials. In: Dahman Y, Kamil A, Baena D (2017) Nanotechnology and functional materials for engineers, pp 47–66. https://doi.org/10.1016/B978-0-323-51256-5.00003-4
89. Metin T, Parlak C, Alver Ö, Tepe M (2022) Hydrogen storage and electronic properties of C20, C15M5 and H2@C15M5 (M = Al, Si, Ga, Ge) nanoclusters. J Mol Struct 1247:131272. https://doi.org/10.1016/J.MOLSTRUC.2021.131272
90. Abid N et al (2022) Synthesis of nanomaterials using various top-down and bottom-up approaches, influencing factors, advantages, and disadvantages: A review. Adv Colloid Interface Sci 300:102597. https://doi.org/10.1016/J.CIS.2021.102597
91. Vijayaram S et al (2023) Applications of green synthesized metal nanoparticles—a review. Biological Trace Element Res 202(1):360–386. https://doi.org/10.1007/S12011-023-03645-9
92. Ghaffari-Moghaddam M, Hadi-Dabanlou R, Khajeh M, Rakhshanipour M, Shameli K (2014) Green synthesis of silver nanoparticles using plant extracts. Korean J Chem Eng 31(4):548–557. https://doi.org/10.1007/S11814-014-0014-6/METRICS
93. Sanchez C, Julián B, Belleville P, Popall M (2005) Applications of hybrid organic–inorganic nanocomposites. J Mater Chem 15(35–36):3559–3592
94. Dey S, Mukherjee S, Saha B et al (2009) Development and implementation of sol-gel coatings for aerospace applications. SAE Technical Paper 2009–01–3208
95. Santos A, Ferreira MG, Montemor MF (2020) Design of corrosion protective and antistatic hybrid sol-gel coatings for aluminium alloys. Coatings 10(5):441
96. Guglielmi M, D'Amico M, Montesperelli G et al (2013) Potentiality of UV-cured hybrid sol–gel coatings for aeronautical applications. Surf Coat Technol 235:553–560
97. Saji VS, Cook R, Subramanian B (2020) Hybrid sol–gel coatings for corrosion mitigation: A critical review. Coatings 10(5):441
98. Lakshmi RV et al (2020) Sol-gel coating for corrosion protection of AA 2024. In: Proceedings of the 31st congress of the international council of the aeronautical sciences (ICAS)
99. Wang B, Zhong S, Lee TL et al (2020) Non-destructive testing and evaluation of composite materials/structures: a state-of-the-art review. Compos Struct 236:111–117
100. Poznyak SK, Zheludkevich ML, Raps D, Gammel F, Yasakau KA, Ferreira MGS (2008) Preparation and corrosion protective properties of nanostructured titania-containing hybrid sol–gel coatings on AA2024. Prog Org Coat 62(3):226–235
101. Rosero-Navarro NC, Pellice SA, Castro Y, Aparicio M, Durán A (2009) Improved corrosion resistance of AA2024 alloys through hybrid organic–inorganic sol–gel coatings produced from sols with controlled polymerisation. Surf Coat Technol 203(15):1897–1903
102. Theodoratou A, Bonnet L, Dieudonne-George P, Massiera G, Etienne P, Robin JJ et al (2018) Vegetable oil hybrid films cross-linked at the air-water interface: formation kinetics and physical characterization. ArXiv preprint arXiv:1811.06903
103. Blohowiak KY, Osborne JH, Seebergh JE (2009) Development and implementation of sol-gel coatings for aerospace applications. SAE Technical Paper 2009–01–3208
104. González C, Vilatela JJ, Molina-Aldareguía JM, Lopes CS, Llorca J (2017) Structural composites for multifunctional applications: current challenges and future trends. ArXiv preprint arXiv:1703.09917

105. Androulidakis C, Kotsidi M, Gorgolis G, Pavlou C, Sygellou L, Paterakis G et al (2021) Multi-functional 2D hybrid aerogels for gas absorption applications. arXiv preprint arXiv: 2109.12829
106. Subasri R, Soma Raju KRC (2019) Multifunctional sol-gel nanocomposite coatings for aerospace, energy, and strategic applications: challenges and perspectives. In: Handbook of advanced ceramics and composites. Springer, pp 1–31
107. Moreau N, Ni L, Croutxé-Barghorn C, Chemtob A (2014) Potentiality of UV-cured hybrid sol–gel coatings for aeronautical and direct-to-metal applications. Surf Coat Technol 239:116–120
108. Subasri R, Soma Raju KRC (2020) Multifunctional sol-gel nanocomposite coatings for aerospace, energy, and strategic applications: challenges and perspectives. In: Mahajan YR, Johnson R (eds) Handbook of advanced ceramics and composites. Springer, pp 1413–1442

Chapter 2
High-performance Polymer Materials for Aeronautical Engineering Applications

M. Prashanth, K. Thavasilingam, L. A. Kumaraswamidhas, Kumaresan Gladys Ashok, and A. Sudhakaran

Abstract This chapter examines the significant potential of the high-performance polymer that is emerging rapidly on several parts of the aerospace components. In this review articles, we begin with a brief overview of the high-performance polymers and the types of high-performance polymers used in aerospace applications. The various fabrication techniques that are involved in the development of HPPs and its composites through different methods are such as fused deposited model (FDM), stereolithography (SLA), selective laser sintering (SLS), and so on. The report further highlights on the portrayal of the various characterizations such as X-ray diffraction, scanning electron microscope, transmission electron microscope, Fourier transform infrared spectroscopy (FTIR) and X-ray photoelectron spectroscopy (XPS), thermogravimetric analysis (TGA), differential scanning calorimetry (DSC), and dynamic mechanical analysis (DMA) of the HPPs nanocomposites fabricated by different techniques. In addition, the high-performance polymers have brought about substantial revolution in many fields such as medical, automotive, aerospace, consumer product, and so on. Thus, this chapter helps the beginners, who can gain a comprehensive understanding of the fundamental principles of nanocomposite fabrication through additive manufacturing techniques.

M. Prashanth (✉) · L. A. Kumaraswamidhas
Department of Mechanical Engineering, Indian Institute of Technology (ISM), Dhanbad, Jharkhand, India
e-mail: prashanthmuralishankar@gmail.com; prashanth.mech@sonatech.ac.in

M. Prashanth
Department of Mechanical Engineering, Sona College of Technology, Salem, TamilNadu, India

K. Thavasilingam (✉) · K. G. Ashok
Department of Mechanical Engineering, Easwari Engineering College, Chennai, Tamil Nadu, India
e-mail: thavasi.k.89@gmail.com

A. Sudhakaran
Department of Civil Engineering,, Indian Institute of Technology, Hyderabad, Hyderabad, India

S. K. Panda (ed.), *Nanocomposites for Defense, Aeronautical and Aerospace Applications*, Advanced Structured Materials 248,
https://doi.org/10.1007/978-3-032-11568-3_2

Keywords High performance polymers · Fabrication techniques · Characterization techniques · Mechanical properties

2.1 Introduction on High-Performance Polymer Materials

Generally, polymer matrix is classified based on the physical, thermal, and mechanical properties. Polymer material is inexpensive to fabricate and demonstrate low thermal and mechanical properties. However, the HPPs exhibit higher cost of fabrication and increase the mechanical and thermal properties. Furthermore, there are other polymers that possess superior properties compared to that of other conventionally available polymers. As shown in Fig. 2.1, polymer is generally sub-divided into several types based on the morphology.

Figure 2.1 represents the sub-divided categories of polymer material based on the morphology with random orientation of the polymer matrix. Polymer matrix contains some highly packed molecular structure that contains appropriate geometries and intermolecular attraction to attain a thermomechanical and chemically favourable structures.

High-performance polymers (HPPs) materials are designed to withstand extreme conditions that include high temperature, aggressive chemicals, and mechanical stresses. High-performance polymers are most widely used in several industries such as oil and gas, aerospace, automobile, and so on. Among these fields, one of the significant areas of advancement for high-performance polymers is aerospace engineering applications. Because of the outstanding strength-to-weight ratio, and its ability to withstand physical, chemical and thermal stresses. In addition, the HPPs material has

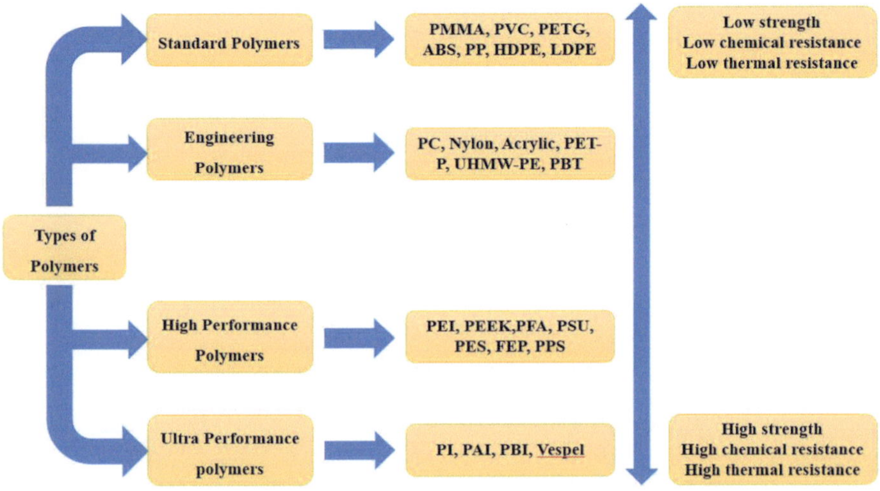

Fig. 2.1 Types of polymers based on chemical thermal and mechanical properties

dimension stability and flexibility that can easily designed, moulded, and fabricated into different complex geometries to satisfy the demand of the industrial applications [1–3].

The fabrication of HPPs structure has a huge interest among the researchers that can have exceptional mechanical, dimensional, and chemical durability of the 3D printed material. One of the most crucial factors in fabricating the HPPs material is bond dissipation energy. If the bond dissipation energy is higher, then the material is very stronger, and it can be used in the harsh environment. However, if the bond dissipation energy is lower, then the material is not feasible to use in the harsh environment. The bond energies of various combination of carbon, hydrogen, and fluorine are as follows: (i) C–C is 83 kcal mol^{-1}, (ii) C=C is 145 kcal mol^{-1}, (iii) C–H is 99 kcal mol^{-1}, and (iv) C–F is 123 kcal mol^{-1} [4, 5]. It is noted that the C=C is harder compared to that of other bond energies. The HPPs bond energies can be improved by adding aromatic components that increase the bond strength because of the resonance stabilization. In addition, the HPPs material with high aromatic content is very difficult to break and does not release volatile substances when they are heated to higher temperature.

One of the major drawbacks of aromatic content on the HPPs material is that the polymer becomes very hard and stiff. In order to overcome this problem, the heterogeneous structure such as nitrogen, oxygen, and sulphur elements is incorporated into the aromatic structure. As a result, the processing of the polymer material becomes easier with decrease in the thermal and chemical resistance due to the bond strength. The factors that contribute to enhance the bonding strength are molecular weight of the polymer, crosslink density, crystallinity, and the incorporation of the reinforcements.

2.1.1 Types of High-Performance Polymers

The use of HPPs in aerospace parts meets the high durability and strength with high quality. Figure 2.2 represents the types of the high-performance polymers. They are: (i) Amorphous high-performance polymers, (ii) Semi-crystalline high-performance polymers, and (iii) Liquid crystalline high-performance polymers: The following sections provide a comprehensive explanation on the types of high-performance polymers [5].

(i) Amorphous high-performance polymers.

Amorphous high-performance polymers are type of polymer material that have distinct crystalline and amorphous regions that have a randomly arranged molecular structure that have a distinct property such as optical transparency, dimensional stability, and uniform mechanical properties. They are generally classified as polysulfone (PSU), polyetherimide (PEI), and polymethyl methacrylate (PMMA), respectively .

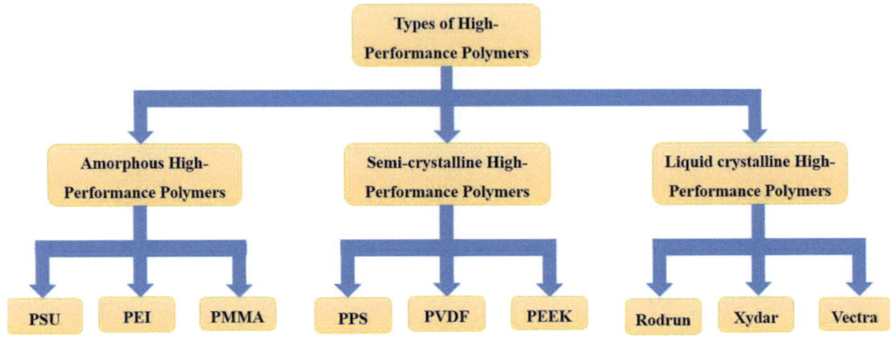

Fig. 2.2 Types of high-performance polymers

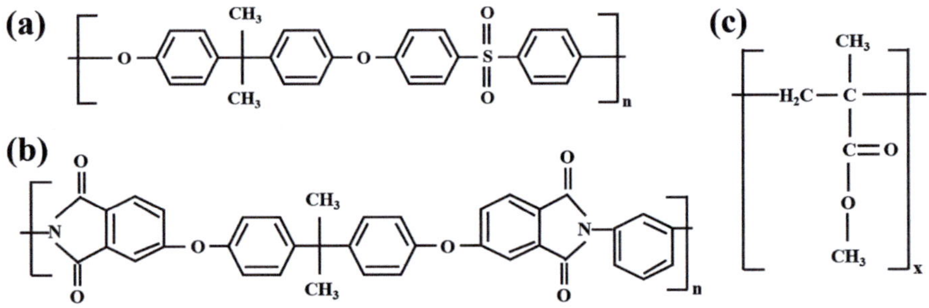

Fig. 2.3 Chemical structure of amorphous high-performance polymers **a** polysulfone, **b** polyether-imide, and **c** polymethyl methacrylate

Polysulfone (PSU)

Polysulfone is a family of amorphous high-performance polymers, which is more transparent and heat resistance thermoplastic polymers due to its outstanding properties such as toughness, rigidity, and stability at high temperature environments. Figure 2.3a represents the chemical structure of the polysulfone polymer material. PSU has one of the highest service temperatures among other HPPs, which has a glass transitions temperature of 185 °C [6]. Additionally, PSU can maintain its mechanical characteristics even at low temperature ranging between − 100 and − 150 °C. PSU is more resistance to oxidizing agents, acids, alkaline, and so on. However, the use of PSU is restricted to certain reagents such as acetone, aromatic hydrocarbons, and dichloromethane. PSU has extremely high flexural modulus and impact resistance in terms of mechanical characteristics due to the combination of polar group of chains. Furthermore, PSU has exceptionally high-temperature dimension stability and its use in aerospace applications involving aircraft interiors, catering trolleys, and precision gears [6, 7].

Polyetherimide (PEI)

Polyetherimide is a type of high-performance amorphous thermoplastic that has excellent heat resistance, strength, and chemical resistance making it a suitable to use in the field of aerospace structural components and interior parts. PEI possess significant mechanical and chemical properties that includes high strength, creep resistance dimension stability, and thermal resistance. Because of this the PEI polymer material is most frequently used in the structural components. PEI is most resistance chemicals such as alcohols, halogenated solvents, and hydrocarbon agents. It also has the ability to retard flame and produce very less smoke. However, it possesses excellent properties, and it still got some drawbacks such as high material and production cost, high processing temperature, and notch sensitive [8, 9]. The molecular structure of the polyetherimide is represented in Fig. 2.3b.

Polymethyl methacrylate (PMMA).

Polymethyl methacrylate is a family of transparent high-performance thermoplastic that has excellent optical clarity, weather resistance, and impact strength. PMMA is employed in several fields such as aircraft canopies and aircraft instrument panels. It has the ability to keep a light beam to reflect within the surfaces [10]. PMMA is fabricated by polymerization of methyl methacrylate (MMA) that undergoes phase transitions to a viscoelastic stage above the glass transition temperature of 105 °C. The main benefit of PMMA is better dimensional stability and adaptation with minor polymer shrinkage. However, PMMA has certain disadvantage high level of residual monomers resulting in strength reduction. PMMA is not chemically resistance to solvents such as acetone and alcohol. In order to overcome these problems, PMMA can be combined with aromatic polymer groups to improve the mechanical and chemical properties of the material [10, 11]. Figure 2.3c indicates the molecular structure of the polymethyl methacrylate polymer.

(ii) Semi-crystalline high-performance polymers.

Semi-crystalline polymer is a type of high-performance polymers that have a combined effect of crystalline and amorphous regions. Semi-crystalline HPPs have the combination of both arranged polymer chains and randomly orientated polymer chains that have a high contribution to rigidity, toughness, impact, and wear resistance. Examples of semi-crystalline HPPs include polyphenylene sulphide (PPS), polyvinylidene fluoride (PVDF), and polyetheretherketone (PEEK), respectively.

Polyphenylene sulphide (PPS)

Polyphenylene sulphide belongs to a semi-crystalline thermoplastic polymer family that has benzene rings connected by atoms. It is well known that PPS is highly resistant to corrosion, and it can withstand medium such as acid, base, and alkaline medium. In addition, PPS has very good thermal and dimensional stability, electrically conductive, and abrasion resistance. However, one of the drawbacks of PPS is that it has high processing temperature which results in high cost, and the material is very brittle. In order to overcome the brittleness failure, the PPS material can

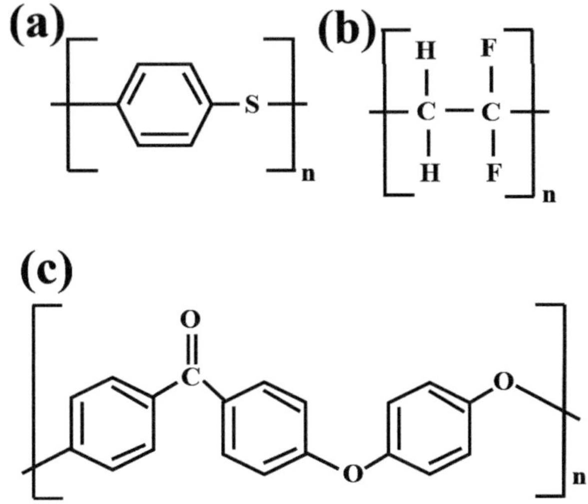

Fig. 2.4 Chemical structure of semi-crystalline high-performance polymers **a** polyphenylene sulphide, **b** polyvinylidene fluoride, and **c** polyetheretherketone

be reinforced with different nanofillers and fibres [12, 13]. The chemical molecular structure of polyphenylene sulphide is presented in Fig. 2.4a.

Polyvinylidene fluoride (PVDF)

PVDF is a non-reactive thermoplastic polymer that is most widely used in the aerospace, chemical, and construction industries. The synthesis of PVDF is carried out through free-radical polymerization of vinylidene fluoride (VDF) monomers. PVDF possess high mechanical strength, better chemical resistant, and thermally stable. PVDF is more expensive, and it gets degraded for exposure at high temperature for 150 °C for longer duration. In order to overcome the degradation, nanocomposites such as graphene nanocomposite and carbon nanotubes can be synthesized along with PVDF to improve the mechanical and electrochemical properties of the material [14, 15]. Figure 2.4b indicates the chemical structure of semi-crystalline high-performance polyvinylidene fluoride polymer.

Polyetheretherketone (PEEK)

Polyetheretherketone is a type of polymer that has both ketone an ether linkage, and it is a best example of aromatic semi-crystalline polymer. PEEK is generally synthesized by polymerization of 4,4′-difluorobenzophenone (DFBP) with 1,4-benzenediol in proper organic solvent conditions. PEEK has a very high mechanical strength, and it is fatigue resistance. In addition, it can withstand chemical attack and flame attack. PEEK has excellent wear properties and good dielectric properties. As PEEK material can withstand high temperature of about 250 °C which is most suitable for aerospace applications that includes flight control system, electrical system, aircraft interiors, and so on. However, the cost of fabricating PEEK is very expensive compared to that of other polymers [16, 17]. The chemical structure of polyetheretherketone is represented in Fig. 2.4c.

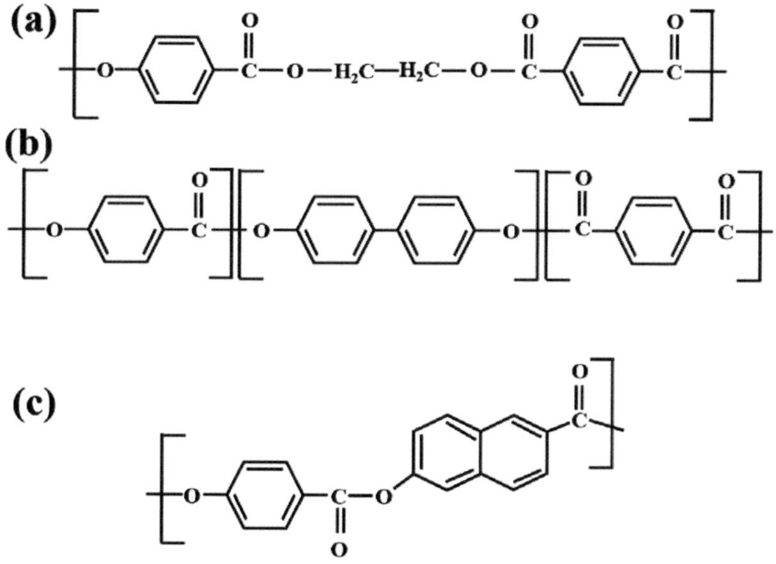

Fig. 2.5 Chemical structure of liquid crystalline high-performance polymers **a** Rodrun, **b** Xydar, and **c** Vectra

(iii) Liquid crystalline high-performance polymers.

Liquid crystalline polymers (LCP) exhibit highly ordered that are partially aromatic polymer. This polymer shows a combination of both molecular structures of solid and liquid states, making it a suitable for aerospace structural components of fuselage and wings. LCP possess excellent mechanical and thermal stability, chemical resistance, and low flammability. LCP-HPPs are generally classified mainly into different types, and they are Rodrun, Xydar, and Vectra, respectively.

Rodrun LC-HPPs

Rodrun is a type of liquid crystalline family that has excellent heat resistance, material flowability, and dimensional stability making it a suitable application for aircraft sockets, connectors, flexible printed circuits, etc. Rodrun is generally composed in the combination of 40% of polyethylene terephthalate and 60% p-hydroxybenzoic acid. It is highly molecular structured polymer that has exceptional properties across several industries. Rodrun possesses low melt viscosity and high strength at elevated temperature, and it is chemically resistant. Rodrun-type polymer can be recycled in the environment condition that could result in the decrease in the mechanical properties of the polymer material [18, 19]. Figure 2.5a represents the chemical molecular structure of Rodrun liquid crystalline.

Xydar LC-HPPs

Xydar is a specialized family of thermotropic liquid crystalline high-performance polymer. This polymer is highly ordered molecular structures with unique molecular alignment that provides excellent properties such as high stiffness, very good strength-to-weight ratio, and dimensional stability. Xydar liquid crystalline possess very high melting and glass transition temperature at about 400 and 280 °C. Xydar is suitable to use in harsh environment, and it is more resistance to acid and bases solvent. This material can be used in lightweight structural components and high-temperature electrical insulation parts. However, one of the major drawbacks of the Xydar polymer is that it is very high cost compared to the other conventional polymer [20, 21]. The thermotropic liquid crystalline of Xydar chemical formula is shown in Fig. 2.5b.

Vectra LC-HPPs

Figure 2.5c represents the chemical structure of Vectra liquid crystalline high-performance polymers. Vectra is a thermotropic liquid crystalline family that offers excellent mechanical, thermal and chemical properties. This Vectra is fabricated by Celanese method that is widely used in the application that requires high strength, high temperature resistance, outstanding chemical properties, and dimensionally stabile material. Vectra exhibit highly ordered molecular structure, even in the melt phase. Vectra high-performance polymer possesses high mechanical strength and wear resistance. It has high glass transition and melting temperature of 280 and 350 °C. It is most widely used in aerospace applications on electrical connectors, sensors housing, and high-performance gears and bushings [22, 23].

2.2 Preparation of High-Performance Polymer Materials

High-performance polymer material is generally classified into several types based on molecular structure, and they are fabricated by different methodologies based on the state of fusion, material distribution, feedstock material, and so on. Our attention is on the most widely used techniques that are exclusive for polymeric material. This sub-division will extensively insight on the various fabrication techniques of high-performance polymeric material.

2.2.1 Fused Deposition Modelling (FDM)

Fused deposition modelling is one of the additive manufacturing techniques. In this technique, the polymer filament is heated and melted on the flat surface on layer-by-layer process. The fabricated sample is then removed from the base plate, and it is made to use for characterization and mechanical testing. The samples were

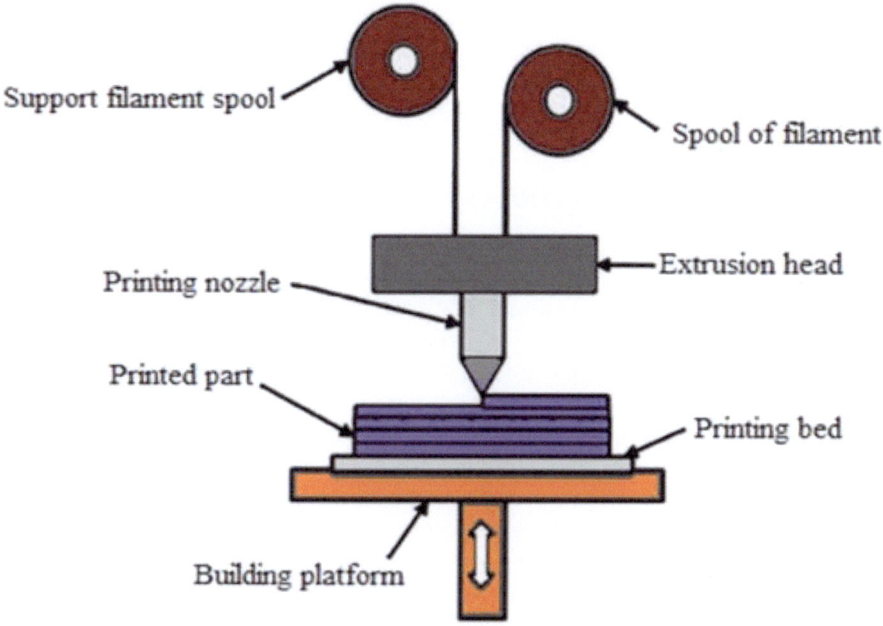

Fig. 2.6 Schematic diagram of the fused deposition model [26]

fabricated by varying several process parameters such as nozzle temperature, bed temperature, nozzle speed, raster angle, layer thickness, and infill pattern. FDM has several advantages, and they are: (i) the ease of maintenance, (ii) relatively low cost compared to that of other fabrication process, (iii) it is easy to store raw materials, (iv) it has free open-source software, and (v) it can be easily relocated to any place which makes suitable for several industrial applications. Figure represents the schematic diagram of the fused deposition modelling [24, 25]. The schematic diagram of the fused deposition model is shown in Fig. 2.6.

The influence of mechanical properties such as tensile, comprehensive, and bending point on the FDM-printed ABS and PEEK fabricated by varying the raster angle and layer thickness. Authors observed higher mechanical properties for a layer thickness of 300 μm and raster angle (0°–90°). It is noted that the mechanical properties obtained for PEEK polymer by FDM method are very high compared to that of ABS polymer by FDM and injection moulded specimen [27]. Rahman et al. [28] studied on the variation of the raster angle at different position of horizontal and vertical printing of CF loaded PEEK by FDM techniques. The tensile results showed that the sample fabricated at horizontal direction with varying raster angle showed better improvement of 10–20% of UTS value compared to that of the tensile sample fabricated in the vertical direction. The bonding between the extruded filaments helps achieve strength through the intermolecular interactions of the polymer chains. Turner et al. [29] investigated on the several parameters that have a significant impact on the fabrication of the polymer material that results in the enhancement of the viscosity, thermal conductivity, heat capacity, and cooling rate. In addition, the author also observed on the interlaminar adhesion on the fabricated samples that undergoes delamination during the shear force between the layers.

2.2.2 *Stereolithography (SLA)*

The schematic diagram of the stereolithography is represented in Fig. 2.7. Stereolithography is also called as vat-photopolymerization. SLA is one of the additive manufacturing techniques that uses UV irradiation in curing the liquid polymer. The UV irradiation develops a cross-linking reaction with the polymer material which helps in generating the free radicals which further reacts to form a reaction through physical and chemical process. The UV irradiation strikes the resin repeatedly prints the samples layer by layer, and solidifies the layers to form a product. The layer thickness of the sample printed is controlled with the help of UV energy and exposure time. SLA techniques work on the polymer resin which is composed on carbon, oxygen, hydrogen, and vinyl group that acts as a multi-functional acrylate monomer. One of the major advantages of the SLA technique is that the samples fabricated will have high resolution and excellent surface finish in compared to that of other AM techniques of polymer [30, 31]. The mechanical properties such as hardness, damping, and compression strength of the 3D polymer material reinforced with different concentration of 0, 0.5, and 1% of boron nitride by SLA process. The 1% of boron nitride shows twice the improvement of the damping force compared to that of the 0.5% and pure. On the other hand, the compression strength improved by 23.8% for 1% of boron nitride, and it is found to be very higher compared to that of the 0.5% boron nitride and pure polymer [33].

Hu et al. [34] investigated on the photosensitive epoxy-acrylic resin reinforced with different weight (0, 2.5, 5, 7.5, and 10%) percentage of SiO_2 resin fabricated through SLA process. The mechanical properties such as tensile strength and bending strength were found to increase above 5 wt% SiO_2 resin due to the good strengthening and toughening effect. In addition, the tribological properties such as friction coefficient and wear mass loss were significantly reduced for above 5 wt% SiO_2 resin. The authors investigated the effect of SiO_2 reinforcement and reported that the SLA specimen containing 10% SiO_2 resin exhibited improved mechanical and tribological properties. Yang et al. [35] studied on the mechanical properties of the fluorapatite (FAp) glass-ceramics fabricated through SLA techniques. Authors have investigated on the variation of the laser power and scanning speed on the mechanical and tribological properties. The result revealed that the laser power and debinding time significantly impact on the microhardness, flexural strength, and elastic modulus of 772.05 Hv, 205.97 MPa, and 97.06 GPa, respectively. In addition, higher mechanical properties show less wear depth and wear volume, and the dominant wear mechanism was observed to be fatigue and adhesive wear.

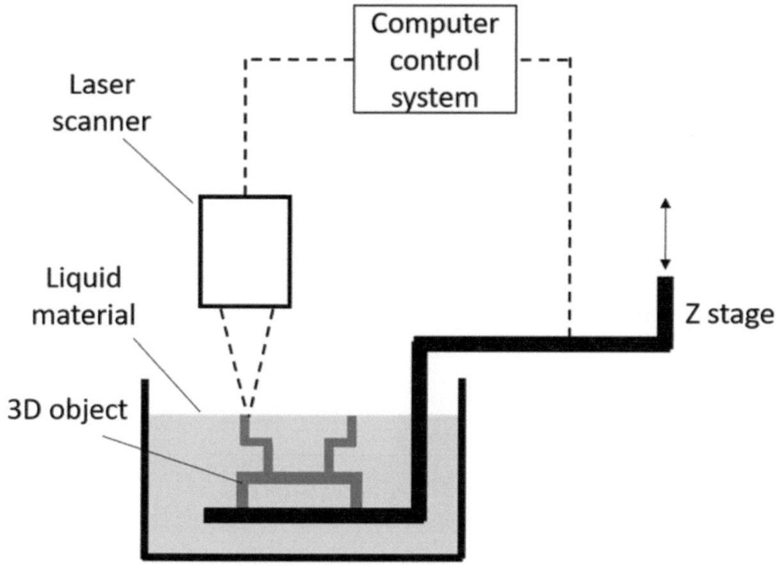

Fig. 2.7 Schematic diagram of the stereolithography [32]

2.2.3 *Selective Laser Sintering (SLS)*

SLS is based on the family of additive manufacturing that uses powder-based system to fabricate the sample with high laser power by melting the powder particles and sintering it to form a required 2-dimensional component. SLS is the most widely used PA11 and PA12 composite material. In addition, SLS also uses polyethylene, polyurethanes, and other thermoplastic materials that have been used to fabricate the various components in automobile, aircraft, and aerospace sectors. SLS techniques can be used to print high melting temperature material that provide very good surface finish of the polymer material. However, the cost of printing the material is very high due to high power laser source [36, 37]. The schematic diagram of the selective laser sintering is represented in Fig. 2.8.

Yuan et al. [39] investigated on the MWCNT/polymer composites fabricated through selective laser sintering on the electrical and thermal properties of the composites. The CNT were added in different ratios of 0.1, 0.5, and 1 wt% to PA12 and PU powders. It is found that the 1 wt% of CNT showed significant results in the electrical properties. However, the thermal properties were not improved significantly. In addition, the microstructure and its corresponding mechanism related to thermal and electrical properties were elucidated. The different combination of the polymer material such as high-performance carbon fibres/polyamide12/epoxy were fabricated by selective laser sintering process. The fabricated sample was investigated through various techniques such as FESEM, FTIR, tensile, and flexural testing of the polymer ternary composite samples. The mechanical properties such as ultimate

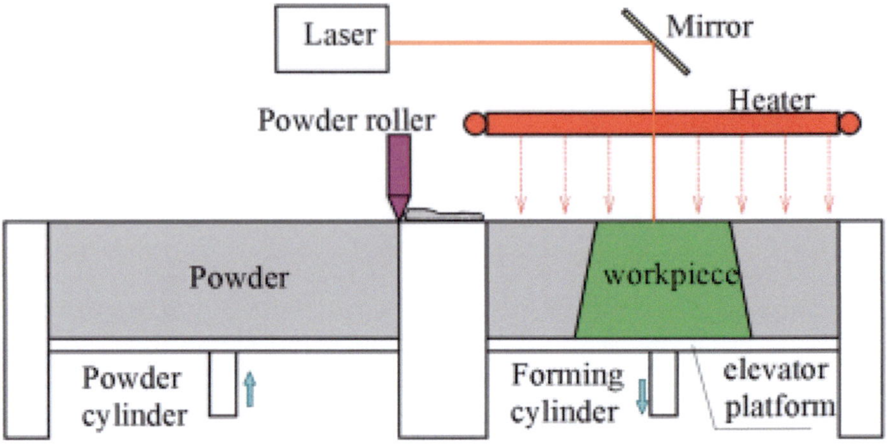

Fig. 2.8 Schematic diagram of the selective laser sintering [38]

tensile strength and flexural strength were found to be 101.3 and 153.43 MPa, respectively. In addition, authors have also made use of the modified Kelly–Tyson model to identify the predicted value, and it is found to be higher than that of the experimental values which was reported by Zhu et al. [40]. Recently several authors [41, 42] have worked on the carbon fibre (CF)-reinforced polymer fabricated through the combination of SLS process, furnace carbonization, and infiltration of epoxy resin. Authors observed that when the CF content is more than 25 Vol.%, the flexural strength was almost reached to 50 MPa which is 1.5 times higher than the pure polymer without CF reinforcement.

2.2.4 Compression Moulding

Compression moulding technique is one of the polymer matrix composite fabrication techniques. This technique gains more popularity due to the fabrication of high strength composite parts which are useful to the aerospace sector. The use of compression moulding process has several advantages such as high production rate, excellent dimensional stability to fabricate complex shapes, very good mechanical properties, better surface finish, and more resistance to flame [43, 44]. Figure 2.9 represents the schematic diagram of the compression moulding of the polymer.

Rangaswamy et al. [46] studied on the investigation of hybrid composite on MWCNT reinforced with E-glass/Kevlar/epoxy composite. The reinforcement fillers with different wt% of 0.3%, 0.6%, and 0.9%, has been investigated on the mechanical properties such as hardness, tensile, and bending strength. The result shows that 0.6% reinforced filler in MWCNT shows better mechanical properties compared to that of the 0.3 and 0.9% of the reinforced filler material. In similar way, the wear

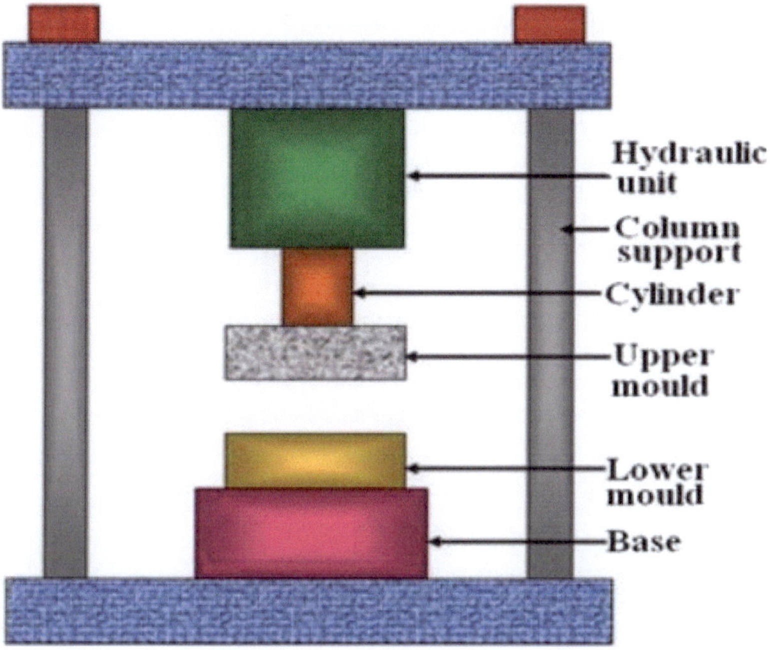

Fig. 2.9 Schematic diagram of the compression moulding [45]

resistance of the 0.6 wt% of the hybrid reinforced filler on MWCNT compared to that of unreinforced filler and 0.3 wt% and 0.9 wt% filler hybrid composites. Mirzazadeh et al. [47] fabricated the high-performance polymer of PVDF reinforced with filler material of sodium–potassium–niobium (KNN) compound. The different weight percentage of both the material of PVDF and KNN were prepared through hot compression moulding process. The thermal stability of the composite increased with increase in the KNN filler material. On the other hand, the yield strength and Young's modulus of the composite show better results compared to that of the pure PVDF polymer. In addition, the KNN with higher filler content in PVDF shows better dielectric properties and notable hysteresis loop. The high-performance thermoplastic PPS reinforced with carbon fibre (CF) used in the aerospace industry for its excellent mechanical and thermal stability. The PPS/CF composites were produced by hot compression moulding process. Authors have fabricated the composite material at different process parameter of pressure cycle, temperature, and plateau time on the thermal and mechanical properties of the composite material. The results of DSC analysis of the composite laminate showed no significant change on the thermal properties for different process parameters. The mechanical properties for lower cooling rate show a very good adhesion bonding of the fibre-matrix interface with enhance mechanical properties reported by the Montagna et al. [48].

2.2.5 Digital Light Processing (DLP)

Digital light processing is one of the promising additive manufacturing techniques where the sample is produced by a bottom-up project in a liquid medium which does not require any support material to improve the layer objects. In this technique, the layer thickness and exposure time has a significant impact on the geometric accuracy and mechanical properties of the final printed samples. DLP process has greater advantages such as rapid fabrication process, more energy efficient, good precision, and wide varieties in shape compared to that of the conventional method [49, 50]. Figure 2.10 represents the schematic diagram of the digital light processing of the polymer material.

Joe et al. [52] studied on the fabrication of 2 wt% of graphene/polyurethane resin using digital light processing technique. The results showed that there is 41% of increase in the tensile strength of the polymer composite than that of pure polymer matrix. In addition, the electric/magnetic fields applied on the DLP process in order to improve the properties and the material to be more bionic which can result in the application of 3D printed composites. Guessasma et al. [53] reported on the 3D printing of the acrylate-modified PLA/lignin blends fabricated through digital light procession examined on the microstructure and mechanical properties of the composites. The varying weight percentage of the lignin ranges from 5 to 30%. The results revealed that the post curing of the polymer composite significantly influenced the tensile properties of the 3D printed. The mechanical properties of polyurethane acrylate resin were investigated through digital light processing technique. In the journal, authors have studied on the different strain rate at quasi-static tensile test. Wang et al. [54] observed that the strain rate has a significant effect of the anisotropy and ductile to brittle transition of the polymer material. Furthermore, the increase

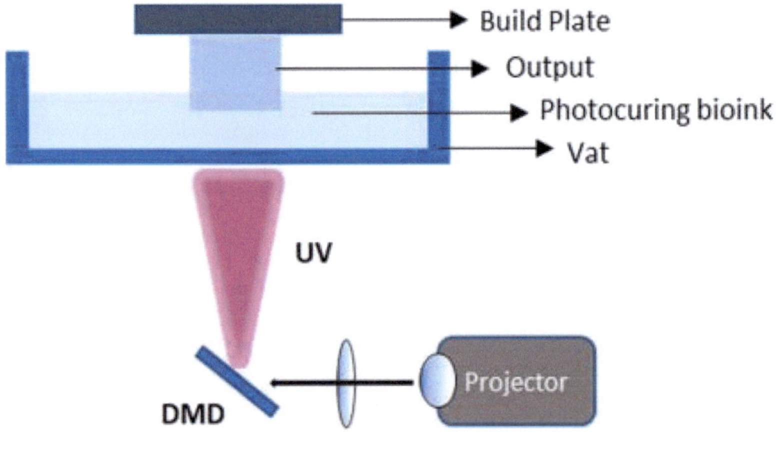

Fig. 2.10 Schematic diagram of the digital light procession [51]

in the strain rate has greatly increased the anisotropy of the material. In addition, the tensile specimen was fabricated on vertical and horizontal direction on the build plate. It is noted that the specimen printed along the vertical direction shows better mechanical properties compared to that of the horizontal direction for different strain rates.

2.3 Characterizations of High-Performance Polymer Materials

To ensure that high-performance polymer (HPP) materials are suitable for demanding aeronautical applications, characterization is an essential first step. Under harsh circumstances, these materials must demonstrate remarkable mechanical strength, thermal stability, structural integrity, and functional performance. A wide range of sophisticated characterization methods are used to assess these characteristics. While X-ray diffraction (XRD) evaluates crystallinity and phase composition, microstructural investigation using scanning electron microscopy (SEM) and transmission electron microscopy (TEM) shows morphological characteristics, porosity, and filler dispersion. Understanding polymer-filler interactions requires knowledge about surface composition and chemical bonding, which can be found in Fourier transform infrared spectroscopy (FTIR) and X-ray photoelectron spectroscopy (XPS). Thermogravimetric analysis (TGA) is used to ascertain thermal deterioration behaviour, while differential scanning calorimetry (DSC) is used to examine melting transitions and glass transition temperatures. Viscoelastic characteristics, including storage and loss moduli, are assessed using dynamic mechanical analysis (DMA) at different frequencies and temperatures. Tensile, flexural, and impact tests measure strength, ductility, and toughness and are used to thoroughly test mechanical performance. Environmental durability is also evaluated by measuring physical attributes like hardness (e.g. Shore-D scale), dimensional stability, and water absorption. Because of their effects on interfacial bonding, load transfer, and fracture mechanics, the incorporation of nanofillers—such as graphene, carbon nanotubes (CNTs), and nano-silica—into polymer matrices makes characterization even more challenging. For example, UHMW-PE fibres improve axial thermal conductivity and mechanical resilience, whereas fumed silica can limit polymer chain mobility, decreasing flexural strength. Deeper understanding of failure processes is made possible by emerging techniques like as 3D X-ray tomography and atomic force microscopy (AFM), which permit bulk and nanoscale structure study, respectively. Because HPPs are used in a wide range of applications, from wearable technology to aerospace components, accurate characterization guarantees the best possible material choice. For instance, PEEK and PTFE composites have better tribological performance than epoxy-based composites, which have great hardness but brittleness. In order to balance strength and toughness, the development of fibre-reinforced polymers (FRPs) containing nanofillers requires customized characterization. In order to advance HPPs in high-stakes industries, a multidisciplinary strategy that combines microscopic, spectroscopic, thermal, and mechanical investigations is essential.

Commercial fabrics composed of ultra-high molecular weight polyethylene (UHMW-PE) fibres are now used in sports equipment and personal body armour due to their exceptional strength, durability, and abrasion resistance. UHMW-PE fibres exhibit exceptionally high axial thermal conductivity and outstanding mechanical properties due to a high degree of polymer chain orientation. However, these materials' potential for usage in flexible and wearable temperature control systems has not been fully explored [55]. Selecting the appropriate materials has proven to be one of the most important aspects of developing equipment for chemical processes. Particularly in the oil and gas industry, non-metals (such advanced and high-performance polymeric materials) have replaced metals in the majority of structural elements, coatings, equipment parts, and the like. Therefore, it is crucial to comprehend the advantages and disadvantages of these polymer materials before they can be employed successfully for a given purpose [4]. The enhanced flexural strength is caused by molecular stretching at the bending strain. However, the flexural strength was decreased by adding fumed silica particles of fine size. When the bending load is applied, the higher particle size of fumed silica relative to the resin molecule usually results in an inaccurate and uneven load transmission. The polymer chain's mobility is restricted by the fumed silica particle that appears in the polymer vacuum. This kind would cause cleavage and raise the stress intensity factor [56]. Polymers, also known as polymeric materials, are used in practically every aspect of our everyday lives and are produced by significant industrial endeavours across the globe. Many of these polymers have been synthesized and will continue to be created. They must therefore be characterized using a variety of analytical tools and techniques. Characterization and categorization of polymers use traditional thermal analysis as well as physico-chemical, mechanical, thermal, and viscoelastic properties [57]. The obtained porous materials were characterized using several techniques (density, SEM, and TEM) to ascertain the size and shape of the porous structure, as well as the cell density. With our approach, a variety of porous materials can be obtained, depending on the parameters of the heat treatment [58].

The coated surfaces were ready for characterization and checked for surface flaws on the mechanical properties of the composites. 86 shore-D is the hardness, 16.74 MPa is the lap shear strength, and 98 MPa is the flexural strength of the pure epoxy composite coating material. Epoxy's higher hardness and poorer flexural and shear strength values result on its high brittle molecular structure [59]. Semiconductor carbon nanotubes (s-CNTs) have emerged as a promising alternative to traditional silicon in ultrascaled field-effect transistors (FETs) due to their exceptional properties. For practical applications, aligned s-CNTs (A-CNTs) are particularly desirable because they can provide a larger driving current and a lower contact resistance than individual s-CNTs or random networks. After conjugated polymer wrapping for the selective separation of s-CNTs, self-assembly techniques are often used to manufacture high-semiconducting-purity A-CNTs. It is still challenging to accurately characterize the extent of removal, despite the fact that several methods have been studied for polymer removal. Conventional techniques such as X-ray photoelectron spectroscopy (XPS) and absorption spectroscopy may not give a sufficient image of the remaining polymer content on A-CNTs due to their inherent detection limits [60]. High-Performance Polymer (HPP)-based composites' were evaluated for their mechanical and tribological properties, wear patterns, and composition for use in friction units. There has been little research on a comparable polyaryletherketone (PAEK) polymer, despite the fact that several of these materials—especially plain polyetheretherketone (PEEK) and PEEK-based composites—are too detailed in the literature. However, polytetrafluoroethylene (PTFE), a well-known solid lubricant filler, is not included in this list of matrix materials. Different tribological test conditions, including the counterpart material and its roughness, temperature and environment, load-speed (P·V) characteristics, and a loading scheme (contact type), are the primary reasons for an effect. A contact scheme of the polymer composites and their counterparts (point, line, and flat ones) is chosen as a crucial criterion (the highest degree of categorization) since it establishes the macro-scale level of the rubbing bodies' tribological interaction. Along with the fundamental mechanisms that determine an increase or reduction in wear resistance, the impact of other parameters is taken into account when interpreting reported values of the coefficient of friction (CoF) and wear rate (WR) [61].

High-performance polymer materials are characterized as shown in Fig. 2.11. These methods include mechanical testing (like tensile testing), heat analysis (like DSC), physical property evaluation (like dimensional analysis and water absorption), and microscopic imaging for structural evaluation. Furthermore, minute characteristics offer high-resolution analysis for precise material characterization. Together, these techniques provide a comprehensive understanding of polymer characteristics, which is essential to maximizing their performance in high-tech fields like aviation [62].

Each material has benefits and drawbacks, as can be deduced from Table 2.1. It follows from this that no single material can provide all the required functionalities. ABS is not appropriate for use in orthoses because, despite its good mechanical properties, it absorbs water. In contrast, PP has the lowest water absorption, a low yield strength, and printing difficulties. In order to find a property that can offer the necessary functionality, it is crucial to compromise on one or two properties. Fiber-reinforced polymer (FRP) composite materials are highly versatile due to their high specific stiffness and high specific strength characteristics. This material's main flaw is its brittleness, which is caused by the resin's low stiffness and fracture toughness. Impact strength, compressive strength, in-plane shear, fracture toughness, and interlaminar strength are among the matrix-dominated metrics that suffer as a result. These limitations can be addressed by filling a FRP composite with nanoparticles. Therefore, the current study emphasizes on investigating how the inclusion of nanofillers affects the mechanical behaviours of glass/epoxy composite materials.

Nanofillers that can react with the resin system in epoxy to greatly enhance the polymer cross-linking web, such as multi-walled carbon nanotubes (MWCNTs), nano-silica (NS), and nano-iron oxide (NFe), were selected. In glass/epoxy composites, nanoparticles in four different weight percentages (0.1, 0.2, 0.5, and 1.0%)

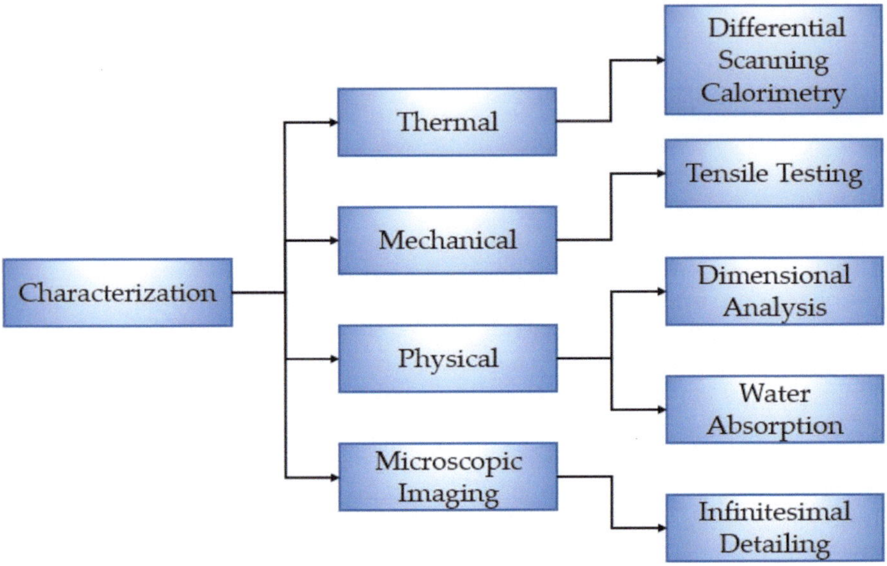

Fig. 2.11 Methodology adopted to characterize various materials [62]

Table 2.1 Mechanical and physical properties measured for different polymers [62]

Material	PLA	ABS	PETG	TPU	PP
Yield strength	25.98	20.70	19.55	6.75	6.58
Tensile strength	30.89	24.99	29.72	28.07	20.23
Elongation at break	5.96	4.07	3.38	332.10	407.99
Water absorption (%)	1.37	1.44	1.59	1.29	0.15
Shrinkage percentage (%)	0.46	0.85	1.18	1.19	2.75
Ease of printing	Easy	Easy	Easy	Difficult	Difficult
Cost (USD)/kg	34.99	34.99	34.99	79.99	59.99

modify the four layers of unidirectional E-glass fibre. For comparison, a sample devoid of nanoparticles was made. Under compressive, tensile, flexural, and impact loads, these samples were mechanically characterized. Figure 2.12 displays the schematic diagram used to produce the modified fibre-reinforced polymer nanocomposites. Nanofillers were dispersed in a resin system in two stages: an hour of bath sonication was followed by an hour of mechanical stirring. Then, using this modified resin system, samples were made by hand lay-up and compression moulding at a pressure of 20 MPa. The samples required for the tensile, compression, bending, and impact tests were cut from a single laminate in accordance with the ASTM (D3039, 695, D790, and D265) standard. The preparation of the different laminates from the different nanoparticles is listed in Table 2.2 according to their weight concentration, size and concentration, and kind.

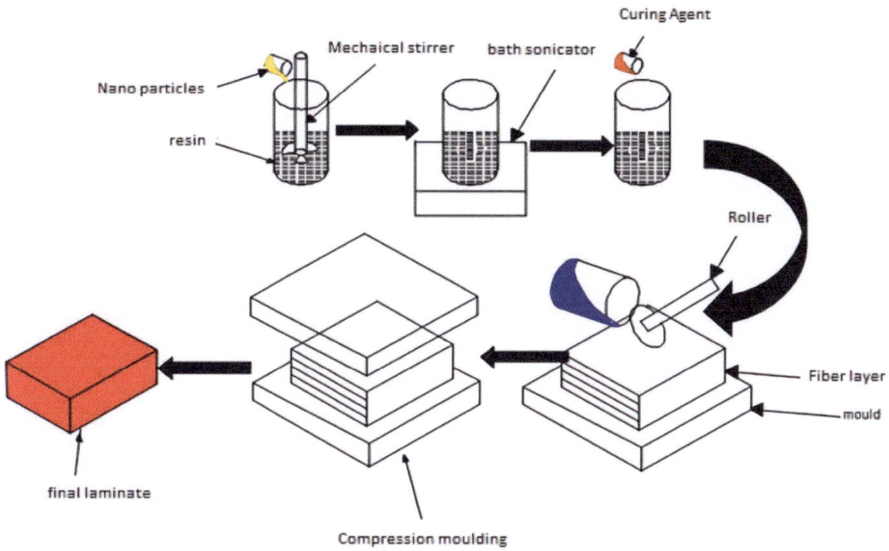

Fig. 2.12 Schematic view of the fabrication of modified nano-polymer composites [63]

Table 2.2 Properties of Lapox [63]

Properties	Value	Properties	Value
Appearance	Clear, viscous liquid	Pot life	6–8 h at 20 °C 5–7 h at 30 °C 3–5 h at 40 °C
Viscosity at 25 °C	9000–12.000 m pas	Tensile strength	70–80 MPa
Specific gravity at 25 °C	1.1–1.2	Elastic modulus in tension	4.0–4.8 GPa
Solubility	30 g/25 mL	Glass transition temperature (DSC)	150–160 °C
Melting point	88–92 °C	Coefficient of linear thermal expansion	45–55 K 10^{-6} K^{-1}

To fully utilize high-performance polymer materials in industrial and aviation applications, their characterization is essential. By employing advanced analytical approaches, researchers may interpret the complicated links between microstructure, composition, and performance. For instance, SEM/TEM imaging supports nanofiller dispersion, whereas XRD and FTIR demonstrate crystallinity and chemical stability. DMA forecasts how polymers will behave under dynamic loads, while thermal investigations (DSC/TGA) make sure they can tolerate operating temperatures. For parts like turbine blades or aeroplane fuselages, mechanical testing like tensile and impact evaluations offers empirical data on durability. The shortcomings of conventional materials (such as water absorption in ABS and brittleness in epoxies) highlight the necessity of hybrid and nanocomposite solutions. For instance, UHMW-PE fibres redefine lightweight armour, and MWCNT-enhanced FRPs show better fracture toughness. Nevertheless, problems still exist, such as the identification of polymer residues on CNTs or the compromises in flexural strength caused by silica additions. In order to speed up material development, future directions include machine learning-driven property prediction and in-situ characterization during additive manufacturing. Ultimately, the combination of multiscale characterization techniques, such as atomic-level spectroscopy and macro-scale mechanical testing, guarantees that HPPs satisfy the exacting requirements of the biomedical, automotive, and aerospace industries. In order to create next-generation polymers that are stronger, lighter, and more versatile than ever before, industries will continue to rely on strong characterization techniques as they push the limits of material science.

2.4 High-Performance Polymer Materials for Aeronautical Engineering Applications

High-performance polymers are being designed to meet the specific requirement owing to their excellent properties, includes less in weight with high strength, good corrosion resistance, long-term durability. High-performance polymers have key advantages compared to that of conventional polymers materials due to its strength properties, weight saving, dimension stability, less thermal expansion, and outstanding fatigue and fracture resistance. Aerospace component should possess properties like low solar absorption, very high thermal emissivity, resistance to radiation, and electrical conductivity. The polymer materials are most widely used in military flight aircraft, many satellite components like equipment panels, support structures, antennas, solar structures, etc. Space shuttle rocket motor cases are made up of graphite reinforced polymer matrix. Selection of proper polymer material is a key role in achieving the desired properties that are required for the aerospace applications. Figure 2.13 represents the different applications of high-performance polymers for aeronautical applications.

2.4.1 Structural Applications

In some structural applications, high performance is the most crucial factor for the aircraft components in addition to being lightweight. Here are some of the components made up of polymer material, and they are: aircraft interiors, coatings, cockpit, polymer casings, and solar array substrate that offers more chemical stability, resistance to fire, low operating cost, and lightweight. In addition, there are few drawbacks in the usage of composites in aerospace structural applications that includes (i) polymers used in circuits, antennas, and so on require higher electrical resistance, (ii)

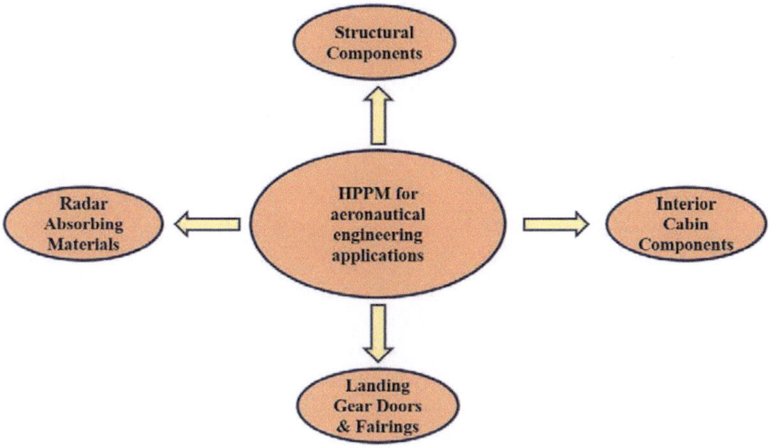

Fig. 2.13 Applications of the high-performance polymers

polymers may reduce thermal conductivity which indulge in more strain on electrical heaters, and (iii) polymer composites are more susceptible to impact, moisture, absorption, deterioration, and ageing over time [64–67]. The various parts of polymers used in aerospace applications are indicated in Fig. 2.14. Dinca et al. [68] reported on the processing of the polymer/clay and polymer/CNT which involves dispersion of the clay or CNT in the polymer matrix material through ultrasonication process and curing it for 24 h at 70 °C to mould the structural component. Polymer/clay and polymer/CNT shows better thermomechanical properties compared to that of pure polymer matrix. In addition, the increase in the weight percentage of the CNT increases the modulus of elasticity and decreases the tensile strength of the material.

Meenakshi et al. [69] investigated on the high performance of 10-dihydro-9-xa-10-phosphaphenanthrene-10-oxide (DOPO)-based nanocomposite reinforced with polyhedral oligomeric silsesquioxanes (POSS). The incorporation of the POSS nanoreinforcement enhanced the mechanical properties of the polymer nanocomposite than that of the matrix material. The enhancement of the properties was mainly because of the size of the POSS reinforcement that has very good interfacial adhesion bonding with the polymer chain matrix material.

2.4.2 Interior Cabin Components

In aircraft interior works, the most common UHPPs used in aircraft interior components include polyetherimide (PEI), polyphenylsulfone (PPSU), and polyvinylidene fluoride (PVDF) especially in the applications of seat frames and panelling, due to exceptional fire resistance and less smoke emissions. Figure 2.15 shows the aerospace interior parts made up on polymer materials.

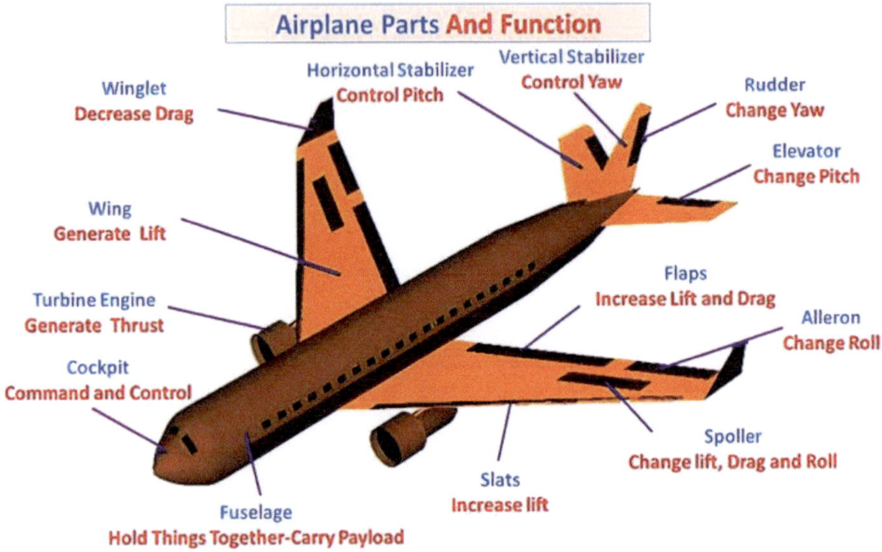

Fig. 2.14 Applications of the high-performance polymers [68]

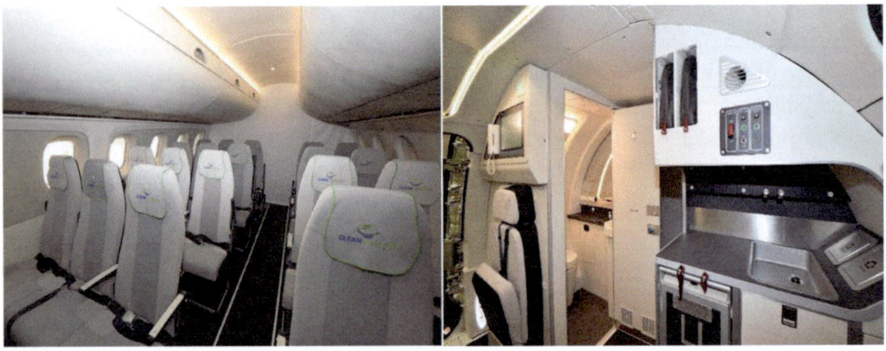

Fig. 2.15 Applications of the high-performance polymers on interior cabin [70]

A high-performance blend of polyetherimide (PEI) and polycarbonate (PC) is manufactured for aircraft applications due to their thermal properties. According to Vīndedze et al. [71], studies of Ultem 9085 exhibit important anisotropy, focusing on tensile and thermal properties varying by printed materials. Whereas these materials meet with aviation standards for the smoke density and toxicity, making it sustainable for aircraft interior components under FAR 25.853. According to studies conducted by Kourtides et al. [72] identify an alternative decorative film for aircraft interior sandwich panels to substitute polyvinyl fluoride (PVF). They estimated several high-performance polymer films, but PEEK establishes outstanding fire-resistant properties, low ignition tendency, and significantly reduces the emission of toxic gases that makes a passenger safe application.

2.4.3 Radar Absorbing Materials (RAM)

The composite materials of conductive polymers are used in the stealth aircraft as radar absorbing materials (RAMs) to reduce the radar wave reflections. The composite is often composed of conductive polymers like polyaniline [73], combined with various fillers of fibre-based carbon materials that provide tailored electrical conduction and permittivity that efficiently absorb and dissolve radar waves. Figure 2.16 indicates the use of RAM on the flights. Proietti et al. [74] and Kim et al. [75] established that PEEK reinforced with graphene nanoplatelets (GNPs) revealed tunable dielectric properties and broad absorption bandwidths, making them effective for RAM integration in aircraft. Moreover, benzoxazine resins, with their low water absorption and high char yield, have also been discovered as RAM matrices, mainly in hybrid sandwich structures, a combination of magnetic fillers like Fe_3O_4 or carbonyl iron for improved absorption performance.

Jayalakshmi et al. [76] studied that the development of radar absorbing structures (RAS) is crucial for refining stealth capabilities in armed aircraft by minimizing radar cross section (RCS). This study highlights that the usage of polymer matrix composites (PMCs), entrenched with radar absorbing materials (RAMs), also the multilayered grid devices such as frequency selective surfaces (FSS) and metamaterials. Carbon-based fillers, like black carbon, carbon fibres, CNTs, and graphene, are used as conductors. Multilayer grid designs and different patterns enable absorption through multiple channels. A review concludes that the development of lightweight,

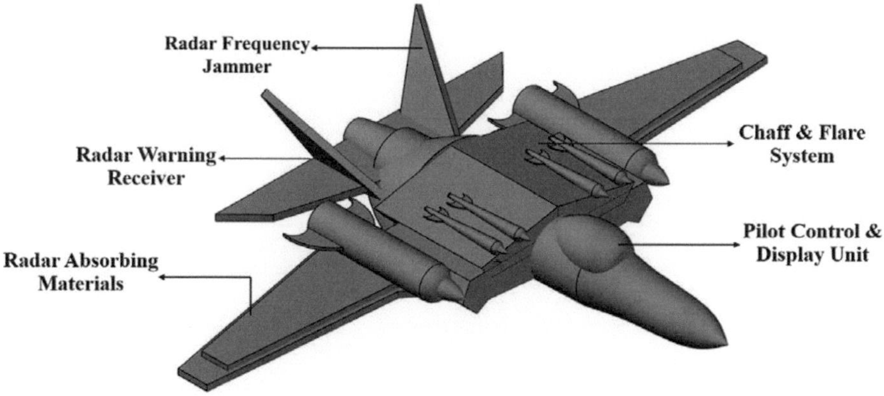

Fig. 2.16 Applications of the high-performance polymers on radar absorbing materials

broadband RAS demands an interdisciplinary approach that involves aeronautical, mechanical, electronics, and materials engineering.

2.4.4 Landing Gear Doors and Fairing

It is a critical path in aerodynamic components, which can be designed to reduce drag and also protect from internal mechanisms and also the fuel efficiency. This kind of structure ensures that it is higher in mechanical loads, impact forces, and mutable in environmental conditions. The aerospace production technology is utilized by ultra-high-performance polymers (UHPPs) and advanced fibre-reinforced polymer matrix composites. The high-performance polymers used in the landing gears are shown in Fig. 2.17.

Zhang et al. [77] developed a continuous carbon-fibre-reinforced magnesium matrix composite (C_f/Mg) by using epoxy resin as a matrix polymer phase, which can focus on the high-impact resistance in aircraft applications such as landing gear doors. According to this research, it reveals that the epoxy-based bonding increases the peak impact loading by 14.5% under repeated impacts. In high-heat zones, polyimide (PI) and PMR-15-based composites have been utilized due to their superior thermal stability and oxidative resistance, sustaining performance at temperatures exceeding 300 °C. As noted by Ghosh and Mittal [78], these materials are ideal for use in fairings located near engine exhaust zones or heat-sensitive landing mechanisms, although they may require careful processing due to issues like brittleness and toxicity in older formulations.

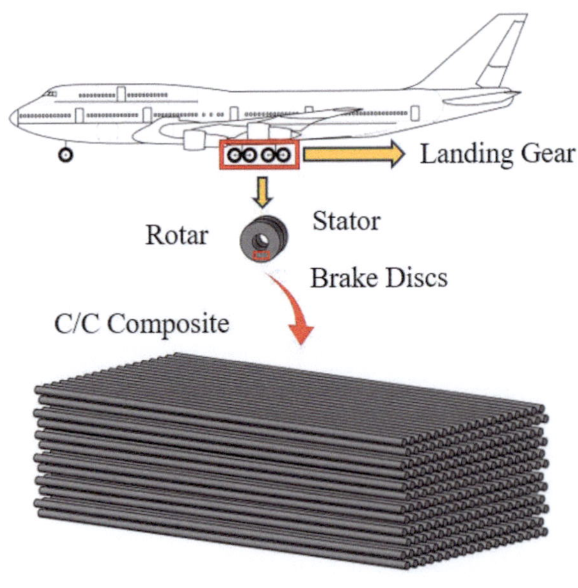

Fig. 2.17 Applications of the high-performance polymers on landing gear doors and fairings

2.5 Conclusion

This chapter provides a comprehensive introductory overview of various high-performance polymer, including their fabrication, characterization and mechanical properties. High-performance polymer alloy and nanocomposites which is one of the novel materials work in extreme conditions, and it is predicted that it will broaden the use of materials to meet the requirement for aerospace applications by different fabrication techniques. High-performance polymers make a good choice to meet the wide range of needs. These polymers can be easily synthesized by different polymerization process, and it can also be combined with other filler material to improve the properties such as mechanical, thermal, chemical, and electrical properties to use them for aerospace applications. However, there are some drawbacks such as structural failure, lightning strike, icing and erosion pose of the aerospace components. To give a clear picture of the state of art, a lot of work have been made to organize and arrange the most pertinent applications and to include illustrations of a few chosen examples from recent studies. There will be more opportunity to use new equipment's and materials to use in the advancement of the fabrication of the aerospace components.

References

1. Hergenrother PM (2003) The use, design, synthesis, and properties of high performance/high temperature polymers: an overview. High Perform Polym 15:3–45. https://doi.org/10.1177/095400830301500101
2. Caldona EB, De Leon ACC, Pajarito BB, Advincula RC (2017) A review on rubber-enhanced polymeric materials. Polym Rev 57:311. https://doi.org/10.1080/15583724.2016.1247102
3. Tsai C-L, Yen H-J, Liou G-S (2016) Highly transparent polyimide hybrids for optoelectronic applications. React Funct Polym 108:2–30. https://doi.org/10.1016/j.reactfunctpolym.2016.04.021
4. De Leon ACC, da Silva ÍG, Pangilinan KD, Chen Q, Caldona EB, Advincula RC (2021) High performance polymers for oil and gas applications. React Funct Polym 162:104878. https://doi.org/10.1016/j.reactfunctpolym.2021.104878
5. de Leon AC, Chen Q, Palaganas NB, Palaganas JO, Manapat J, Advincula RC (2016) High performance polymer nanocomposites for additive manufacturing applications. React Funct Polym 103:141–155. https://doi.org/10.1016/j.reactfunctpolym.2016.04.010
6. Chukov DI, Tcherdyntsev VV, Stepashkin AA, Zadorozhnyy MY (2023) Structure, thermal, and mechanical behavior of the Polysulfone solution impregnated unidirectional carbon Fiber yarns. Polymers 15:4601. https://doi.org/10.3390/polym15234601
7. Gargol M, Podko'scielna, B. (2022) Synthesis and characterization of polymeric blends containing Polysulfone based on cyclic bisphenol. Polymers 14:3148. https://doi.org/10.3390/polym14153148
8. Gao M, Zhu Y, Yan J, Wu W, Wang B (2022) Micromechanism study of molecular compatibility of PVDF/PEI blend membrane. Membranes 12:809. https://doi.org/10.3390/membranes12080809
9. Alvaredo-Atienza Á, Chen L, San-Miguel V, Ridruejo Á, Fernández-Blázquez JP (2020) Fabrication and characterization of PEEK/PEI multilayer composites. Polymers 12:2765. https://doi.org/10.3390/polym12122765

10. Rahman F, Carbaugh DJ, Wright JT, Rajan P, Pandya SG, Kaya S (2020) A review of polymethyl methacrylate (PMMA) as a versatile lithographic resist—with emphasis on UV exposure. Microelectron Eng 224:111238. https://doi.org/10.1016/j.mee.2020.111238
11. Weilong W, Ouyang Q, He L, Huang Q (2022) Optical and thermal properties of polymethyl methacrylate (PMMA) bearing phenyl and adamantyl substituents. Colloids Surf A Physicochem Eng Asp 653:130018. https://doi.org/10.1016/j.colsurfa.2022.130018
12. Won JS, Kwon M, Lee JE, Lee JM, Kwak TJ, Lee SG (2022) Polyphenylene sulfide-modified carbon fiber reinforced high-strength composites via electrophoretic deposition. J Sci Adv Mater Devices 7:100456. https://doi.org/10.1016/j.jsamd.2022.100456
13. Chen G, Mohanty AK, Misra M (2021) Progress in research and applications of Polyphenylene Sulfide blends and composites with carbons. Compos Part B 209:108553. https://doi.org/10.1016/j.compositesb.2020.108553
14. Wang D, Li K, Teo WK (1999) Preparation and characterization of polyvinylidene fluoride (PVDF) hollow fiber membranes. J Membr Sci 163:211–220. https://doi.org/10.1016/S0376-7388(99)00181-7
15. Dallaev R, Pisarenko T, Sobola D, Orudzhev F, Ramazanov S, Trˇcka T (2022) Brief review of PVDF properties and applications potential. Polymers 14:4793. https://doi.org/10.3390/polym14224793
16. Liu G, Hu N, Huang J, Tu Q, Xu F (2024) Experimental investigation on the mechanical and dynamic thermomechanical properties of polyether ether ketone based on fused deposition modeling. Polymers 16:3007. https://doi.org/10.3390/polym16213007
17. Sastri VR (2010) High-temperature engineering thermoplastics: polysulfones, polyimides, polysulfides, polyketones, liquid crystalline polymers, and fluoropolymers. Plast Des Libr:175–215. https://doi.org/10.1016/B978-0-8155-2027-6.10008-X
18. Incarnato L, Motta O, Acierno D (1998) Thermal behaviour of the PET/Rodrun 3000 system. Polymer 39:5085–5091. https://doi.org/10.1016/S0032-3861(97)10058-1
19. Saengsuwan S, Bualek-Limcharoen S, Mitchell GR, Olley RH (2003) Thermotropic liquid crystalline polymer (Rodrun LC5000)/polypropylene in situ composite films: rheology, morphology, molecular orientation and tensile properties. Polymer 44:3407–3415. https://doi.org/10.1016/S0032-3861(03)00244-1
20. Banerjee S, Kar KK (2020) Introduction to liquid crystalline polymers. In: Polymers and polymeric composites: a reference series. Springer, Berlin, Heidelberg. https://doi.org/10.1007/978-3-642-37179-0_49-1
21. Chung T-S, Xing J (2000) Studies on the phase transition and thermal stability of Xydar and Zenite series liquid crystalline polymers. Polym Eng Sci 40:841–856. https://doi.org/10.1002/pen.11213
22. Pisitsak P, Magaraphan R (2009) Influences of a liquid crystalline polymer, Vectra A950, on crystallization kinetics and thermal stability of poly (Trimethylene terephthalate). J Therm Anal Calorim 95:661–666. https://doi.org/10.1007/s10973-008-9214-9
23. Georgiev G, Cebe P, Capel M (2005) Effects of nematic polymer liquid crystal on crystallization and structure of PET/Vectra blends. J Mater Sci 40:1141–1152. https://doi.org/10.1007/s10853-005-6930-4
24. Mohan N, Senthil P, Vinodh S, Jayanth N (2017) A review on composite materials and process parameters optimisation for the fused deposition modelling process. Virtual Phys Prototyp 12:47–59. https://doi.org/10.1080/17452759.2016.1274490
25. Komarnicki P (2017) A review of fused deposition modeling process models. In: Rusinski E, Pietrusiak D (eds) Proceedings of the 13th international scientific conference: computer aided engineering, Springer International Publishing, Cham, pp 241–247. https://doi.org/10.1007/978-3-319-50938-9_25
26. Mogan J, Harun WSW, Kadirgama K, Ramasamy D, Foudzi FM, Sulong AB, Tarlochan F, Ahmad F (2023) Fused deposition modelling of polymer composite: a progress. Polymers 15:28. https://doi.org/10.3390/polym15010028
27. Wenzheng W, Geng P, Li G, Zhao D, Zhang H, Zhao J (2015) Influence of layer thickness and raster angle on the mechanical properties of 3D-printed PEEK and a comparative mechanical study between PEEK and ABS. Materials 8:5834–5846. https://doi.org/10.3390/ma8095271

28. Rahman KM, Letcher T, Reese R (2015) Mechanical properties of additively manufactured peek components using fused filament fabrication. In: ASME international mechanical engineering congress and exposition, proceedings (IMECE). https://doi.org/10.1115/IMECE2015-52209
29. Turner BN, Strong R, Gold SA (2014) A review of melt extrusion additive manufacturing processes: I process design and modeling. Rapid Prototyp J 20:192–204. https://doi.org/10.1108/RPJ-01-2013-0012
30. Nagarajan B, Mertiny P, Qureshi AJ (2020) Magnetically loaded polymer composites using stereolithography—material processing and characterization materials today. Communications 25:101520. https://doi.org/10.1016/j.mtcomm.2020.101520
31. Choong YYC, Maleksaeedi S, Eng H, Wei J, Su P-C (2017) 4D printing of high-performance shape memory polymer using stereolithography. Mater Des 126:219–225. https://doi.org/10.1016/j.matdes.2017.04.049
32. Huang J, Qin Q, Wang J (2020) A review of stereolithography: processes and systems. Process 8:1138. https://doi.org/10.3390/pr8091138
33. Bustillos J, Montero-Zambrano D, Loganathan A, Boesl B, Agarwal A (2017) Stereolithography-based 3D printed photosensitive polymer/ boron nitride nanoplatelets composites. Polym Compos. https://doi.org/10.1002/pc.24662
34. Haixia H, Zhang L, Wang Z, Ruan X, Liu J (2024) Mechanical and tribological performances of epoxy-acrylate resin nanocomposites production via stereolithography. Mater Lett 360:135782. https://doi.org/10.1016/j.matlet.2023.135782
35. Yang B, Wang S, Wang G, Yang X (2021) Mechanical properties and wear behaviors analysis of fluorapatite glass-ceramics based on stereolithography 3D printing. J Mech Behav Biomed Mater 124:104859. https://doi.org/10.1016/j.jmbbm.2021.104859
36. Bai J, Goodridge R, Yuan S, Zhou K, Chua C, Wei J (2015) Thermal influence of CNT on the polyamide 12 nanocomposite for selective laser sintering. Molecules 20:19041. https://doi.org/10.3390/molecules201019041
37. Yuan S, Bai J, Chua CK, Wei J, Zhou K (2016) Material evaluation and process optimization of CNT-coated polymer powders for selective laser sintering. Polymers 8:370. https://doi.org/10.3390/polym8100370
38. Zheng X, Meng J, Liu Y (2022) Strain rate dependence of compressive mechanical properties of polyamide and its composite fabricated using selective laser sintering under saturated-water conditions. Micromachines 13:1041. https://doi.org/10.3390/mi13071041
39. Yuan S, Zheng Y, Chua CK, Yan Q, Zhou K (2018) Electrical and thermal conductivities of MWCNT/polymer composites fabricated by selective laser sintering. Compos Part A 105:203–213. https://doi.org/10.1016/j.compositesa.2017.11.007
40. Zhu W, Yan C, Shi Y, Wen S, Liu J, Wei Q, Shi Y (2016) A novel method based on selective laser sintering for preparing high-performance carbon fibres/polyamide12/epoxy ternary composites. Sci Rep 6:33780. https://doi.org/10.1038/srep33780
41. Bourell D, Leu M-C, Chakravarthy K, Guo N, Alayavalli K (2011) Graphite-based indirect laser sintered fuel cell bipolar plates containing carbon fiber additions. CIRP Ann-Manuf Technol 60:275–278. https://doi.org/10.1016/j.cirp.2011.03.105
42. Guo N, Leu MC (2012) Effect of different graphite materials on the electrical conductivity and flexural strength of bipolar plates fabricated using selective laser sintering. Int J Hydrog Energy 37:3558–3566. https://doi.org/10.1016/j.ijhydene.2011.11.058
43. Park CH, Lee WI (2012) Compression moulding in polymer matrix composites. In: Suresh G, Advani Suresh G, Kuang TH (eds) Manufacturing techniques for polymer matrix composites (PMCs). Woodhead Publishing, pp 47–94. https://doi.org/10.1533/9780857096258.1.47
44. Kim HS, Chang SH (2019) Simulation of compression moulding process for long-fibre reinforced thermoset composites considering fibre bending. Compos Struct 230:111514. https://doi.org/10.1016/j.compstruct.2019.111514
45. Sathish T, Mohanavel V, Raja T, Djearamane S, Velmurugan P, Nasif O, Alfarraj S, Wong LS, Manikandan V, Ravichandran M (2021) Influence of compression Molding process parameters in mechanical and tribological behavior of hybrid polymer matrix composites. Polymers 13:4195. https://doi.org/10.3390/polym13234195

46. Rangaswamy H, Harsha HM, Chandrashekarappa MPG, Pimenov DY, Giasin K, Wojciechowski S (2021) Experimental investigation and optimization of compression moulding parameters for MWCNT/ glass/kevlar/epoxy composites on mechanical and tribological properties. J Mater Res Technol 15:327–341. https://doi.org/10.1016/j.jmrt.2021.08.037
47. Mirzazadeh Z, Sherafat Z, Bagherzadeh E (2021) Physical and mechanical properties of PVDF/ KNN composite produced via hot compression molding. Ceram Int 47:6211–6219. https://doi.org/10.1016/j.ceramint.2020.10.199
48. Montagna LS, de Melo Morgado GF, Marini J, Guimar˜aes A, Passador FR, Rezende MC (2023) Mechanical properties of carbon fiber/PPS laminates processed by hot compression molding under different consolidation cycles. Polym Compos 44: 3358–3371. https://doi.org/10.1002/pc.27326
49. Patel DK, Sakhaei AH, Layani M, Zhang B, Ge Q, Magdassi S (2017) Highly stretchable and UV curable elastomers for digital light processing based 3D printing. Adv Mater 29:1606000. https://doi.org/10.1002/adma.201606000
50. Rehbein T, Johlitz M, Lion A, Sekmen K, Constantinescu A (2021) Temperature-and degree of cure-dependent viscoelastic properties of photopolymer resins used in digital light processing. Prog Addit Manuf 6:743–756. https://doi.org/10.1007/s40964-021-00194-2
51. Sultan MT, Lee OJ, Lee JS, Park CH (2022) Three-dimensional digital light-processing bioprinting using silk fibroin-based bio-ink: recent advancements in biomedical applications. Biomedicine 10: 3224. https://doi.org/10.3390/biomedicines10123224
52. Joo H, Cho S (2020) Comparative studies on polyurethane composites filled with polyaniline and graphene for DLP-type 3D printing. Polymers 12:67. https://doi.org/10.3390/polym12010067
53. Guessasma S, Stephant N, Durand S, Belhabib S (2024) Digital light processing route for 3D printing of acrylate-modified PLA/lignin blends: microstructure and mechanical performance. Polymers 16:1342. https://doi.org/10.3390/polym16101342
54. Wang Y, Li X, Chen Y, Zhang C (2021) Strain rate dependent mechanical properties of 3D printed polymer materials using the DLP technique. Addit Manuf 47:102368. https://doi.org/10.1016/j.addma.2021.102368
55. Candadai AA, Nadler EJ, Burke JS, Weibel JA, Marconnet AM (2021) Thermal and mechanical characterization of high-performance polymer fabrics for applications in wearable devices. Sci Rep 11:8705. https://doi.org/10.1038/s41598-021-87957-7
56. Thavasilingam K, Kumar AS, Khan MA, Devanand S, Giridharan K (2021) Effect of fumed silica in rice bran wax-epoxy coating on aluminum substrate: mechanical, thermal, and water absorption properties. Biomass Convers Biorefinery 13(2021):4229–4240. https://doi.org/10.1007/s13399-021-01889-w
57. Lin S (2016) An overview of advanced hyphenated techniques for simultaneous analysis and characterization of polymeric materials. CritAl Rev Solid State Mater Sci 41:482–530. https://doi.org/10.1080/10408436.2016.1186598
58. Merlet S, Marestin C, Schiets F, Romeyer O, Mercier R (2007) Preparation and characterization of Nanocellular poly(phenylquinoxaline) foams. A new approach to Nanoporous high-performance polymers. Macromolecules 40:2070–2078. https://doi.org/10.1021/ma062259t
59. Thavasilingam K, Senthil Kumar A, Sakthimurugan D, Meenatchisundaram P (2023) Synthesis and characterization of fumed silica dispersed bees/carnauba wax-epoxy biocomposite coating: a comparative study. Biomass Convers Biorefinery 13:7375–7383. https://doi.org/10.1007/s13399-023-04112-0
60. Bai L, Lin Y, Chen X, Yin H, Jin C, Wang Y, Zhang Z, Peng L, Liang X, Cao Y (2024) Achieving high-performance polymer-wrapper-free aligned carbon nanotube field-effect transistors through degradable polymer wrapping and efficient removal techniques. ACS Nano 18:23392–23402. https://doi.org/10.1021/acsnano.4c06700
61. Panin SV, Alexenko VO, Buslovich DG (2022) High performance polymer composites: a role of transfer films in ensuring tribological properties—a review. Polymers 14:975. https://doi.org/10.3390/polym14050975

62. Mian SH, Nasr EA, Moiduddin K, Saleh M, Alkhalefah H (2024) An insight into the characteristics of 3D printed polymer materials for orthoses applications: experimental study. Polymers 16:403. https://doi.org/10.3390/polym16030403
63. Rajhi AA (2022) Mechanical characterization of hybrid Nano-filled glass/epoxy composites. Polymers 14:4852. https://doi.org/10.3390/polym14224852
64. Peng HX (2011) Polyurethane nanocomposite coatings for aeronautical applications. In: Leng J, Lau AKT (eds) Multifunctional polymer nanocomposites. CRC Press, Boca Raton, pp 337–387. https://doi.org/10.1201/b10462-9
65. Mahieux CA (2006) Environmental degradation in industrial composites. Elsevier, Chicago
66. Martin R (2008) Ageing of composites. Woodhead Publishing, Cambridge
67. Bafakeeh OT, Shewakh WM, Abu-Oqail A, Abd-Elaziem W, Abdel Ghafaar M, Abu-Okail. M (2021) Synthesis and characterization of hybrid Fiber-reinforced polymer by adding ceramic nanoparticles for aeronautical structural applications. Polymers 13:4116. https://doi.org/10.3390/polym13234116
68. Dinca I, Ban C, Stefan A, Pelin G (2012) Nanocomposites as advanced materials for aerospace industry. INCAS Bull 4:73–83. https://doi.org/10.13111/2066-8201.2012.4.4.6
69. Meenakshi KS, Sudhan EPJ, Kumar SA, Umapathy MJ (2011) Development and characterization of novel DOPO based phosphorus tetraglycidyl epoxy nanocomposites for aerospace applications. Prog Org Coat 72:402–409. https://doi.org/10.1016/j.porgcoat.2011.05.013
70. Reichherzer A, Norrefeldt V, Müller B, Straninger D, Herbig B, Euhus A, Bellmann M, Peter Vink Y, Song NM (2025) Evaluation of an on-ground regional passenger cabin demonstrator. Eng Proc 90:100. https://doi.org/10.3390/engproc2025090100
71. Vīndedze E, Glaskova-Kuzmina T, Dejus D, Jātnieks J, Sevcik S, Bute I, Sevcenko J, Stankevich S, Gaidukovs S (2025) Effects of printing orientation on the tensile, Thermophysical, smoke density, and toxicity properties of Ultem® 9085. Polymers 17:145. https://doi.org/10.3390/polym17020145
72. Kourtides DA (1983) Fire resistant films for aircraft applications. J Fire Sci 1:200–220. https://doi.org/10.1177/073490418300100306
73. Luiza de Castro F, Rezende MC, Multilayer radar absorbing material processing by using polymeric nonwoven and conducting polymer, Mater Res 11 (2008) 245–249. https://doi.org/10.1590/S1516-14392008000300003
74. Proietti A, Rinaldi A, Tamburrano A, De Bellis G, Sarto MS (2016) Wideband radar absorbing panels with lossy multilayer graphene and carbon nanofiber-based coating. In: IEEE international symposium on electromagnetic compatibility (EMC), pp 239–244. https://doi.org/10.1109/ISEMC.2016.7571651
75. Kim S-H, Lee S-Y, Zhang Y, Park S-J, Junwei G (2023) Carbon-based radar absorbing materials toward stealth technologies. Adv Sci 10:2303104. https://doi.org/10.1002/advs.202303104
76. Jayalakshmi CG, Inamdar A, Anand A, Kandasubramanian B (2019) Polymer matrix composites as broadband radar absorbing structures for stealth aircrafts. J Appl Polym Sci 136:47241. https://doi.org/10.1002/app.47241
77. Zhang H, Zhao P, Li J, Cao J (2024) Impact response characteristics and damage mechanism of continuous carbon-fiber-reinforced magnesium matrix composite materials. J Reinf Plast Compos. https://doi.org/10.1177/07316844241256157
78. Ghosh M (2018) Polyimides: fundamentals and applications. CRC press, p 912. https://doi.org/10.1201/9780203742945

Chapter 3
Mechanical Behavior for Hybrid Metal Nanoparticles-Based Thermoplastic Laminates for Aeronautical Applications

Mrinal, Neha Govind, and Anu Radha Pathania

Abstract The creation of hybrid metal nanoparticle-based thermoplastic laminates represents a significant leap in materials research, especially for aeronautical applications. These laminates include nanoparticles such as silver, copper, or aluminum into thermoplastic matrices, providing increased mechanical properties like as tensile strength, toughness, and durability while retaining the lightweight features required for aviation. The introduction of nanoparticles improves load transfer efficiency, reduces micro-crack propagation, and increases fatigue and impact stress resistance, all of which are critical for aircraft structural components. Furthermore, the adjustable thermal and electrical properties of metal nanoparticles aid in effective heat dissipation and electromagnetic interference (EMI) shielding, hence improving the safety and efficiency of modern airplanes. This research examines recent advances in hybrid metal nanoparticle-reinforced thermoplastic laminates, including their mechanical properties, production processes, and possible applications in aerospace. The findings show that these advanced composites have the potential to make aeronautical structures lighter, stronger, and more reliable, hence improving overall aircraft performance and sustainability. This abstract discusses recent advances in the creation of hybrid metal nanoparticle-based thermoplastic laminates, with an emphasis on their mechanical performance under various loading circumstances. Key aspects influencing mechanical performance, such as nanoparticle concentration, distribution, and matrix compatibility, are examined, as well as fabrication techniques such as compression molding, injection molding, and additive manufacturing that have been modified for these hybrid materials.

Keyword Hybrid · Thermoplastic · Aeronautical · Electromagnetic · Sustainability · Fabrication · Mechanical performance

Mrinal · N. Govind · A. R. Pathania (✉)
Department of Chemistry, University Institute of Sciences, Chandigarh University, Gharuan, Punjab, India
e-mail: anuthakur2014@gmail.com; anu.radha40256@paruluniversity.ac.in

A. R. Pathania
Department of Chemical Sciences, Parul Institute of Applied Sciences, Parul University, Limda, Vadodara, Gujrat, India

S. K. Panda (ed.), *Nanocomposites for Defense, Aeronautical and Aerospace Applications*, Advanced Structured Materials 248,
https://doi.org/10.1007/978-3-032-11568-3_3

3.1 Introduction

Nanoparticles are essential materials, despite their nanoscale size. Their spectrum of nanoscale sizes allows the relationships between different constituents to be identified. The size of the nanoparticle determines the categories into which the nanocomposite materials are separated. Because of the material's dimensional properties, the link between volume and surface area is especially important. Nanomaterials play an important role in the field of contemporary nanomaterials and are becoming increasingly popular among researchers due to their degradability, biocompatibility, and economic benefits [1]. Polymer-oriented nanoparticles are regarded biodegradable and biocompatible, and their fascinating features make them viable choices for a variety of applications (shown in Fig. 3.1). Materials have distinct characteristics at the nanoscale because of their higher surface-to-volume ratio [2]. The material's reactivity and interaction with its surroundings are improved by its large surface area, which is advantageous for uses including sensing and catalysis. Additionally, quantum phenomena seen in bulk forms are absent in nanomaterials. For example, size-dependent electrical and optical characteristics of semiconductor nanoparticles, often known as quantum dots, enable customizable performance. Several sectors are changing due to nanomaterials. They are employed in medicine for imaging, diagnosis, and medication delivery [3]. They improve transistor and sensor performance in electronics. They enhance the energy efficiency of solar cells and batteries and are essential for environmental uses including air and water filtering. Top-down techniques that reduce bigger materials into nanoscale particles, such lithography and ball milling, can be used to create nanomaterials. On the other hand, atomic or molecular precursors are used in bottom-up techniques including chemical vapor deposition (CVD), sol-gel processing, and chemical reduction to create nanoparticles. Advanced materials known as nanoparticle composites mix nanoparticles with a matrix material to improve the materials' characteristics and functionality [4]. These composites provide materials with better mechanical, thermal, electrical, and chemical properties than conventional composites by utilizing the special qualities of nanoparticles, such as their large surface area and quantum effects.

Nanocomposites made with metallic, nonmetallic, and polymer-oriented materials offer amazing advantages, like surface deformities and other material properties [5]. These composites are unique due to their increased surface area, significant mechanical properties, resistance to wear and scratches, and improved optical properties. Dimensional quantities, quality, resemblance to the microstructural lattice, and the dispersion of nanomaterials are the characteristics that determine the selection of nanocomposites. Previous studies have discovered that evenly distributed oxide components in nanoparticles enhanced the mechanical characteristics, thermal stability, and fire resistance of the composites over micro-composites [6]. Nanoparticles play a significant influence in influencing the molecular structural properties of the polymer matrix. They have a greater impact on the molecular structure of the material, which directly determines its behavior. Nanoparticle scattering might

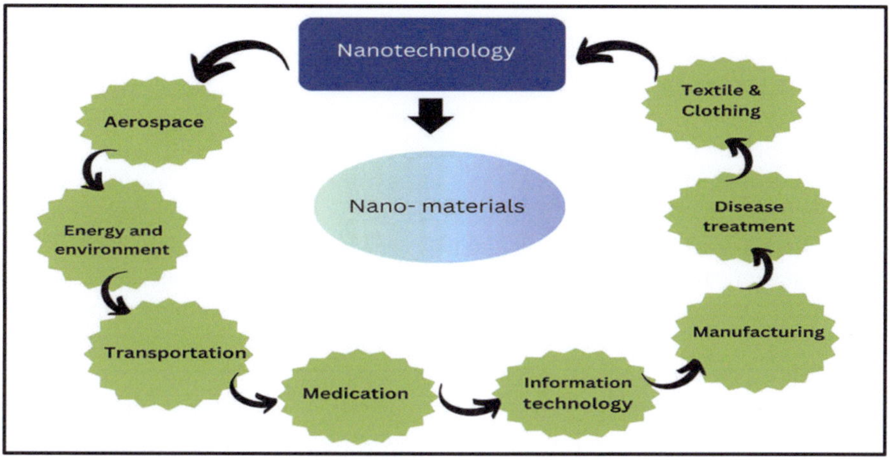

Fig. 3.1 Various application of nanomaterials

be regarded as a very fine phase at the nano level. The morphological characteristics of a substance helps explain why nanoparticles accumulate on its surface [7]. By improving mechanical qualities, thermal stability, fatigue and creep resistance, wear and corrosion resistance, and offering novel functions, nanoparticles have the potential to revolutionize the aerospace sector. The development of next-generation aerospace technologies, which will result in safer, more effective, and longer-lasting airplanes and spacecraft, depends heavily on their capacity to increase material performance while reducing weight [8]. Composite materials' mechanical strength and stiffness can be significantly increased by adding nanoparticles, such as metal or carbon nanotubes. These nanoparticles strengthen the material at the microscopic level when incorporated into matrices like metals or polymers, creating composites that are stronger and lighter. This is especially helpful for aeronautical components, where increasing performance and fuel efficiency require reducing weight without sacrificing strength. Adding nanoparticles can also increase a material's toughness, increasing its resistance to damage and impact. For example, adding nanoparticles to aerospace composites can improve their ability to absorb and release impact energy, lowering the risk of catastrophic failure and raising the safety [9]. By improving a material's thermal stability, nanoparticles help it endure the high temperatures seen in aerospace applications. Ceramic nanoparticles, for instance, can be utilized to strengthen the thermal resistance of aircraft parts that are subjected to high temperatures in engines or during atmospheric re-entry. This preserves the structural integrity of important components and aids in the prevention of thermal damage.

In certain applications, nanoparticles can also be designed to increase thermal conductivity. High thermal conductivity nanoparticles, for example, can be utilized in thermal management systems to effectively dissipate heat and prevent the overheating of delicate components [10]. Nanoparticles help materials have better fatigue resistance, which is important for materials that are subjected to cyclic loads. Because operating circumstances vary often, aerospace components are subjected to repetitive stress. Nanoparticles assist to distribute these stresses more evenly throughout the material, lowering the possibility of fatigue-related failure. Another property of nanoparticles is their ability to increase a material's resistance to gradual deformation over time under continuous tension. This feature is especially crucial for high-temperature applications where materials must withstand extended loads. Nanoparticles contribute to the long-term preservation of the material's mechanical characteristics and shape by strengthening the matrix [11]. Materials that are subjected to friction and abrasion need to have improved wear resistance, which is what nanoparticles do. For instance, incorporating ceramic nanoparticles into composites or coatings can assist minimize wear and increase the longevity of moving components like landing gear and engines. Aerospace components are shielded against deterioration by exposure to chemicals, moisture, and other corrosive substances using nanoparticles, which can help improve corrosion resistance. In hostile conditions, this is essential for preserving the integrity of the components [12]. Materials with self-healing properties that automatically fix small damages can be developed thanks to nanoparticles. By using this technology, coatings and composites may automatically patch up fractures and other flaws, increasing the dependability and lifespan of aerospace parts [13].

Additionally, nanoparticles can be used to make anti-icing coatings, which prevent ice from developing on critical surfaces such as engines and wings. In cold weather, this improves operational effectiveness and safety. Nanoparticles play a significant role in improving the shielding properties against electromagnetic interference (EMI) that is necessary to safeguard sensitive electronics in aerospace systems from electromagnetic interference [14]. The ability of nanoparticles to reduce material weight without sacrificing strength or durability is one of its most important benefits. This is important for aircraft applications since reduced weight results in higher payload capacities and better fuel economy per gram. Composites reinforced with nanoparticles enable the design of lightweight structures that function well under the demanding conditions of aeronautical operations [15].

The aerospace industry is very much interested in the materials that offer superior performance. Most commonly aluminum and titanium alloys have been the backbone of aeronautical structures for a long time due to their favorable strengths. However, the constant push for fuel efficiency and performance improvements has catalyzed the exploration of advanced materials, among which thermoplastic laminates reinforced with nanoparticles have emerged as promising candidates [16]. Material required for aerospace structure must have higher strength and rigidity to retain mechanical properties. The unique combination of composites makes their environment ideal to the space applications [17]. Thermoplastic composites are progressively being recognized for their potential in aerospace applications due to their unique characteristics. Unlike thermosetting polymers, thermoplastics can be reshaped and reprocessed, which offers significant advantages in terms of manufacturing flexibility and recyclability. Additionally, thermoplastic composites generally exhibit higher toughness and impact resistance, which are critical for the safety and durability of aerospace components [18].

Thermoplastic matrix composite laminates reinforced with continuous fibers have shown great promise as materials for present and future airplane components. Thermoplastic laminates are superior to thermoset fiber-reinforced composite laminates and offer several advantages. They are simpler to handle since they do not require complex chemical reactions or lengthy curing processes. Additionally, thermoplastic laminates are easily recyclable and do not need refrigeration for storage, resulting in virtually infinite shelf life [19]. Moreover, thermoplastic composites have extremely low amounts of moisture uptake, meaning their mechanical properties are less susceptible to degradation in a humid and warm environment. Furthermore, they offer higher damage tolerance characteristics and greater repairability potential compared to thermosetting matrix laminates. Unlike thermosetting resins, thermoplastics can be remelted after they are formed. They can also be joined using various assembly processes [20]. Joining is crucial in the manufacturing of composite structures in the marine, automotive, and aerospace industries. Mechanical fastening and adhesive bonding are commonly used to assemble both metal and composite components. The extensive research conducted in the field of composites over the past few years has greatly influenced the ability to meet the demands posed by the structural engineering community [21]. The two main categories of polymer matrix composites are thermoset matrix composites and thermoplastic matrix composites and elastomers can also include. Thermoset polymers go through a chemical curing process that causes polymer chains to cross-link, forming an often-irreversible network. Because of the durability that these cross-linked bonds give, thermosets do not melt or soften considerably when exposed to heat after they have cured [22]. In contrast, when heated, thermoplastic polymers will melt and soften. These polymers might be semi-crystalline, having sections of organized molecular chains, or amorphous, having randomly arranged molecular chains. While polymer matrix composites include a wide variety of reinforcement materials, continuous and discontinuous (chopped) fibers are the most used. Particles and flakes are two other types of reinforcements. It is possible to weave fabrics or arrange continuous fibers in a unidirectional strand configuration [23]. On the other hand, discontinuous fibers can be made with different lengths, widths, or diameters and are usually orientated arbitrarily inside the matrix. Glass, aramid, and carbon fibers are often utilized fiber types in polymer matrix composites. Furthermore, jute, flax, palm, and kenaf are examples of natural fibers that are utilized in certain applications as reinforcement materials [24]. The usefulness and adaptability of polymer matrix composites are enhanced by these materials.

Advanced composite materials called thermoplastic laminates are renowned for their superior mechanical strength, impact resistance, and thermal stability, which makes them perfect for high-performance uses in sectors like the automotive and aerospace industries [25]. These laminates, which are reinforced with glass or carbon fibers, have excellent fatigue resistance, a high strength-to-weight ratio, and resistance to environmental deterioration, chemicals, and moisture. Their capacity to be recycled and reshaped increases their manufacturing flexibility and sustainability, establishing them as a vital component for strong and lightweight parts in harsh situations [26]. Polymer matrix composites (shown in Fig. 3.2) and carbon-carbon matrix composites are the two categories into which organic matrix composites (OMCs) fall. Polymer composites are the most widely used and prevalent type of OMC. Polymeric composites are suitable for a variety of structures because to their favorable mechanical qualities, which include low weight, high stiffness, and high-temperature resistance. Although polymer composites have lesser strength and stiffness than metal and ceramic composites, this disadvantage is sometimes overcome by reinforcing them with additional materials [27]. Furthermore, polymer composites are processed at low temperatures and pressures, and the equipment required for manufacture is simple. As a result, researchers have been increasingly interested in polymer matrix composites for a variety of applications. Composites are desirable because their combined qualities outperform those of their individual components. Thermoplastic materials are solid at normal temperature and must be heated to soften or melt before shaping, after which they are cured to achieve the desired shape. These substances can be recycled. Not recyclable, thermosetting polymers, on the other hand, can exist in both liquid and solid phases. Thermoplastics are held together by covalent connections, but thermosetting polymers have cross-linked bonds, making them stronger and stiffer than thermoplastics [28]. Thermoplastic materials include polyamides, polyoxymethylene (POM), polycarbonates (PC), and acrylics. Thermosetting materials include epoxies (EP), thermoset polyamides, amino plastics, and polyester thermosets. Natural fibers are derived from animal (protein) and plant (cellulose) sources, as well as mineral-based resources. Natural fibers outperform other fibers in terms of physical and mechanical qualities [29]. Their advantages, including as lightweight, high strength, low cost, and corrosion resistance, make them more appealing for industrial applications. However, natural fibers have some disadvantages, such as high-water absorption, high anisotropy, less homogeneity than synthetic fibers, and the creation of tensions between fibers in a composite. The chemical composition of fiber materials plays a crucial role in their properties and varies depending on the fiber source. This composition differs not only between different types of fibers but also within the same family [30]. The performance of these fibers is largely dependent on their aspect ratio and cellulose crystallinity. Previous studies have indicated that these fibers are biodegradable due to their chemical constituents. For example, UV degradation caused by lignin and hemicellulose contributes to thermal and biodegradation, as well as water absorption [31].

The main components of natural fibers are cellulose, hemicellulose, and lignin. The percentages of these components vary based on development conditions. Understanding the behavior of these fibers requires examining the structure of their components. Cellulose, one of the primary building blocks of natural fibers, constitutes about 50% of plants and impacts moisture absorption [32]. Hemicellulose, the second major component, follows cellulose and is less significant as a filler material. Lignin, composed of phenylpropane derivatives, is an amorphous natural polymer that controls fluid transport and binds hard cells in the plant. Acting as a cementing material, lignin influences the structure, properties, and morphology of natural fibers. Despite having several advantages over other modern fibers, natural fibers are not without flaws. They have strong polar connections, which can affect their bonding ability inside a polymer matrix. To address this condition, chemical treatment is

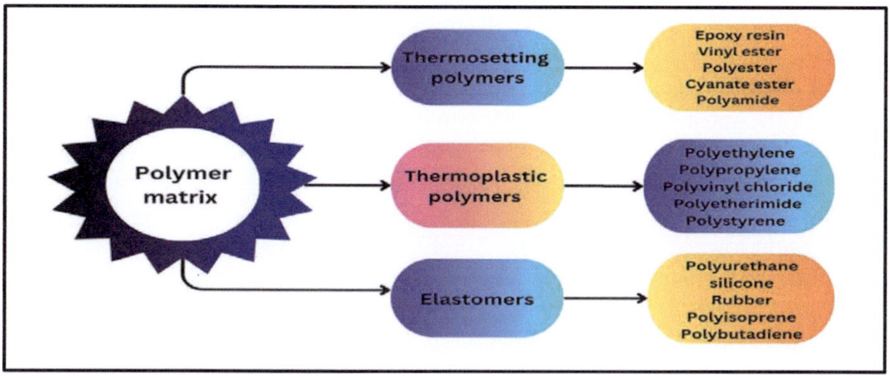

Fig. 3.2 Classification of different polymer matrix

frequently required. This treatment strengthens the composite material by increasing the contact between the fiber and the matrix material during the curing process. Furthermore, it improves fiber surface quality and decreases moisture absorption by the composite material [33]. Numerous man-made or synthetic fibers have been introduced into contemporary research networks due to their exceptional qualities. Glass and carbon fibers are the most utilized fillers in hybrid composites. These fibers are made using polymerization techniques, which include connecting monomers to produce a large chain [34].

In aerospace applications, most engineering composite materials consist of continuous carbon or glass fiber reinforcing a polymeric matrix. Because of its excellent strength-to-weight ratio and ability to improve mechanical and thermal properties by changing its form, orientation, and composition, carbon fiber is frequently utilized to create composites for high-end applications [35]. Carbon fiber is distinguished by its exceptional strength, high modulus, low thermal expansion, superior creep resistance, and resistance to corrosion. Currently, carbon fiber-reinforced plastics are increasingly replacing metallic components in high-temperature and corrosion-resistant applications. Among the composites used in aerospace and structural applications, carbon fiber-reinforced epoxies are preferred for their properties such as high failure strength, low creep, high modulus, and relatively low cost [36]. However, epoxy-based composites have notable drawbacks, including brittleness, poor resistance to crack initiation and growth, high moisture absorption, and limited shelf life. Thermosetting composites face limitations regarding storage, hydrothermal aging, insufficient toughness, and processing constraints due to lengthy and strict multistep processing requirements [37]. Hybrid nanocomposites combine multiple types of nanoparticles within a single matrix to exploit the synergistic effects of different reinforcing materials. This approach can lead to materials with tailored properties that are not achievable with single-type nanoparticle reinforcements. For instance, combining carbon nanotubes with metal nanoparticles can result in a composite that benefits from the high strength and electrical conductivity of carbon nanotubes and the mechanical reinforcement provided by metal nanoparticles [38].

As a result of the development of energy-saving tactics and emission reduction rules, automakers have looked into various ways to satisfy consumer demands and fuel economy requirements without sacrificing performance. Reducing weight is one of the most effective measures to reduce energy consumption and emissions, especially CO_2 emissions [39]. Both fiber-reinforced polymer composites (FRP) and advanced metal alloys are competing materials for automobiles that have the ability to save weight without sacrificing vehicle performance or safety, which would lower the overall cost. Automakers aim to replace traditional metals with innovative materials to achieve tight performance and weight requirements, resulting in improved fuel efficiency and lower production costs without compromising safety [40]. Natural fiber polymer composites are continual, renewable and offer several interests, including low cost, low density, biodegradability, improved thermal stability, tool wear resistance, and fewer health issues. Although biofiber composites are more environmentally friendly and economically favorable compared to traditional synthetic fiber composites. These composites also face challenges when using untreated fibers, which can lead to increased moisture absorption and poor thermal properties, resulting in issues such as weak matrix-fiber adhesion and uneven fiber distribution during composite fabrication [41].

Natural fibers are classified based on their sources, which can be plants, animals, or minerals. The characteristics of hybrid composites depend on their individual sustainable components and the fiber/matrix interfacial bonding. Natural cellulosic plant fibers in these composites range in size from micro to centimeters [42]. Synthetic fibers possess superior mechanical properties and damping resistance, whereas natural fibers do not. Hybrid composites (HCs) are multifunctional materials used in advanced structural applications where multiple benefits from constituent materials are needed. HCs are favorable and well-designed materials that incorporate one type of filler or reinforcement in a blend of two similar or dissimilar polymer matrices more than one filler in a single polymer resin, or a combination of these methods. The properties of HCs are influenced by the weight, geometry, and specific applications in various structural engineering components [43].

Hybrid fiber metal laminates (FML) mechanical performance remains a serious challenge in automotive practice applications due to their interaction and interfacial bonding behavior (shown in Fig. 3.3). It becomes more complicated due to the diversity of material selection, the designability of structures, the selectivity of production processes, etc. As a result, additional information is required for the desired applications through ongoing research of many elements that influence [44]. FMLs for the automobile industry must be able to handle unexpected challenges including drop-offs, vibration, and collision. However, FMLs' interfacial bonding limits their potential applications. To lengthen the service life of FMLs, researchers have focused on improving interfacial bonding. Previous research suggests that strengthening the interfacial bonding of FMLs through fiber treatment and metal surface modification improves mechanical performance [45].

When compared to a full-metal material, FMLs in the automobile industry have the ability to achieve comparable structural qualities at a weight reduction of around 25%. They are also ideal for mass production. For the automobile industry, researchers have

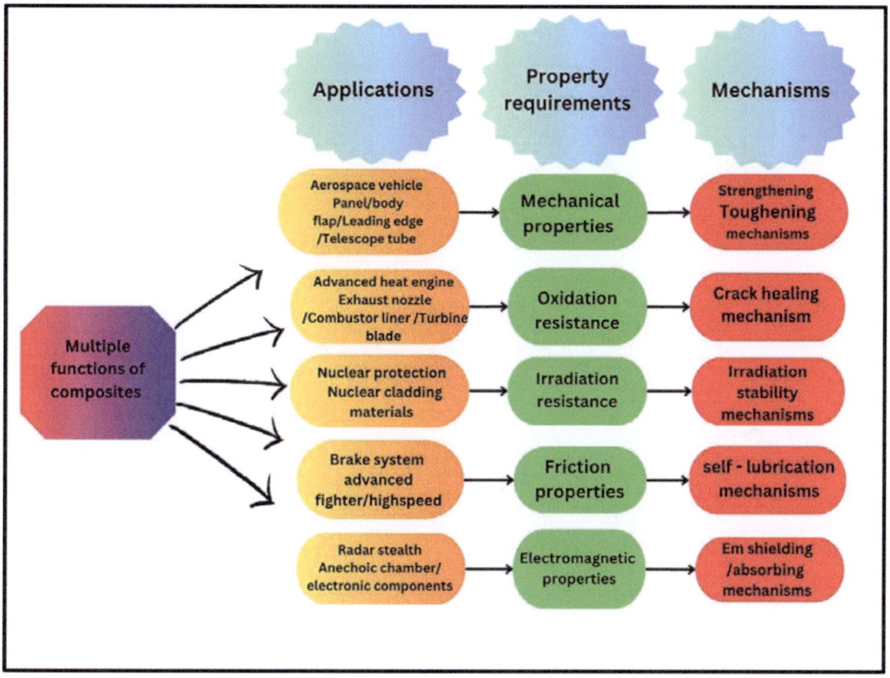

Fig. 3.3 Multiple functions of hybrid metal composites

been investigating new materials, cutting-edge processing techniques, and unique FML architectures nonstop in recent years. Nanocomposites are gaining popularity among researchers in the applied materials science sector as a potential solution to classic composite difficulties. Nanocomposites are created by combining a polymeric matrix and nanosized fillers to improve material properties. The presence of nanoparticles in nanocomposites alters their chemical, electrical, mechanical, and physical properties significantly. Nanoparticles added to a conventional laminate changed its behavior but did not change its mass. Hybrid nanocomposites use nanofiller in biofiber-reinforced polymer composites [46].

Vehicle designers must understand composite fabrication techniques and their impact on structural development. Hybrid composite has a unique and typical working mechanism (shown in Fig. 3.4). Designers must understand how fiber reinforcements disperse within a composite after production. Initially, there was a significant disparity between vehicle designers' and composite manufacturers' needs, leading to material waste and manufacturing delays.

Fiber metal laminates (FMLs) are hybrid composites composed of metal and fiber-reinforced polymer layers. Many academics have studied how adding nanoparticles to polymeric composites can improve their mechanical properties. However, the impact of nanoparticles on the mechanical properties of FMLs and adhesively bonded joints (ABJs) between metal sheets and polymers is still not fully known. We gathered

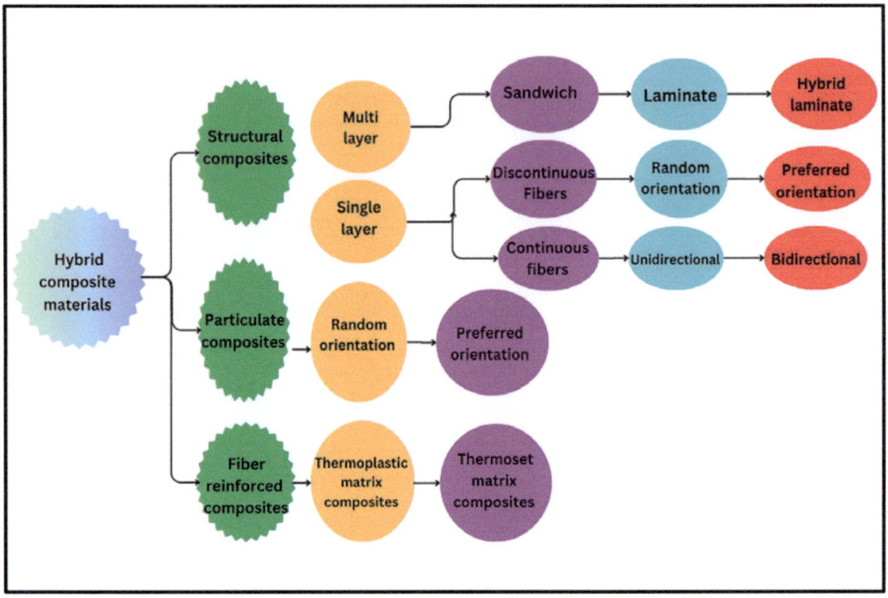

Fig. 3.4 Working mechanism of hybrid composite materials

recent works on the impact of nanoparticles on the properties of FMLs and ABJs between metal sheets and polymers, drawing on our expertise in the subject [47]. Most studies have focused on the effects of carbon nanotubes (CNTs) on the characteristics of FMLs and ABJs produced between metal sheets. Some researchers have examined the impact of various characteristics on FMLs and adhesive joints reinforced with CNTs, including metal sheet surface shape and specimen size. However, there is limited research on the characteristics of FMLs and ABJs between metal sheets when other nanofillers are added. Most studies only focus on the concentration of nanofillers [48]. Other nanoparticles, including as graphene, clay, TiO_2, $Al2O_3$, SiO_2 nanoparticles, and so on, must be studied to see how they affect the mechanical behavior of FMLs and ABJs between metal sheets. Until far, the effectiveness of nanofillers in enhancing the characteristics of FMLs and ABJs between metal sheets has only been studied at room temperature and in dry conditions. Future studies should examine the effects of hygrothermal aging and thermal/cooling settings on the mechanical characteristics of FMLs and ABBs reinforced with nanofillers [49].

3.1.1 Preparation Methodology

To improve material qualities for high-performance applications, metallic nanoparticles and thermoplastic matrices are integrated in a complex process called hybrid metal nanoparticle-based thermoplastic laminate preparation. The first step in the process is material selection. Based on the required mechanical, thermal, and chemical qualities, a suitable thermoplastic polymer, such as polycarbonate (PC), polyamide (PA), polyetheretherketone (PEEK), or polyimide (PI), is selected [50]. The specific enhancements required, such as increased conductivity, strength, or catalytic capabilities, are taken into consideration when choosing metal nanoparticles, such as those made of silver, gold, copper, or metallic oxides (such as titanium dioxide or zinc oxide). The following stage entails creating the metal nanoparticles using either physical or chemical techniques, such as liquid laser ablation or chemical reduction or sol-gel processing. To make sure the nanoparticles have the necessary size, shape, and crystalline structure, methods such as transmission electron microscopy (TEM), scanning electron microscopy (SEM), and X-ray diffraction (XRD) are employed to explain the particles [51]. After production, a uniform distribution of the metal nanoparticles is achieved by dispersing them into a solvent to form a nanoparticle dispersion using methods like ultrasonic sonication. To get a uniform blend of nanoparticles inside the polymer matrix, this dispersion is subsequently combined with a solution of the thermoplastic polymer dissolved in an appropriate solvent. Casting the nanoparticle-polymer solution onto a mold or flat surface to generate thin films is the first step in the laminate preparation process. Then, using thermal processing techniques like hot pressing or compression molding, these films are stacked and consolidated [52]. The temperature is carefully regulated to guarantee correct bonding without deteriorating the polymer or nanoparticles. Following manufacture, the laminated material is cooled, shaped, and tested using a variety of characterization methods. To verify the even dispersion of nanoparticles and assess the improvement in characteristics, mechanical testing (e.g., tensile, flexural, impact), thermal analysis (e.g., differential scanning calorimetry (DSC), thermogravimetric analysis (TGA)), and microscopy are used. To evaluate the laminate's performance in areas like temperature cycling, wear resistance, and electromagnetic interference (EMI) shielding, application-specific tests are carried out. Advanced materials with improved mechanical, thermal, and functional qualities are produced because of this thorough process, making them appropriate for demanding aerospace applications [53].

3.2 Hybrid Metal Nanoparticles

Thermoplastic laminates based on hybrid metal nanoparticles provide numerous benefits in aeronautical applications, including solving some of the trickiest problems in aerospace engineering. The mechanical strength, thermal management, corrosion resistance, electrical conductivity, and optical features of these laminates are improved by fusing the special qualities of metal nanoparticles with the flexibility and processability of thermoplastics. Because of these qualities, they are essential for creating next-generation aerospace materials that will make airplanes safer, more dependable, more efficient. Different metal laminates have different percentage of involvement in the production of components (shown in Fig. 3.5) [54].

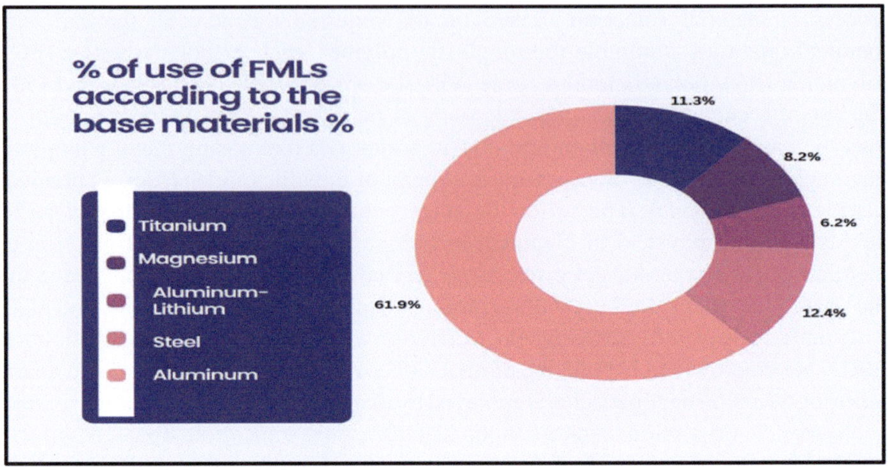

Fig. 3.5 Percentage of fiber metal laminates in formation of different materials

3.2.1 Silver Laminates

Based on silver nanoparticles laminates because of their special qualities, which include strong electrical conductivity, antibacterial activity, and outstanding reflectivity, silver nanoparticles are frequently employed in thermoplastic composites. These fiber metal laminates are very helpful for shielding against electromagnetic interference (EMI), conductive coatings, and flexible circuits in electronic applications. Silver's exceptional electrical conductivity, the highest among all metals, makes it particularly advantageous when incorporated into thermoplastic matrices [55]. The addition of silver nanoparticles can greatly improve the electrical conductivity of the composite, making it well-suited for electronic applications like conductive films, sensors, and circuits. Moreover, silver nanoparticles are highly effective as antimicrobial agents, offering robust protection against a broad spectrum of bacteria, viruses, and fungi. This makes them particularly valuable in applications where hygiene and sterility are essential, such as in medical devices and food packaging [56]. In addition to these properties, silver nanoparticles possess distinctive optical characteristics, including surface plasmon resonance (SPR). By adjusting the size and shape of the nanoparticles, these optical properties can be fine-tuned, making silver nanoparticles ideal for use in optical sensors and imaging technologies [57].

Since silver is the most conductive metal, silver nanoparticles are perfect for applications where electrical conductivity is crucial. This characteristic is used in the aerospace sector to create lightweight, highly conductive materials for electronic systems, sensors, and wiring. Conductive Coatings: To guarantee dependable electrical connections and signal integrity, coatings based on silver nanoparticles are applied to the surfaces and components of aircraft. EMI Electromagnetic interference (EMI) shielding materials, which shield delicate electronic equipment on airplanes from erratic electromagnetic fields, contain silver nanoparticles. This is especially crucial for avionics, where shielding is required to avoid interfering with systems that are used for navigation, communication, and control [58]. Effective heat management is essential in aircraft applications to avoid overheating of electrical systems and components. Because of their excellent thermal conductivity, silver nanoparticles are essential for creating materials that dissipate heat effectively. They are included, for example, in thermally conductive adhesives that facilitate heat transmission while bonding components together, keeping electrical equipment in airplanes from overheating. Furthermore, thermal interface materials (TIMs), which are positioned between heat-generating components and heat sinks to improve thermal conductivity and guarantee efficient heat dispersion, employ silver nanoparticles [59].

The potent antibacterial qualities of silver nanoparticles are widely recognized; they work well against a variety of bacteria, viruses, and fungi. This makes them very useful in the aerospace sector, where it is crucial to keep everything clean, notably in passenger cabins, crewed spacecraft, and medical transport aircraft. For instance, surfaces of medical equipment and airplane cabins are coated with coatings based on silver nanoparticles to prevent the growth and spread of dangerous bacteria. Moreover, air filtration systems employ silver nanoparticles to improve their capacity to eradicate or neutralize airborne infections, guaranteeing a safer atmosphere for travelers and workers. Surface plasmon resonance (SPR), one of the special optical characteristics of silver nanoparticles, is used in the creation of sophisticated sensors for structural health monitoring in airplanes [60]. These characteristics make it possible to develop optical sensors that are especially useful for monitoring the structural integrity of crucial aircraft parts. These sensors can identify even minute variations in strain, temperature, or pressure. Furthermore, early wear or damage in aircraft structures can be detected because to the great sensitivity of silver nanoparticle-based sensors, which makes maintenance much more efficient and lowers the chance of catastrophic failure [61]. Silver nanoparticles can be added to composite materials to increase electrical and thermal conductivity as well as improve mechanical characteristics including tensile strength and durability. These nanoparticles are employed in aerospace applications to strengthen polymer matrices and produce composite materials with electrical conductivity and mechanical strength. They are also used in multifunctional composites, which offer adaptable solutions for a range of aeronautical requirements by combining electrical or thermal conductivity with structural reinforcement [62].

3.2.2 Aluminum Laminates

Because of their great strength and light weight, aluminum laminates are used in the fabrication of many structural parts of aircraft. The lightweight and durable fuselage panels made possible using these laminates significantly reduce weight and increase fuel efficiency. Aluminum laminates are also employed in wing structures, where they strengthen structural integrity and improve aerodynamic performance, guaranteeing the aircraft's overall efficacy and safety. Heat shields and thermal protection systems use aluminum laminates to regulate high temperatures and protect delicate components [63]. These laminates provide strong protection from the extreme heat experienced during engine operations or atmospheric re-entry. Furthermore, aluminum laminates are employed in thermal barrier coatings to protect components from extreme heat, guaranteeing their longevity and peak functionality. Applications such as electronic systems and heat exchangers that demand efficient electrical and thermal conductivity use aluminum laminates. These laminates effectively divert heat from important parts and systems in heat exchangers. They are also ideal for use in cooling systems and electronic enclosures because to their high electrical and thermal conductivity, which guarantees the best possible performance and thermal management [64].

To achieve the best possible balance between strength and weight, lightweight composites composed of aluminum laminates are used in aeronautical engineering. These composites include panels made of aluminum laminate that are mixed with other materials to provide lightweight, robust panels that are utilized in a variety of aircraft constructions. Furthermore, sandwich structures employ aluminum laminates in conjunction with a core material to achieve improved mechanical performance and notable weight savings [65]. Excellent corrosion resistance is provided by aluminum laminates, which is crucial for preserving the integrity of aviation parts exposed to hostile conditions. They are employed in the production of corrosion-resistant panels that shield chemically and moisture-exposed sections from deterioration, prolonging the life of the constituent parts. Furthermore, aluminum laminates can be used to cover other materials to protect them and increase their resistance to corrosion [66].

3.2.3 Copper Laminates

Due to the excellent heat management capabilities, copper laminates are widely used in aircraft applications. Because of their excellent thermal conductivity, they can effectively drain heat from important parts, avoiding overheating and preserving system dependability. Here, heat sinks using copper laminates are used to absorb and transport heat from high-power electronic components, preventing thermal breakdowns and guaranteeing stable operation [67]. To increase thermal conductivity and facilitate heat dispersion, they are also utilized in thermal interface materials (TIMs), which are positioned between heat-generating components and cooling systems. Apart from its potential for managing heat, copper laminates play a critical role in guaranteeing dependable electrical connectivity in-flight systems. They provide a stable and conductive foundation for electronic components, making them perfect for use in printed circuit boards (PCBs) for aircraft electronics due to their superior electrical conductivity [68]. To ensure reliable and effective electrical connections, they are also used in the creation of electrical contacts and connectors. Lightweight composite materials also incorporate copper laminates, which provide a compromise between mechanical strength and decreased weight. These composites are employed in many structural applications where weight reduction and durability are essential [69]. For instance, composite panels that are lightweight and robust enough for use in airplane constructions are created by combining copper laminates with other materials. They are also employed in sandwich constructions, where the combination of core materials and copper layers improves mechanical performance and further reduces weight. Furthermore, copper laminates offer superior resistance to corrosion, a crucial attribute for preserving the integrity of aerospace parts subjected to severe environmental circumstances. They serve as coatings to stop underlying materials from corroding, especially in places where there is a lot of moisture and chemical exposure. Additionally, copper laminates safeguard numerous components, guaranteeing their long-term performance and durability [70].

3.2.4 Titanium Laminates

The aerospace engineering community is beginning to recognize titanium laminates more and more because to their exceptional properties, which include resistance to corrosion, thermal stability, and a high strength-to-weight ratio. These laminates, which are made of titanium and other materials alternating layers, have several advantages for different aerospace applications. Titanium laminates are used extensively in structural components because of their strength and light weight. They reduce total weight and greatly improve performance and durability. For instance, titanium laminates are utilized to make fuselage panels that are robust and lightweight, improving both structural integrity and fuel efficiency. They also increase the wing structures' longevity and performance, which makes them perfect for harsh aerospace conditions [71]. Titanium laminates are used in thermal management because of their exceptional resistance to high temperatures and thermal stability. In heat shields, they insulate aircraft parts from high temperatures—especially during high-speed or re-entry operations. Furthermore, titanium laminates are added to thermal barrier coatings to shield aircraft components from extremely hot conditions and lengthen their lifespan. Because of their exceptional resistance to corrosion, titanium laminates are a good choice for applications that are subjected to challenging environmental conditions. They are employed in the production of corrosion-resistant components, especially those that are exposed to chemicals or high levels of moisture. Moreover, titanium laminates are used as protective coatings to strengthen other materials' resistance to deterioration and corrosion [72]. Furthermore, well-known for their superior resistance to impact, titanium laminates are appropriate for uses involving strong impact forces. They improve safety for a variety of aircraft systems by serving as impact resistance in armor plates. Titanium laminates are also used in impact-resistant panels that must endure strong impacts to maintain the structural integrity of important parts [73].

3.3 Aeronautical Application of Hybrid Metal Nanoparticle-Based Thermoplastic Laminates

Hybrid metal nanoparticle-based thermoplastic laminates (HMNTLs) combine the best qualities of thermoplastic polymers and metal nanoparticles, making them especially useful for usage in airplanes (shown in Fig. 3.6). Below is a detailed discussion of the various uses for these laminates in the aerospace industry:

3.3.1 Fuselage Panels

1. High Strength-to-Weight Ratio:

Fuselage panels are an essential part of an aircraft's body, and the weight and fuel efficiency of the aircraft are directly impacted by their performance. Because HMNTLs are designed to have a high strength-to-weight ratio, they can bear a lot of strain and stress without gaining extra weight. This characteristic is essential for maximizing fuel efficiency without compromising the aircraft's structural integrity.

2. Durability:

Because of their resilience, HMNTLs are especially helpful in the fuselage, where they must withstand exposure to the elements, pressure variations, and temperature swings. The incorporation of metal nanoparticles into the thermoplastic matrix augments the laminate's resilience to abrasion, therefore prolonging the fuselage panels' lifespan [74].

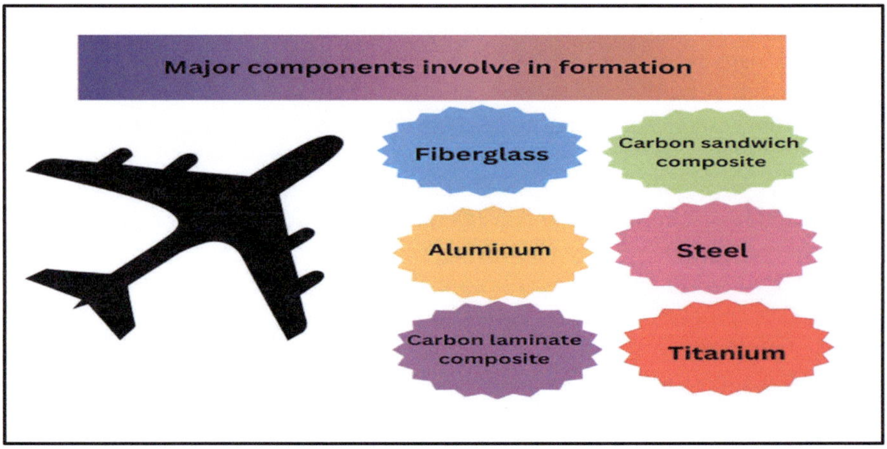

Fig. 3.6 Components involve in preparation of aerospace materials

3.3.2 *Impact Resistance*

1. Bird Strike Resistance:

In aircraft, bird strikes are a frequent hazard. One important benefit of using HMNTLs to reduce this risk is their impact resistance. By absorbing and dispersing impact energy, these laminates lessen the chance that crucial aircraft parts may sustain damage. This increase in impact resistance guarantees that structural integrity is preserved in unplanned crashes, which enhances aircraft safety overall.

2. Safety Enhancement:

The increased impact resistance also guards against additional in-flight dangers that could jeopardize the aircraft's safety, like hail and debris. One important component in increasing the overall safety of the aircraft is HMNTLs' capacity to absorb these impacts without suffering severe deformation or failure [75].

3.3.3 *Wing Structure*

1. Load Bearing:

The principal load-bearing components of an airplane are its wings, which are subjected to a great deal of stress during flight due to weight distribution and aerodynamic forces. Because of its superior fatigue resistance and high tensile strength, HMNTLs are perfect for use in wing structures. The wings will retain their structural integrity over time because to these laminates' ability to withstand the high loads and stresses involved in take-off, flight, and landing.

2. Fatigue Resistance:

The materials used in wing structures must be fatigue resistant because they will be loaded and unloaded repeatedly during flight. Because HMNTLs are more fatigue resistant than other materials, they can withstand repeated cycles without degrading, extending the service life of the wing structures [76].

3.3.4 *Thermal Management*

1. Enhanced Thermal Conductivity:

Aerodynamic friction and system functions, particularly at high altitudes, expose aircraft components to high temperatures. Because of HMNTLs' improved thermal conductivity, heat is dispersed more efficiently, reducing the risk of hotspots and structural damage. This characteristic keeps the aircraft's integrity intact by guaranteeing that its parts stay within safe temperature ranges.

2. Structural Integrity:

Especially in hot conditions, efficient thermal control is essential to preserving the aircraft's structural integrity. Because of HMNTLs' capacity to conduct and disperse heat, materials experience less thermal stress, which prevents thermal wear and extends the life of important components [77].

3.3.5 Interior Components

1. Weight Reduction:

A fundamental goal in aerospace design is to reduce an aircraft's overall weight, as this has a direct impact on fuel economy and operational costs. HMNTLs are naturally lightweight, making them ideal for interior components including panels, seats, and walls. The usage of these laminates allows for significant weight savings while maintaining the strength and durability required for interior constructions.

2. Fuel Efficiency:

The weight reduction achieved through the usage of HMNTLs helps to increase fuel efficiency because lighter aircraft require less energy to maintain flight. This reduction in fuel usage not only saves operational costs, but it also benefits the environment by lowering the aircraft's carbon imprint [78].

3.3.6 Engine Components

1. Heat Resistance:

Engine components are subjected to intense temperatures during operation, hence materials utilized in these areas must be highly heat resistant. HMNTLs are designed to endure high temperatures, making them acceptable for use in components near the engine compartment. The laminates' capacity to maintain structural integrity under heat stress ensures that engine components perform dependably even in the most extreme conditions.

2. Vibration Damping:

Engines cause large vibrations, which can impair the performance and longevity of adjacent components. HMNTLs offer intrinsic toughness that helps to dampen these vibrations, decreasing wear and tear on engine components. This vibration dampening capability improves the overall dependability and durability of the engine systems [79].

3.3.7 Shielding against Electromagnetic Interference (EMI)

1. Electrical Conductivity:

EMI can interfere with the operation of sensitive electronic equipment on board an aircraft. The metal nanoparticles in HMNTLs have great electrical conductivity, making these laminates ideal EMI shielding. These laminates avoid electromagnetic interference, ensuring the smooth operation of avionics and other essential electronic devices [80].

2. Lightning Strike Protection:

Lightning strikes on aircraft can cause structural components and onboard systems to be damaged. Because of their increased electrical conductivity, HMNTLs can better disperse the energy from lightning strikes, preserving both the aircraft's structure and its sensitive electronic equipment from harm [81].

3.3.8 Aerodynamic Surfaces

1. Smooth Surface Finishes:

Hybrid metal nanoparticle-based thermoplastic laminates can be made with smooth surface finishes, which benefit aerodynamic surfaces such as control surfaces and fairings by lowering drag and boosting fuel efficiency.

2. Shape Retention:

The high stiffness and strength of HMNTLs guarantee that aerodynamic surfaces retain their shape and performance characteristics across a wide range of flight conditions. This shape preservation is critical to maintaining the aircraft's aerodynamic efficiency and stability throughout its operating life [82].

3.3.9 Structural Health Monitoring

1. Embedded Sensors:

Metal nanoparticles found in HMNTLs can be used to embed sensors directly into laminates. These sensors allow for real-time monitoring of crucial components' structural health, including stress, strain, and potential flaws. This functionality enables for more efficient maintenance plans and improves overall safety by detecting faults before they cause component failure.

2. Efficient Maintenance:

Structural health monitoring with embedded sensors enables predictive maintenance, which addresses possible faults before they become significant. This method decreases downtime, lowers maintenance costs, and ensures that the aircraft performs optimally throughout its service life [83].

3.3.10 Thermal Insulation

1. Fire Resistance:

Fire resistance is an important safety property in aerospace materials, especially in components that are exposed to high temperatures or fire dangers. HMNTLs can be designed to provide increased fire resistance, lowering the danger of fire spread and protecting both the aircraft structure and its occupants in the case of a fire.

2. Thermal Insulation:

In addition to fire resistance, HMNTLs provide thermal insulation, which helps maintain acceptable cabin temperatures and protects temperature-sensitive components. This heat insulation capability is critical for guaranteeing passenger comfort and the safe operation of electronic devices [84].

3.4 Methodology for Manufacturing of Different Metal Composites for Aerospace Application

The process of shot peening forming (SPF) involves interacting with hard balls to cause localized sheet deformation. The kinetic energy of these balls' high-speed impacts with the FML surface is converted into a localized plastic strain in the material, altering the panel's overall curvature. Elastic-plastic properties characterize the balls' influence on the workpiece surface. The SPF technology was first applied to the formation of metallic sheets in aircraft. SPF was applied to the formation of fiber metal laminates (FMLs), which are innovative materials that combine metal and composite layers for increased strength and reduced weight, as the technique gained popularity and research yielded encouraging results.

X-ray computed tomography is another crucial instrument used in the aerospace production process (XCT). Since XCT is a non-destructive procedure, the substance being tested is not harmed. It is employed to track the whole production process in real time and identify flaws. By precise X-ray imaging, engineers may view inside

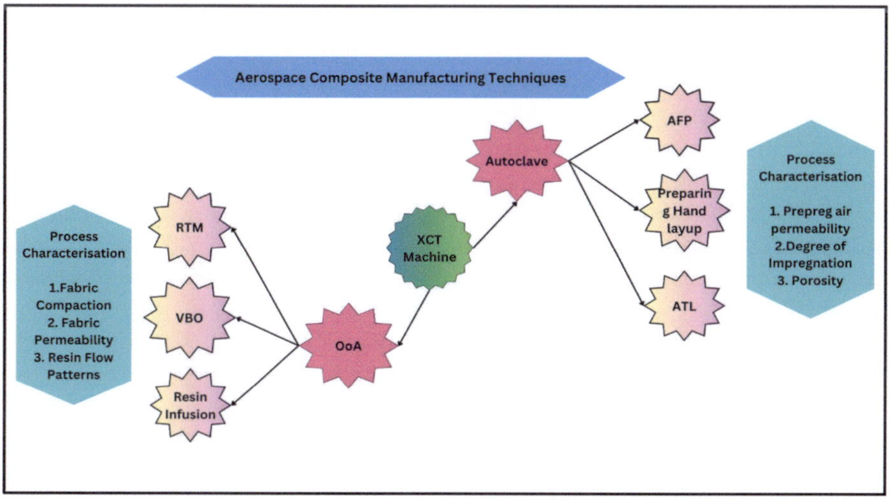

Fig. 3.7 Techniques of manufacturing metal-based composites

structures without having to break them open thanks to XCT technology. This technique is perfect for evaluating the quality of aerospace composites, which are materials created by combining various materials. It is also highly fast and exact [85]. Aerospace composites are manufactured using XCT, particularly in autoclave (a high-pressure, high-temperature oven) and out-of-autoclave techniques. They discovered that more effective production procedures for the upcoming generation of composite aircraft parts might result from fusing cutting-edge XCT modeling with existing aerospace manufacturing techniques. This implies that XCT may contribute to the development of increasingly stronger and more efficient airplane parts in the future. RTM is a process for producing laminates in closed forms that involves packing a tightly closed mold with sandwich material or reinforcing fibers before injecting catalyzed resin into it. Like the primary mold, the counter-mold is constructed. Its steel reinforcements and stiffeners enable the utilization of high injection pressure [86]. These techniques (shown in Fig. 3.7) are involved in manufacturing metal-based composites utilized in aerospace component designing.

3.5 Conclusion

In conclusion, hybrid metal nanoparticle-based thermoplastic laminates have enormous potential to enhance the mechanical properties of materials utilized in aviation applications. The addition of metal nanoparticles, such as silver, copper, or titanium, to thermoplastic matrices improves the material's mechanical properties, including tensile strength, impact resistance, and fatigue life. This improvement is mostly due to the homogeneous dispersion of nanoparticles, which act as reinforcement inside the matrix and aid in stress distribution under mechanical loads. Furthermore, hybrid composites have high thermal stability, making them ideal for the harsh thermal conditions seen in aircraft structures. Metal nanoparticles help to improve thermal conductivity, allowing for better heat management and lowering the risk of material failure due to overheating. This thermal attribute is critical for the continued operation of electrical systems and structural components in aviation conditions. The application of nanoparticles prolongs the material's life and dependability by strengthening the laminate's resistance to environmental deterioration, such as oxidation and corrosion. Nonetheless, maintaining uniform dispersion of nanoparticles and preventing agglomeration pose significant hurdles that may put in danger the performance of the laminate. All things considered; thermoplastic laminates based on hybrid metal nanoparticles offer an effective way for aviation engineers to create lightweight, high-performing materials. To fully realize the promise of nanoparticles for commercial aviation applications, future studies should concentrate on enhancing the distribution of the particles, increasing manufacturing cost-efficiency, and assessing their long-term durability.

References

1. Hassan T, Salam A, Khan A, Khan SU, Khanzada H, Wasim M, et al. (2021) Functional nanocomposites and their potential applications: a review. J Polym Res 28(2):36
2. Banik BL, Fattahi P, Brown JL (2016) Polymeric nanoparticles: the future of nanomedicine. Wiley Interdiscip Rev Nanomed Nanobiotechnol 8(2):271–299
3. Wang Y, Herron N (1991) Nanometer-sized semiconductor clusters: materials synthesis, quantum size effects, and photophysical properties. J Phys Chem 95(2):525–532
4. Tamirat Y (2017) The role of nanotechnology in semiconductor industry: review article. J Mater Sci Nanotechnol 5:202
5. Kashfipour MA, Mehra N, Zhu J (2018) A review on the role of interface in mechanical, thermal, and electrical properties of polymer composites. Adv Compos Hybrid Mater 1:415–439
6. Zhang H, Zhu H, Xu C, Li Y, Liu Q, Wang S, Yan S (2022) Effect of nanoparticle size on the mechanical properties of polymer nanocomposites. Polymer 252:124944
7. Ondreas F, Lepcio P, Zboncak M, Zarybnicka K, Govaert LE, Jancar J (2019) Effect of nanoparticle organization on molecular mobility and mechanical properties of polymer nanocomposites. Macromolecules 52(16):6250–6259
8. Ramdani N (ed) (2019) Nanotechnology in aerospace and structural mechanics. IGI Global
9. Ashraf MA, Peng W, Zare Y, Rhee KY (2018) Effects of size and aggregation/agglomeration of nanoparticles on the interfacial/interphase properties and tensile strength of polymer nanocomposites. Nanoscale Res Lett 13:1–7
10. Zhou K, Xie F, Wu X, Wang S (2019) Fabrication of high temperature oxidation resistance nanocomposite coatings on PEO treated TC21 alloy. Materials 13(1):11
11. Hassanzadeh-Aghdam MK, Ansari R, Mahmoodi MJ, Darvizeh A (2018) Effect of nanoparticle aggregation on the creep behavior of polymer nanocomposites. Compos Sci Technol 162:93–100
12. Gu Y, Xia K, Wu D, Mou J, Zheng S (2020) Technical characteristics and wear-resistant mechanism of nano coatings: a review. Coatings 10(3):233
13. Kausar A, Ahmad I, Maaza M, Bocchetta P (2023) Self-healing nanocomposites—advancements and aerospace applications. J Compos Sci 7(4):148

14. Xie Q, Hao T, Zhang J, Wang C, Zhang R, Qi H (2019) Anti-icing performance of a coating based on nano/microsilica particle-filled amino-terminated PDMS-modified epoxy. Coatings 9(12):771
15. Arepalli S, Moloney P (2015) Engineered nanomaterials in aerospace. MRS Bull 40(10):804–811
16. Trzepieciński T, Najm SM, Sbayti M, Belhadjsalah H, Szpunar M, Lemu HG (2021) New advances and future possibilities in forming technology of hybrid metal–polymer composites used in aerospace applications. J Compos Sci 5(8):217
17. Bhat A, Budholiya S, Aravind Raj S, Sultan MTH, Hui D, Md Shah AU, Safri SNA (2021) Review on nanocomposites based on aerospace applications. Nanotechnol Rev 10(1):237–253
18. Barile M, Lecce L, Iannone M, Pappadà S, Roberti P (2020) Thermoplastic composites for aerospace applications. In: Revolutionizing aircraft materials and processes. Springer International Publishing, Cham, pp 87–114
19. Costa APD, Botelho EC, Costa ML, Narita NE, Tarpani JR (2012) A review of welding technologies for thermoplastic composites in aerospace applications. J Aerosp Technol Manag 4:255–265
20. Hou M, Ye L, Mai YW (1997) Manufacturing process and mechanical properties of thermoplastic composite components. J Mater Process Technol 63(1–3):334–338
21. Alagesan PK (2020) Recent advances of hybrid fiber composites for various applications. In: Hybrid fiber composites: materials, manufacturing, process engineering, pp 381–404
22. Sudhin AU, Remanan M, Ajeesh G, Jayanarayanan K (2020) Comparison of properties of carbon fiber reinforced thermoplastic and thermosetting composites for aerospace applications. Mater Today Proc 24:453–462
23. Blumentritt BF, Vu BT, Cooper SL (1975) Mechanical properties of discontinuous fiber reinforced thermoplastics. II. Random-in-plane fiber orientation. Polym Eng Sci 15(6):428–436
24. Srinivas K, Naidu AL, Bahubalendruni MR (2017) A review on chemical and mechanical properties of natural fiber reinforced polymer composites. Int J Perform Eng 13(2):189
25. Xavier SF (ed) (2022) Thermoplastic polymer composites: processing, properties, performance, applications and recyclability. Wiley
26. Dharan CKH (1975) Fatigue failure in graphite fibre and glass fibre-polymer composites. J Mater Sci 10:1665–1670
27. Mallick PK (2007) Fiber-reinforced composites: materials, manufacturing, and design. CRC press
28. Lila MK, Komal U, Chaitanya S (2016) Secondary processing of polymer matrix composites: challenges and opportunities. In: International conference on latest development in material, manufacturing and quality control, MMQC-2016
29. Wang X, Xie J, Zhang H, Zhang W, An S, Chen S, Luo C (2016) Determining the lignin distribution in plant fiber cell walls based on chemical and biological methods. Cellulose 26:4241–4252
30. Praveena BA, Buradi A, Santhosh N, Vasu VK, Hatgundi J, Huliya D (2022) Study on characterization of mechanical, thermal properties, machinability and biodegradability of natural fiber reinforced polymer composites and its applications, recent developments and future potentials: a comprehensive review. Mater Today Proc 52:1255–1259
31. Wang X, Chang CH, Jiang J, Liu Q, Liao YP, Lu J, Li L, Liu X, Kim J, Ahmed A, Nel AE (2019) The crystallinity and aspect ratio of cellulose nanomaterials determine their pro-inflammatory and immune adjuvant effects in vitro and in vivo. Small 15(42):1901642
32. Chokshi S, Parmar V, Gohil P, Chaudhary V (2022) Chemical composition and mechanical properties of natural fibers. J Nat Fibers 19(10):3942–3953
33. Nurazzi NM, Asyraf MRM, Fatimah Athiyah S, Shazleen SS, Rafiqah SA, Harussani MM, Kamarudin SH, Razman MR, Rahmah M, Zainudin ES, Ilyas RA (2021) A review on mechanical performance of hybrid natural fiber polymer composites for structural applications. Polymers 13(13): 2170
34. Rajpurohit A, Joannès S, Singery V, Sanial P, Laiarinandrasana L (2020) Hybrid effect in in-plane loading of carbon/glass fibre based inter-and intraply hybrid composites. J Compos Sci 4(1):6

35. Mrazova M (2013) Advanced composite materials of the future in aerospace industry. Incas Bull 5(3):139
36. Ahmad F, Awadh MA, Abas M, Noor S, Hameed A (2022) Optimization of carbon fiber reinforced plastic curing parameters for aerospace application. Appl Sci 12(9):4307
37. Zimmerman MC, Scalzo HL, Parsons JR, Torop AH, Lin TS (1991) Effects of environmental exposure on carbon polysulphone composites. Biomaterials 12(4):424–430
38. Jin C, Wu Q, Yang G, Zhang H, Zhong Y (2021) Investigation on hybrid nanofluids based on carbon nanotubes filled with metal nanoparticles: stability, thermal conductivity, and viscosity. Powder Technol 389:1–10
39. Rüdele K, Wolf M (2023) Identification and reduction of product carbon footprints: case studies from the Austrian automotive supplier industry. Sustainability 15(20):14911
40. DjafariPetroudy SR (2017) 3-physical and mechanical properties of natural fibers. Adv High Strength Nat Fibre Compos Constr:59–83
41. Mohanty AK, Misra M, Hinrichsen G (2000) Biofibres, biodegradable polymers and biocomposites: an overview. Macromol Mater Eng 276:1–24
42. Gholampour A, Ozbakkaloglu T (2020) A review of natural fiber composites: properties, modification and processing techniques, characterization, applications. J Mater Sci 55:829–892
43. Mohd Bakhori SN, Hassan MZ, Mohd Bakhori N, Jamaludin KR, Ramlie F, Md Daud MY, Abdul Aziz SA (2022) Physical, mechanical and perforation resistance of natural-synthetic fiber interply laminate hybrid composites. Polymers 14(7):1322
44. Ding Z, Wang H, Luo J, Li N (2021) A review on forming technologies of fibre metal laminates. Int J Light Mater Manuf 4(1):110–126
45. Xiao H, Sultan MTH, Shahar FS, Gaff M, Hui D (2023) Recent developments in the mechanical properties of hybrid fiber metal laminates in the automotive industry: a review. Rev Adv Mater Sci 62(1):20220328
46. Khalid MY, Al Rashid A, Arif ZU, Sheikh MF, Arshad H, Nasir MA (2021) Tensile strength evaluation of glass/jute fibers reinforced composites: an experimental and numerical approach. Results Eng 10:100232
47. Eslami-Farsani R, Aghamohammadi H, Khalili SMR, Ebrahimnezhad-Khaljiri H, Jalali H (2022) Recent trend in developing advanced fiber metal laminates reinforced with nanoparticles: a review study. J Ind Text 51(5_suppl):7374S–7408S
48. Ebrahimnezhad-Khaljiri H, Eslami-Farsani R, Banaie KA (2017) The evaluation of the thermal and mechanical properties of aramid/semi-carbon fibers hybrid composites. Fibers Polym 18:296–302
49. Saleh M, Tooski MY, Khorshidvand AR, Javadi M (2022) Experimental investigation on mechanical behavior of fiber–metal laminates reinforced by hybrid carbon nanotubes and nanosilica under quasi-static pressure load. Proc Inst Mech Eng C J Mech Eng Sci 236(13):7319–7329
50. Trautmann M, Mrzljak S, Walther F, Wagner G (2020) Mechanical properties of thermoplastic-based hybrid laminates with regard to layer structure and metal volume content. Meta 10(11):1430
51. Khan I, Saeed K, Khan I (2019) Nanoparticles: properties, applications and toxicities. Arab J Chem 12(7):908–931
52. Mbuya TO, Mwema FM, Shagwira H (2021) Lightweight polymer–nanoparticle-based composites. CRC Press
53. Barra G, Guadagno L, Raimondo M, Santonicola MG, Toto E, Vecchio Ciprioti S (2023) A comprehensive review on the thermal stability assessment of polymers and composites for aeronautics and space applications. Polymers 15(18):3786
54. Yang P (2023) Nanomaterials in aerospace: advancements, applications, and the path forward. Highlights Sci Eng Technol 73:116–121
55. Pal S, Tak YK, Song JM (2007) Does the antibacterial activity of silver nanoparticles depend on the shape of the nanoparticle? A study of the gram-negative bacterium Escherichia coli. Appl Environ Microbiol 73(6):1712–1720

56. Rauwel P, Rauwel E, Ferdov S, Singh MP (2015) Silver nanoparticles: synthesis, properties, and applications. Adv Mater Sci Eng 2015:1–2
57. Mayer KM, Hafner JH (2011) Localized surface plasmon resonance sensors. Chem Rev 111(6):3828–3857
58. Mishra RK, Thomas MG, Abraham J, Joseph K, Thomas S (2018) Electromagnetic interference shielding materials for aerospace application: a state of the art. In: Advanced materials for electromagnetic shielding: fundamentals, properties, and applications, pp 327–365
59. Yu H, Li L, Zhang Y (2012) Silver nanoparticle-based thermal interface materials with ultra-low thermal resistance for power electronics applications. Scr Mater 66(11):931–934
60. Rycenga M, Cobley CM, Zeng J, Li W, Moran CH, Zhang Q, Qin D, Xia Y (2011) Controlling the synthesis and assembly of silver nanostructures for plasmonic applications. Chem Rev 111(6):3669–3712
61. Marques C, Leal-Júnior A, Kumar S (2023) Multifunctional integration of optical fibers and nanomaterials for aircraft systems. Materials 16(4):1433
62. Dermanaki Farahani R, Gagne M, Klemberg-Sapieha JE, Therriault D (2016) Electrically conductive silver nanoparticles-filled nanocomposite materials as surface coatings of composite structures. Adv Eng Mater 18(7):1189–1199
63. Hombergsmeier E (2006, Sept) Development of advanced laminates for aircraft structures. In: Proceedings of 25th international congress of the aeronautical sciences, pp 3–8
64. Padture NP, Gell M, Jordan EH (2002) Thermal barrier coatings for gas-turbine engine applications. Science 296(5566):280–284
65. Reyes G, Kang H (2007) Mechanical behavior of lightweight thermoplastic fiber–metal laminates. J Mater Process Technol 186(1–3):284–290
66. Twite RL, Bierwagen GP (1998) Review of alternatives to chromate for corrosion protection of aluminum aerospace alloys. Prog Org Coat 33(2):91–100
67. Bui VP, Thitsartarn W, Liu EX, Chuan JYC, Chua EK (2015) EM performance of conductive composite laminate made of nanostructured materials for aerospace application. IEEE Trans Electromagn Compat 57(5):1139–1148
68. Raghunath CR, Raghunath U, Yeshaswini HS, Rashmi H (2013) Development of new laminates for high performance interconnection technology in avionics applications. Circuit World 39(1):22–30
69. Roudini G, Ghahfarokhi AR, Behzadmehr A (2017, May) Copper laminated composites reinforced with Al2O3 nanoparticles by suspension method and hot pressing. In: IOP conference series: materials science and engineering, vol 201, no 1, p 012026. IOP Publishing
70. Nasir MA, Isaac S, Pasha RA, Wakeel A, Rashid TU (2018) Mechanical characterization of copper nanoparticles-based glass fibre reinforced polymer composites. Tech J 23(02):35–39
71. Nassir NA, Birch RS, Cantwell WJ, Sierra DR, Edwardson SP, Dearden G, Guan ZW (2020) Experimental and numerical characterization of titanium-based fibre metal laminates. Compos Struct 245:112398
72. Miller JL, Progar DJ, Johnson WS, St. Clair, T. S. L. (1995) Preliminary evaluation of hybrid titanium composite laminates. J Adhes 54(1–4):223–240
73. Jakubczak P (2019) The impact behaviour of hybrid titanium glass laminates—experimental and numerical approach. Int J Mech Sci 159:58–73
74. Shagwira H, Mwema FM, Mbuya TO (2021) Lightweight polymer–nanoparticle-based composites: an overview. Nanomater Nanocomposites:31–50
75. Algarni A (2018) Enhancing the mechanical properties of aerospace fiber reinforced polymer composite materials using nanoparticles. King Abdulaziz University
76. Gibson RF (2010) A review of recent research on mechanics of multifunctional composite materials and structures. Compos Struct 92(12):2793–2810
77. Ergün RK, Adin H (2023) Investigation of effect of nanoparticle reinforcement woven composite materials on fatigue behaviors. Iran J Sci Technol Trans Mech Eng 47(2):729–740
78. Singha NR, Chattopadhyay PK, Karmakar M, Mondal H (2020) Hybrid thermoplastic and thermosetting composites. In: Lightweight polymer composite structures. CRC Press, pp 131–192

79. Pathak AK, Dhakate SR (2022) Carbon nanomaterial-carbon fiber hybrid composite for lightweight structural composites in the aerospace industry: synthesis, processing, and properties. In: Advanced composites in aerospace engineering applications. Springer International Publishing, Cham, pp 445–470
80. Kireitseu MV (2006, Jan) Damping properties/dynamics of nanoparticle-reinforced material. In: CANEUS: MNT for aerospace applications, vol 42541, pp 15–20
81. Hemath M, MavinkereRangappa S, Kushvaha V, Dhakal HN, Siengchin S (2020) A comprehensive review on mechanical, electromagnetic radiation shielding, and thermal conductivity of fibers/inorganic fillers reinforced hybrid polymer composites. Polym Compos 41(10):3940–3965
82. Kensrud JR (2010) Determining aerodynamic properties of sports balls in situ. Doctoral dissertation, Washington State University
83. Ferreira PM, Machado MA, Carvalho MS, Vidal C (2022) Embedded sensors for structural health monitoring: methodologies and applications review. Sensors 22(21):8320
84. Ali A, Andriyana A (2020) Properties of multifunctional composite materials based on nanomaterials: a review. RSC Adv 10(28):16390–16403
85. Balasubramanian K, Sultan MT, Rajeswari N (2018) Manufacturing techniques of composites for aerospace applications. In: Sustainable composites for aerospace applications. Woodhead Publishing, pp 55–67
86. Dogea R, Yan XT, Millar R (2022) Additive manufacturing process design for complex aircraft components. Int J Adv Manuf Technol 123(11):4195–4211

Chapter 4
Development and Characterization of Novel DOPO-Based Phosphorus Tetraglycidyl Epoxy Composites for Aerospace Applications

Kamla Rawat and Mukesh C. Joshi

Abstract This chapter delves into the development and characterization of innovative nanocomposites made from DOPO (9,10-dihydro-9-oxa-10-phosphaphenanthrene-10-oxide) and phosphorus tetraglycidyl epoxy for aerospace use. The research addresses the urgent need for advanced materials with improved fire resistance and mechanical properties in aerospace engineering. It starts by explaining the synthesis of phosphorus tetraglycidyl epoxy resin and its modification with DOPO flame retardant additives. Comprehensive characterization of these materials is performed using Fourier-transform infrared spectroscopy (FTIR), scanning electron microscopy (SEM), and thermogravimetric analysis (TGA) to reveal the structural changes and interactions within the nanocomposite system. The chapter then assesses the fire retardancy and thermal stability of the nanocomposites through cone calorimetry and differential scanning calorimetry (DSC), showing promising results with reduced flammability and enhanced thermal resistance compared to traditional epoxy resins. Mechanical tests, including tensile, flexural, and impact assessments, indicate that incorporating DOPO and phosphorus tetraglycidyl epoxy significantly boosts the nanocomposites' mechanical performance, making them viable for structural aerospace applications. In conclusion, this chapter not only demonstrates the successful synthesis and characterization of DOPO-based phosphorus tetraglycidyl epoxy nanocomposites but also highlights their potential as advanced materials for improving safety and performance in aerospace applications.

Keywords Novel DOPO · Phosphorus tetraglycidyl · Epoxy composites · Aerospace applications · Cone calorimetry · Differential scanning calorimetry ·

K. Rawat
Department of Chemistry, School of Chemical and Life Sciences, Jamia Hamdard, New Delhi, India

M. C. Joshi (✉)
Department of Chemistry, Kirori Mal College University of Delhi, Delhi, India
e-mail: mukeshjoshi21@kmc.du.ac.in

S. K. Panda (ed.), *Nanocomposites for Defense, Aeronautical and Aerospace Applications*, Advanced Structured Materials 248,
https://doi.org/10.1007/978-3-032-11568-3_4

Flammability · Thermogravimetric analysis · Fourier-transform infrared spectroscopy

4.1 Introduction

4.1.1 Overview of Epoxy Resins in Aerospace Applications

Epoxy resins are widely recognized for their exceptional mechanical properties, chemical resistance, and ease of processing, making them a preferred choice in the aerospace industry [1, 2]. These thermosetting polymers are commonly used in the fabrication of composite materials, coatings, adhesives, and structural components. The versatility of epoxy resins allows them to be tailored for specific applications, providing the necessary strength and durability required in aerospace engineering [3, 4]. Their ability to form strong adhesive bonds and maintain structural integrity under extreme conditions makes them indispensable in the construction of aircraft, spacecraft, and related components.

4.1.2 Importance of Fire Resistance and Mechanical Properties

In aerospace applications, the importance of fire resistance cannot be overstated. Aircraft and spacecraft are subjected to stringent safety regulations to prevent onboard fires, which can have catastrophic consequences. Materials used in these environments must exhibit high thermal stability and low flammability to enhance passenger and crew safety [5, 6]. Additionally, they should produce minimal smoke and toxic gases during combustion to improve survivability in case of a fire incident.

Alongside fire resistance, the mechanical properties of aerospace materials are crucial. These materials must support significant mechanical loads and resist impacts, fatigue, and harsh environmental conditions throughout their service life. High tensile strength, flexibility, and impact resistance are essential to withstand the rigorous demands of flight [7, 8]. Furthermore, lightweight materials are preferred to enhance fuel efficiency, performance, and payload capacity. The dual requirement for superior fire resistance and mechanical robustness necessitates the development of advanced composite materials that can meet these stringent criteria.

Conventional fire retardant materials exhibit several significant drawbacks that emphasize the need for better alternatives. A major concern is their high toxicity, particularly in halogenated compounds, which release harmful gases like hydrogen chloride (HCl) and dioxins during combustion. These toxic byproducts present serious health risks and environmental hazards. Moreover, many of these fire retardants persist in the environment, leading to bioaccumulation and ecological damage, with some contributing to ozone layer depletion.

In addition to toxicity, conventional fire retardants often require high loading levels—typically 20–30 wt%—to achieve effective flame retardancy. This high filler content negatively affects the mechanical properties of materials, resulting in reduced strength, flexibility, and durability. The excessive incorporation of fire retardants can also lead to deterioration in mechanical properties, such as increased brittleness and lower tensile strength, which is particularly problematic in applications requiring both fire resistance and structural integrity, like aerospace and construction. Furthermore, many conventional fire retardants exhibit poor thermal stability at high temperatures, compromising their effectiveness in critical conditions. Processing challenges also arise due to the high filler content, which can cause uneven dispersion and weak spots in the final material.

Given these limitations, there is a pressing need for better alternatives, such as DOPO-based PTE composites. These substitutes offer safer solutions to address growing concerns over toxic emissions and environmental damage. They do not release harmful substances during combustion and have a minimal environmental impact. Improved mechanical performance is also essential, especially in industries like aerospace, where materials must retain both fire resistance and structural strength. Substitutes that require lower loadings while maintaining mechanical integrity are highly desirable. Additionally, materials with higher thermal stability offer improved performance in extreme environments, making them more reliable for high-temperature applications. Lower filler loadings also simplify processing, ensuring more consistent material quality and reducing production costs.

DOPO-based phosphorus tetraglycidyl epoxy (PTE) composites offer several key advantages. They provide superior fire resistance at lower loadings, making them highly efficient compared to conventional fire retardants that often require higher amounts. These composites also reduce toxicity, as they are halogen-free and do not release harmful gases during combustion. Moreover, DOPO-based PTE composites maintain excellent mechanical integrity, a crucial factor for aerospace applications where both fire resistance and structural strength are required. Hence, DOPO-based PTE composites offer a balanced combination of enhanced fire resistance, mechanical strength, and environmental safety, making them a superior alternative to conventional fire retardant materials. Table 4.1 shows the comparison of DOPO-based (PTE) composites with conventional fire retardant materials, highlighting key advantages.

Table 4.1 Comparison of DOPO-based phosphorus tetraglycidyl epoxy (PTE) composites with conventional fire retardant materials

S. No.	Property	DOPO-based PTE composites	Conventional fire retardant materials	References
1	Fire resistance	High fire resistance due to phosphorus-based char formation	Moderate to high, often requiring higher loadings for effectiveness	[9–12]
2	Mechanical properties	Retains excellent mechanical strength and toughness	May suffer from reduced mechanical performance due to high additive content	[13–15]
3	Thermal stability	Enhanced thermal stability at elevated temperatures	Lower thermal stability, particularly in halogenated retardants	[13, 16, 17]
4	Toxicity of combustion products	Low toxicity (halogen-free, phosphorus-based system)	Higher toxicity due to halogenated compounds releasing toxic gases (e.g., HCl)	[12]
5	Environmental impact	Environmentally friendly, halogen-free	Potential environmental concerns, especially with halogenated materials	[12]
6	Loading requirement	Lower loading levels (often less than 10 wt%) needed for effective fire retardancy	Higher loadings (20–30 wt%) are typically needed for sufficient fire retardancy	[13]
7	Processing ease	Easy to incorporate into epoxy matrices	May cause processing difficulties due to high filler content	[13, 18]

4.1.3 Significance of Phosphorus-Based Additives like DOPO (9,10-Dihydro-9-Oxa-10-Phosphaphenanthrene-10-Oxide)

Phosphorus-based additives, such as DOPO, play a vital role in enhancing the fire resistance of epoxy resins without compromising their mechanical properties. DOPO, in particular, is effective due to its unique chemical structure, which allows it to act both as a reactive and additive flame retardant [19, 20]. When incorporated into epoxy resins, DOPO improves fire resistance by promoting the formation of a protective char layer during combustion. This char layer acts as a barrier, reducing the release of flammable gases and heat, thereby slowing down the spread of fire.

Moreover, DOPO enhances the mechanical properties of epoxy nanocomposites. It contributes to better dispersion of nanoparticles within the resin matrix, resulting in a more uniform and robust composite material. This improved dispersion leads to enhanced tensile strength, flexibility, and toughness, making the material suitable for aerospace structural applications [3, 21]. Additionally, DOPO's environmental benefits, such as reduced toxicity and compliance with regulatory standards, make it a preferable choice over traditional halogenated flame retardants [22]. In summary, the incorporation of DOPO-based phosphorus additives in epoxy resins addresses the critical need for materials with both enhanced fire resistance and superior mechanical properties in aerospace applications. This chapter explores the development and characterization of novel DOPO-based phosphorus tetraglycidyl epoxy nanocomposites, highlighting their potential to meet the demanding requirements of modern aerospace engineering.

4.2 Background and Literature Review

4.2.1 Epoxy Resin Basics and their Properties

Epoxy resins are thermosetting polymers formed by the reaction of an epoxide “resin” with a polyamine “hardener.” They are known for their excellent adhesion, chemical resistance, mechanical strength, and electrical insulating properties [23, 24]. Epoxy resins cure through a process of cross-linking, which forms a three-dimensional network that imparts rigidity and durability to the material. These properties make epoxy resins highly suitable for a wide range of industrial applications, including coatings, adhesives, and composite materials, especially in the aerospace sector [25, 26].

In aerospace applications, epoxy resins are often used in composite materials, where they act as the matrix binding reinforcing fibers such as carbon or glass. This combination yields materials with outstanding strength-to-weight ratios, essential for reducing the overall weight of aircraft and spacecraft while maintaining structural integrity. The ability to tailor the curing process and the formulation of epoxy resins allows for customized properties, such as enhanced toughness or flexibility, making them versatile materials for various aerospace components. The preparation and structure of polyphosphate flame retardant-epoxy resin is shown in Figs. 4.1 and 4.2.

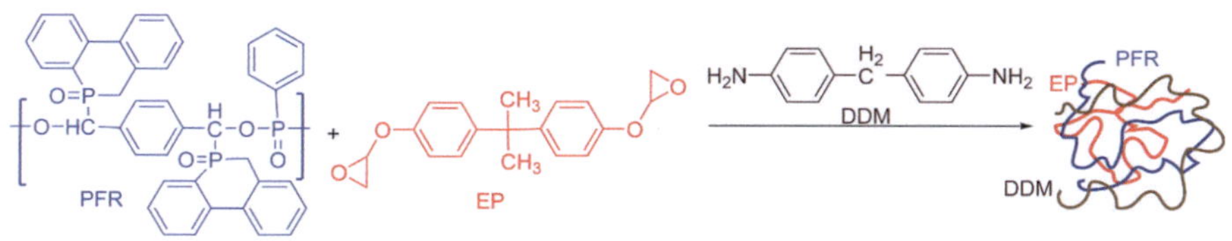

Fig. 4.1 Preparation of polyphosphate flame retardant-epoxy resin (PFR-EP) semi-interpenetrating polymer network (SIPN) composites

Fig. 4.2 Structures for Al (MEP) and Al (DEP)

4.2.2 Challenges in Aerospace Applications

The aerospace industry demands materials that can withstand extreme conditions, including high mechanical stresses, wide temperature variations, and potential exposure to harsh chemicals and UV radiation. Two critical challenges in aerospace materials are fire safety and mechanical requirements. The risk of fire poses a significant threat to aircraft safety. Materials used in aerospace applications must exhibit high fire resistance to prevent ignition and slow down the spread of fire. They must also produce minimal smoke and toxic fumes to protect passengers and crew in the event of a fire. Achieving these properties without compromising other material characteristics is a complex challenge.

Aerospace materials must possess high mechanical strength to endure the loads and stresses encountered during flight, including takeoff, landing, and turbulence. They must also be resistant to impact damage from foreign objects and have good fatigue resistance to withstand repeated stress cycles over the aircraft's operational life. Additionally, lightweight materials are crucial for improving fuel efficiency and performance, necessitating a balance between strength and weight.

4.2.3 Previous Research on DOPO-Based Additives in Epoxy Nanocomposites

Research into phosphorus-based flame retardants, particularly DOPO (9,10-dihydro-9-oxa-10-phosphaphenanthrene-10-oxide), has shown significant promise in enhancing the fire resistance of epoxy resins. DOPO is a highly effective flame retardant due to its ability to promote char formation and reduce flammable gas release during combustion. Studies have demonstrated that incorporating DOPO into epoxy resins can significantly improve their thermal stability and flame retardancy [27]. DOPO-based additives have been shown to act synergistically with other fire retardant agents, further enhancing the overall fire performance of the nanocomposites. For example, combining DOPO with other phosphorus compounds or metal oxides can result in advanced materials with superior fire resistance and mechanical properties [28–30].

Research has also explored the impact of DOPO on the mechanical properties of epoxy nanocomposites, revealing that DOPO can improve tensile strength, flexibility, and impact resistance by promoting better dispersion of reinforcing nanoparticles within the epoxy matrix. Several studies have focused on the synthesis and characterization of DOPO-modified epoxy resins, using techniques such as Fourier-transform infrared spectroscopy (FTIR), scanning electron microscopy (SEM), and thermogravimetric analysis (TGA) to understand the structural changes and interactions within the nanocomposite system. These analyses have provided valuable insights into the mechanisms by which DOPO enhances both fire resistance and mechanical performance.

In summary, previous research has established the effectiveness of DOPO as a flame retardant in epoxy nanocomposites, highlighting its potential to address the critical challenges of fire safety and mechanical performance in aerospace applications. This chapter builds on these findings by exploring the development and characterization of novel DOPO-based phosphorus tetraglycidyl epoxy nanocomposites, aiming to further enhance their suitability for aerospace engineering.

4.3 Various Synthesis Methods

4.3.1 Phosphorus Tetraglycidyl Epoxy Resin

The synthesis of phosphorus tetraglycidyl epoxy resin begins with the careful selection of starting materials and catalysts to ensure high yield and optimal properties. The process involves reacting a phosphorus-containing compound, such as tetrakis(4-hydroxyphenyl) ethane, with epichlorohydrin in the presence of a base like sodium hydroxide. This reaction forms glycidyl ether groups, resulting in the desired epoxy resin. The reaction is conducted under controlled temperature and pH conditions to ensure complete conversion and prevent side reactions. After the reaction, the mixture is neutralized and washed to remove unreacted materials and byproducts. Purification of the resulting resin is typically achieved through vacuum distillation or solvent extraction, yielding high-purity phosphorus tetraglycidyl epoxy resin suitable for further modification.

4.3.2 Incorporation of DOPO and Characterization of the Additive

Incorporating DOPO (9,10-dihydro-9-oxa-10-phosphaphenanthrene-10-oxide) into the synthesized epoxy resin is a crucial step to enhance the material's fire resistance and mechanical properties. Initially, the phosphorus tetraglycidyl epoxy resin is dissolved in an appropriate solvent to facilitate mixing. DOPO is then gradually

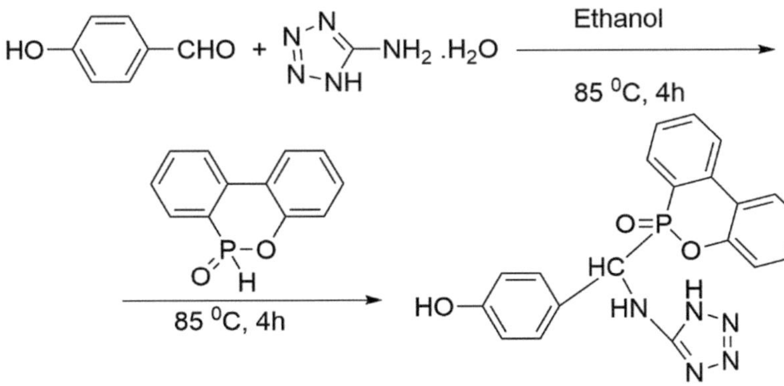

Fig. 4.3 Synthesis route of DOPO-based tetrazole derivative (ATZ)

added to this solution under continuous stirring to ensure homogeneous distribution. Thorough mixing is essential to achieve uniform dispersion of DOPO within the resin matrix. Characterization of the DOPO-modified epoxy resin involves several analytical techniques. (FTIR) is used to identify functional groups and confirm the incorporation of DOPO by analyzing characteristic absorption bands. (SEM) provides detailed images of the nanocomposite's surface morphology, revealing the dispersion of DOPO within the resin. (TGA) measures the thermal stability and decomposition behavior of the modified resin, indicating the effectiveness of DOPO in enhancing fire resistance. The synthesis route of DOPO-based tetrazole derivative is depicted in Fig. 4.3.

4.3.3 *Nanocomposite Preparation Methods*

The preparation of DOPO-based phosphorus tetraglycidyl epoxy nanocomposites involves blending, curing, and post-processing steps to achieve the desired material properties [31]. Initially, the DOPO-modified epoxy resin is mixed with reinforcing agents such as carbon fibers, glass fibers, or nanoparticles to form a homogeneous blend. Mechanical stirring or ultrasonic dispersion techniques are used to ensure uniform distribution of the reinforcing agents within the resin matrix. The blended mixture is then combined with a suitable hardener to initiate the cross-linking reaction. The mixture is poured into molds or onto substrates as required and cured under controlled temperature conditions, typically following a step-wise heating protocol to ensure complete cross-linking and optimal mechanical properties. After curing, the nanocomposites are removed from the molds and may undergo additional finishing processes such as trimming or surface treatment. Further characterization tests, including mechanical testing (tensile, flexural, impact resistance) and fire performance evaluations (cone calorimetry, DSC), are conducted to assess the final properties of the nanocomposites, ensuring they meet the stringent requirements for

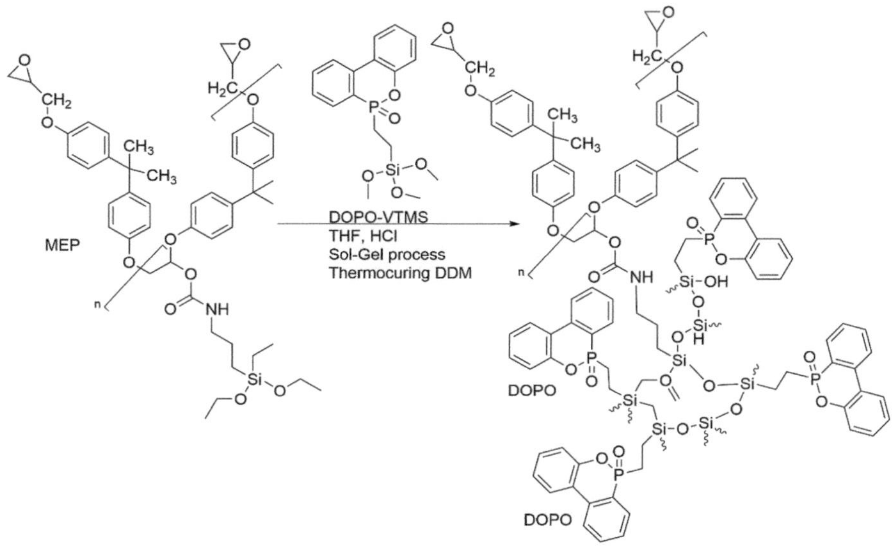

Fig. 4.4 Sol-gel synthesis route of the organic/inorganic epoxy hybrids by using DOPO-vinyl trimethoxy silane (DOPO-VTMS). MEP: modified epoxy resin. THF (tetrahydrofuran) and HCl (hydrogen chloride), reaction solvent and catalyst, respectively

aerospace applications. The Sol-gel synthesis route of the organic/inorganic epoxy hybrids by using DOPO-vinyl trimethoxy silane (DOPO-VTMS). MEP: modified epoxy resin is depicted in Fig. 4.4.

4.4 Characterization Techniques

4.4.1 Thermal Analysis for Thermal Stability

Thermal stability is a critical attribute for materials used in aerospace applications, and it is extensively evaluated using thermal analysis techniques such as (TGA) and differential scanning calorimetry (DSC). TGA measures the weight loss of a material as a function of temperature, providing insights into its thermal decomposition behavior. During TGA, the DOPO-based phosphorus tetraglycidyl epoxy nanocomposites are subjected to a controlled heating program, and the weight change is recorded. This analysis reveals the thermal degradation temperatures and the stability of the nanocomposites under high-temperature conditions, which is essential for determining their suitability for aerospace environments.

DSC, on the other hand, measures the heat flow into or out of a material as it is heated, cooled, or held at a constant temperature. This technique is used to identify the glass transition temperature (T_g), melting point, and crystallization behavior of the nanocomposites. The T_g is particularly important as it indicates the temperature range within which the material transitions from a hard, glassy state to a more flexible, rubbery state. By determining the T_g and other thermal transitions, DSC provides valuable information about the thermal stability and operational temperature range of the nanocomposites.

4.4.2 Mechanical Testing for Performance Evaluation

The mechanical properties of DOPO-based phosphorus tetraglycidyl epoxy nanocomposites are evaluated through various mechanical tests to ensure they meet the performance requirements for aerospace applications. Tensile testing measures the material's response to uniaxial tensile stress, providing data on tensile strength, elongation at break, and Young's modulus. These parameters are crucial for understanding the material's ability to withstand stretching forces and its overall stiffness.

Flexural testing assesses the material's behavior under bending loads, determining the flexural strength and flexural modulus. This test is essential for evaluating the performance of the nanocomposites in applications where they will experience bending or flexural stresses. Impact testing evaluates the material's resistance to sudden and intense forces, measuring properties such as impact strength and toughness. This test is particularly important for aerospace materials, which must resist damage from impacts during flight operations or accidental collisions with debris. By conducting these mechanical tests, the overall performance of the nanocomposites is comprehensively assessed, ensuring that they possess the necessary strength, flexibility, and durability for use in demanding aerospace environments.

4.4.3 Fire Performance Evaluation

Fire performance evaluation is critical for materials intended for aerospace applications, where fire safety is of utmost importance. Cone calorimetry is a widely used technique to assess the fire behavior of materials. This test measures parameters such as heat release rate (HRR), total heat released (THR), time to ignition (TTI), and smoke production. By subjecting the DOPO-based phosphorus tetraglycidyl epoxy nanocomposites to a controlled heat flux, cone calorimetry provides a detailed profile of their flammability and combustion characteristics. Reduced HRR and THR, along with delayed TTI, indicate improved fire resistance, making the material safer for aerospace use.

The limiting oxygen index (LOI) test is another important fire performance evaluation method. LOI measures the minimum concentration of oxygen required to sustain combustion of the material. A higher LOI value indicates better flame retardancy, as the material requires a higher oxygen concentration to burn. This test provides a quick and reliable assessment of the inherent flame retardant properties of the nanocomposites.

Together, these fire performance evaluations provide a comprehensive understanding of the flame retardancy and thermal stability of the DOPO-based phosphorus tetraglycidyl epoxy nanocomposites. The combination of thermal analysis, mechanical testing, and fire performance evaluation ensures that the developed materials meet the stringent safety and performance standards required for aerospace applications, thereby enhancing both safety and performance in this critical field.

4.5 Structure-Property Behavior

4.5.1 Morphological Characterization of Nanocomposites

The morphological characterization of the DOPO-based phosphorus tetraglycidyl epoxy nanocomposites is conducted using (SEM) and transmission electron microscopy (TEM). SEM provides detailed images of the surface morphology, revealing the dispersion and distribution of DOPO and other reinforcing agents within the epoxy matrix. The images show a uniform dispersion of nanoparticles, indicating successful integration of DOPO into the resin. This uniformity is critical for achieving consistent mechanical properties and enhanced performance.

TEM offers further insights by providing high-resolution images of the internal structure of the nanocomposites. TEM analysis confirms the nanoscale distribution of DOPO and other additives, highlighting the intimate interaction between the epoxy matrix and the additives. The combination of SEM and TEM characterization demonstrates that the synthesis and preparation methods yield nanocomposites with a well-dispersed morphology, essential for optimizing their thermal and mechanical properties.

4.5.2 Thermal Stability and Degradation Behavior

The thermal stability and degradation behavior of the DOPO-based nanocomposites are evaluated using (TGA) and (DSC). TGA results show that the nanocomposites exhibit enhanced thermal stability compared to unmodified epoxy resins. The onset of thermal degradation occurs at higher temperatures, and the rate of weight loss during decomposition is significantly reduced. This improvement is attributed to the presence of DOPO, which promotes the formation of a protective char layer that acts as a barrier, slowing down the thermal degradation process.

DSC analysis further supports these findings by revealing an increased glass transition temperature (T_g) for the nanocomposites. The higher T_g indicates that the modified epoxy resin can maintain its structural integrity at elevated temperatures, making it suitable for high-temperature aerospace applications. The combined TGA and DSC results demonstrate that the incorporation of DOPO significantly enhances the thermal stability and degradation resistance of the epoxy nanocomposites.

4.5.3 Mechanical Properties Enhancement with DOPO

Mechanical testing of the DOPO-based phosphorus tetraglycidyl epoxy nanocomposites shows significant improvements in tensile, flexural, and impact properties. Tensile tests reveal increased tensile strength and Young's modulus, indicating that the material can withstand higher loads and stresses without deformation. The improved tensile properties are attributed to the uniform dispersion of DOPO and the resulting enhanced interaction between the epoxy matrix and the reinforcing agents. Flexural testing results show a notable increase in flexural strength and modulus, suggesting that the nanocomposites exhibit better performance under bending stresses. This enhancement is crucial for aerospace components that experience significant flexural loads during operation.

Impact testing demonstrates increased impact strength and toughness, indicating that the nanocomposites can absorb and dissipate energy more effectively during sudden impacts. The presence of DOPO contributes to this improvement by promoting better energy transfer and reducing the likelihood of catastrophic failure. Overall, the mechanical testing results confirm that the incorporation of DOPO enhances the mechanical properties of the epoxy nanocomposites, making them suitable for demanding aerospace applications.

4.5.4 Fire Resistance Improvement and Mechanisms

The fire resistance of the DOPO-based phosphorus tetraglycidyl epoxy nanocomposites is evaluated using cone calorimetry and limiting (LOI) tests. Cone calorimetry results show a significant reduction in (HRR) and (THR) compared to unmodified epoxy resins. The (TTI) is also delayed, indicating improved resistance to ignition. These results suggest that the DOPO-modified nanocomposites are less flammable and release less heat during combustion, enhancing fire safety.

The LOI test results further confirm the improved flame retardancy of the nanocomposites. The higher LOI values indicate that a greater concentration of oxygen is required to sustain combustion, demonstrating the material's enhanced flame retardant properties. The improved fire resistance is attributed to the char-forming mechanism of DOPO. During combustion, DOPO promotes the formation of a stable char layer on the material's surface, acting as a physical barrier that prevents the release of flammable gases and reduces heat transfer.

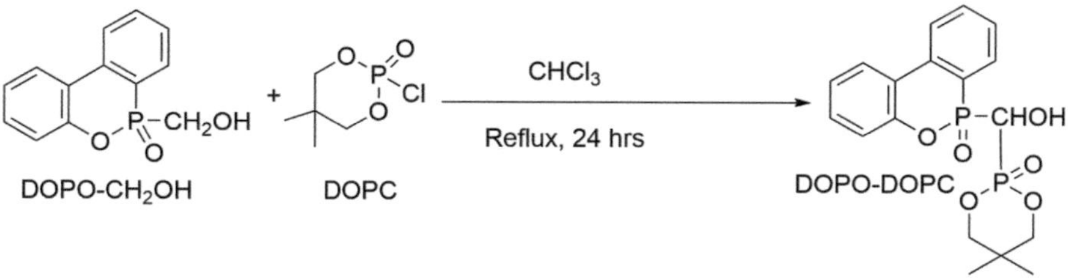

Fig. 4.5 Synthesis route and chemical structure of DOPO-DOPC flame retardant additive. The chemical structures for both reactants, 6*H*-dibenz[c,e][1, 2] oxaphosphorin-6-methanol, 6-oxide (DOPO-CH_2OH) and 2-chloro-5,5-dimethyl-1,3,2-dioxaphosphinane-2-oxide (DOPC), are also reported

The literature results demonstrate that the incorporation of DOPO into phosphorus tetraglycidyl epoxy nanocomposites significantly enhances their morphological, thermal, mechanical, and fire-resistant properties [32]. These advancements make the nanocomposites highly suitable for aerospace applications, addressing the critical needs for improved safety and performance in this demanding field. The synthesis route and chemical structure of DOPO-DOPC and DOPO-VTMS/TGIC-KH flame retardant additive are shown in Figs. 4.5 and 4.6.

4.6 Applications in Aerospace

4.6.1 Compatibility with Aerospace Structural Materials

The DOPO-based phosphorus tetraglycidyl epoxy nanocomposites exhibit excellent compatibility with various aerospace structural materials, making them highly suitable for integration into advanced aerospace applications [33]. Their enhanced thermal stability and mechanical properties complement the high-strength, lightweight materials commonly used in aerospace, such as carbon fiber and glass fiber composites. The improved adhesion and uniform dispersion of the nanocomposites within these structural materials ensure that they maintain structural integrity under extreme conditions, including high mechanical stresses and elevated temperatures. This compatibility facilitates the use of DOPO-based nanocomposites as matrix resins or coatings, where they can effectively bind and reinforce other materials, enhancing overall performance and reliability.

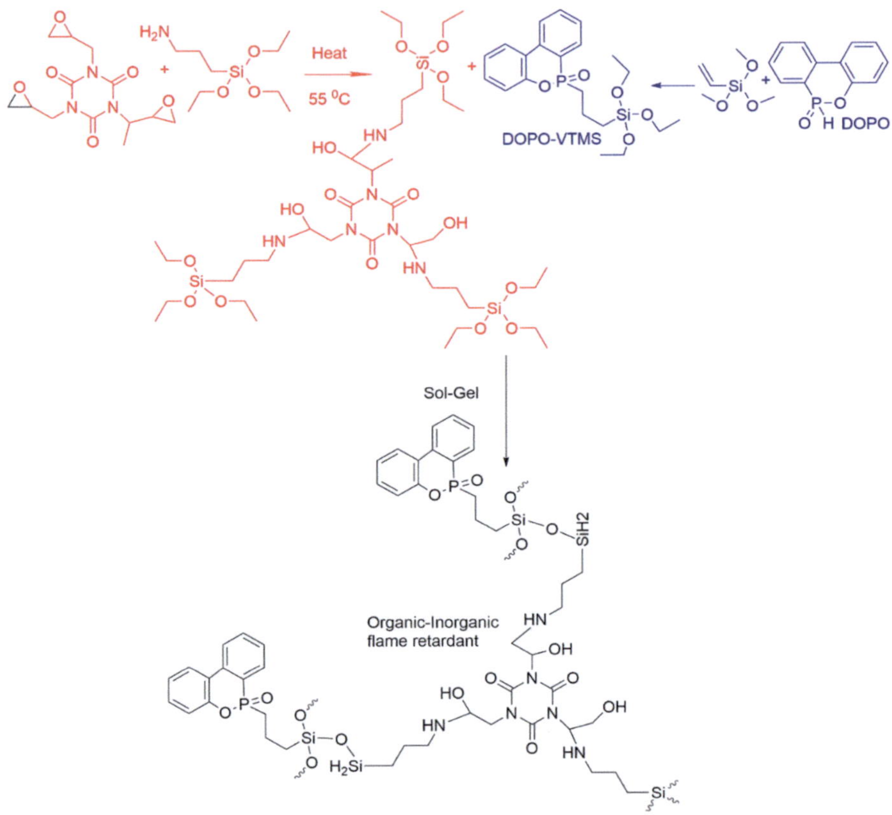

Fig. 4.6 Synthesis route and chemical structure of DOPO-VTMS/TGIC-KH flame retardant additive. DOPO-vinyl trimethoxy silane (DOPO-VTMS) was left to react with a silane modified product (TGIC-KH) based on 1,3,5-triglycidyl isocyanurate and 3-triethoxysilylpropylamine

4.6.2 Potential Applications in Aircraft Interiors and Components

The enhanced fire resistance and mechanical properties of DOPO-based phosphorus tetraglycidyl epoxy nanocomposites open up several potential applications within aircraft interiors and structural components. In aircraft interiors, these nanocomposites can be used for manufacturing lightweight, fire-resistant panels and components, such as seat frames, cabin partitions, and overhead storage bins [34, 35]. Their improved thermal stability ensures that they can withstand the varying temperature conditions experienced during flight, while their fire-resistant properties contribute to overall cabin safety. The chemical structure of flame retardant additives and synthesis routes of each epoxy system is depicted in Fig. 4.7.

In terms of structural components, the nanocomposites are suitable for use in parts that require high strength and durability, such as wing and fuselage components.

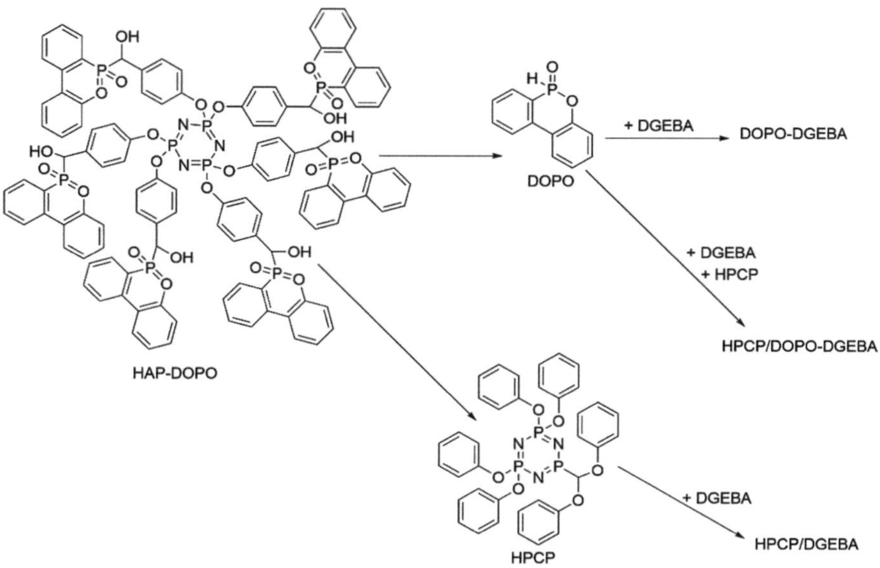

Fig. 4.7 Chemical structure of flame retardant additives and synthesis routes of each epoxy system

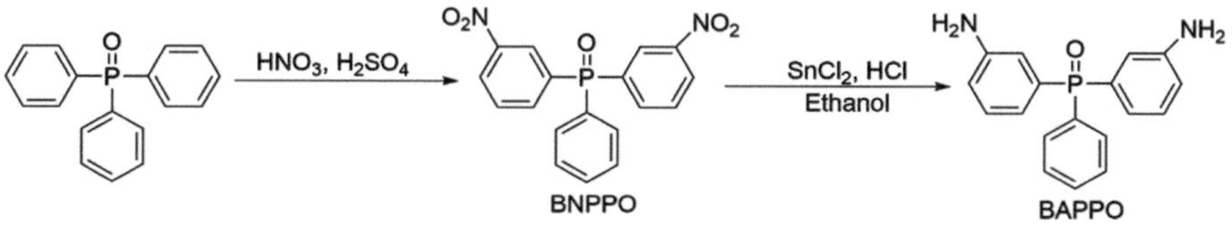

Fig. 4.8 Synthesis of phosphorus diamine

Their ability to absorb and dissipate impact energy makes them ideal for areas subject to mechanical stresses or potential impacts from foreign objects. Additionally, the reduced flammability and smoke generation characteristics enhance safety in critical aerospace applications, where failure of materials can have severe consequences. Figures 4.8, 4.9, and 4.10 represent the synthesis of phosphorus diamine, phosphorus-containing tetraglycidyl epoxy (TG-DOPO) and phosphorus-containing diglycidyl epoxy resin, respectively.

4.6.3 Future Directions and Challenges

Looking ahead, there are several future directions and challenges for the development and application of DOPO-based phosphorus tetraglycidyl epoxy nanocomposites in aerospace. One of the key areas for future research is the optimization of the nanocomposite formulation to further enhance performance attributes while minimizing costs

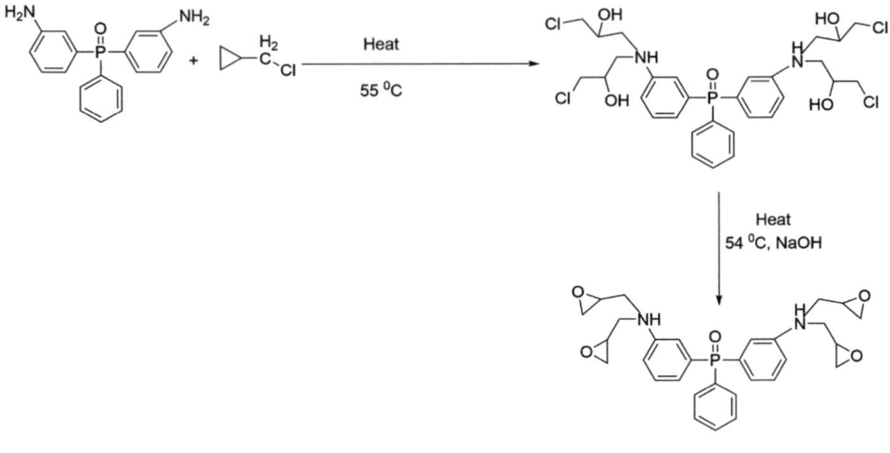

Fig. 4.9 Synthesis of phosphorus-containing tetraglycidyl epoxy (TG-DOPO)

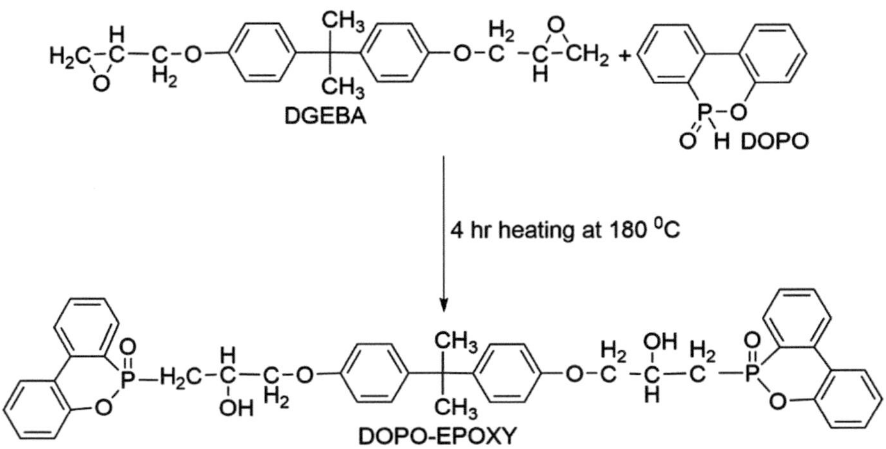

Fig. 4.10 Synthesis of phosphorus-containing diglycidyl epoxy resin

[36, 37]. This may involve exploring different ratios of DOPO to epoxy resin, as well as incorporating additional additives or fillers to achieve desired properties.

Another challenge is to ensure the long-term stability and performance of the nanocomposites under the harsh conditions of aerospace environments [38, 39]. Further studies are needed to assess the durability and aging effects of the nanocomposites over extended periods of use. This includes evaluating their resistance to environmental factors such as UV radiation, humidity, and chemical exposure. Additionally, the scaling up of the synthesis and manufacturing processes for these nanocomposites presents another challenge. Developing cost-effective and scalable production methods while maintaining the high quality and consistency of the nanocomposites is crucial for their widespread adoption in the aerospace industry.

Addressing these challenges will require ongoing collaboration between material scientists, engineers, and industry stakeholders to refine the nanocomposite formulations and production techniques. By overcoming these obstacles, DOPO-based phosphorus tetraglycidyl epoxy nanocomposites have the potential to significantly enhance safety, performance, and efficiency in aerospace applications, paving the way for more advanced and reliable aerospace materials in the future.

4.7 Conclusion

This chapter has presented a comprehensive exploration of the development and characterization of novel DOPO-based phosphorus tetraglycidyl epoxy nanocomposites tailored for aerospace applications. The synthesis process for the phosphorus tetraglycidyl epoxy resin and its subsequent modification with DOPO flame retardant additives has been detailed. Characterization techniques, including SEM, TEM, TGA, DSC, and mechanical testing, have demonstrated that the incorporation of DOPO significantly enhances the thermal stability, mechanical properties, and fire resistance of the epoxy nanocomposites. The improved performance metrics reveal a substantial advancement over conventional epoxy resins, making these nanocomposites highly suitable for demanding aerospace environments.

The key insights from this research highlight the successful integration of DOPO into phosphorus tetraglycidyl epoxy resins and its impact on material properties [40, 41]. The morphological characterization confirms the uniform dispersion of DOPO within the epoxy matrix, which contributes to the improved mechanical performance observed. Thermal analysis results indicate that the nanocomposites exhibit superior thermal stability and degradation resistance, essential for high-temperature aerospace applications. Furthermore, fire performance evaluations demonstrate enhanced flame retardancy and reduced heat release, addressing critical safety concerns in aerospace materials.

These findings contribute significantly to the field by providing a detailed understanding of how DOPO modifies epoxy resin properties, offering a pathway to develop advanced materials with optimized performance for aerospace use. The research not only fills gaps in the current knowledge of flame retardant epoxy composites but also sets a precedent for future developments in high-performance aerospace materials.

Implications for Aerospace Materials Development

The implications of this research for aerospace materials development are substantial. The enhanced fire resistance and mechanical properties of DOPO-based phosphorus tetraglycidyl epoxy nanocomposites make them ideal candidates for a range of aerospace applications, from structural components to interior fittings. Their ability to withstand extreme conditions while maintaining safety and performance standards aligns with the increasing demand for advanced materials in the aerospace industry.

Moreover, the insights gained from this study can guide further research and development efforts aimed at optimizing material formulations and production processes. By addressing the challenges related to long-term stability and cost-effective manufacturing, the aerospace industry can leverage these nanocomposites to achieve safer, more efficient, and reliable components.

In conclusion, the development and characterization of DOPO-based phosphorus tetraglycidyl epoxy nanocomposites represent a significant advancement in aerospace materials science. The successful integration of flame retardant additives and the resultant improvements in performance underscore the potential of these materials to enhance safety and functionality in aerospace applications, paving the way for continued innovation in this critical field.

References

1. Wazalwar R, Sahu M, Raichur AM (2021) Mechanical properties of aerospace epoxy composites reinforced with 2D nano-fillers: current status and road to industrialization. Nanoscale Adv 3(10):2741–2776
2. Hamerton I, Kratz J (2018) The use of thermosets in modern aerospace applications. In: Thermosets. Elsevier, pp 303–340
3. Balakrishnan P, John MJ, Pothen L, Sreekala MS, Thomas S (2016) Natural fibre and polymer matrix composites and their applications in aerospace engineering. In: Advanced composite materials for aerospace engineering. Woodhead Publishing, pp 365–383
4. Bello SA, Agunsoye JO, Hassan SEA, Kana MZ (2015) Epoxy resin based composites, mechanical and tribological properties: a review. Tribol Ind 37(4):500
5. Yang Y, Chen W, Li Z, Huang G, Wu G (2022) Efficient flame retardancy, good thermal stability, mechanical enhancement, and transparency of DOPO-conjugated structure compound on epoxy resin. Chem Eng J 15(450):138424
6. Ciesielski M, Burk B, Heinzmann C, Döring M (2017) Fire-retardant high-performance epoxy-based materials. In: Novel fire retardant polymers and composite materials. Woodhead Publishing, pp 3–51
7. Singh J, Srivastawa K, Jana S, Dixit C, Ravichandran S (2024) Advancements in lightweight materials for aerospace structures: a comprehensive review. Acceleron Aerosp J 2(3):173–183
8. Gohardani O, Elola MC, Elizetxea C (2014) Potential and prospective implementation of carbon nanotubes on next generation aircraft and space vehicles: a review of current and expected applications in aerospace sciences. Prog Aerosp Sci 70:42–68
9. Ciprioti SV, Raimondo M, Vertuccio L, Calabrese E, Aliberti F, Guadagno L (2024) Multi-technique characterization and thermal degradation study of epoxy modified resins designed for multifunctional applications. J Therm Anal Calorim 1–15
10. Javed A. Development of flame retardant cotton fabrics with multifunctional properties
11. Li P, Wang J, Wang C, Xu C, Ni A (2024) The flame retardant and mechanical properties of the epoxy modified by an efficient DOPO-based flame retardant. Polymers 16(5):631
12. Shen R, Quan Y, Wang Q (2022) Thermal stability and flame retardancy of epoxy/synthetic fiber composites. In: Handbook of epoxy/fiber composites. Springer, Singapore, pp 1–35
13. Li J, Cao Q, Weng Z (2023) Producing N-containing chemicals from biomass for high performance thermosets. In: Production of N-containing chemicals and materials from biomass. Springer Nature Singapore, Singapore, pp 271–303
14. Wan J, Zhao J, Zhang X, Fan H, Zhang J, Hu D, Wang DY (2020) Epoxy thermosets and materials derived from bio-based monomeric phenols: transformations and performances. Prog Polym Sci 108:101287

15. Zhao W, Liu J, Wang S, Dai J, Liu X (2024) Bio-based thermosetting resins: from molecular engineering to intrinsically multifunctional customization. Adv Mater 2311242
16. Chen Z, Yu T, Yang Z, Wei Z, Li Y, Yang W, Yu T (2024) Superior mechanical behavior and flame retardancy FRP via a distribution controllable 1D/2D hybrid nanoclay synergistic toughening strategy. Engineering
17. Sangwan A, Malik P, Gupta R, Ameta RK, Mukherjee TK (2021) Nanocomposites: preparation, characterization, and applications. In: Nanotechnology. Jenny Stanford Publishing, pp 201–247
18. Maâti H, Amadine O, Sair S, Abouelhrouz S, Ouadil B, Mahi H, Zahouily M (2023) Carbon nanotubes particles: processing, mechanical properties and application. In: Mechanics of nanomaterials and polymer nanocomposites. Springer, Singapore, pp 19–49
19. Liu Q, Wang D, Li Z, Li Z, Peng X, Liu C, Zhang Y, Zheng P (2020) Recent developments in the flame-retardant system of epoxy resin. Materials 13(9):2145
20. Liu S, Fang Z, Yan H, Wang H (2016) Superior flame retardancy of epoxy resin by the combined addition of graphene nanosheets and DOPO. RSC Adv 6(7):5288–5295
21. Joshi M, Chatterjee U (2016) Polymer nanocomposite: an advanced material for aerospace applications. In: Advanced composite materials for aerospace engineering. Woodhead Publishing, pp 241–264
22. Yang Y, Wang DY, Jian RK, Liu Z, Huang G (2023) Chemical structure construction of DOPO-containing compounds for flame retardancy of epoxy resin: a review. Prog Org Coat 1(175):107316
23. Gonçalves FA, Santos M, Cernadas T, Alves P, Ferreira P (2022) Influence of fillers on epoxy resins properties: a review. J Mater Sci 57(32):15183–15212
24. Mohan P (2013) A critical review: the modification, properties, and applications of epoxy resins. Polym-Plast Technol Eng 52(2):107–125
25. Chang L, Zhang Z, Ye L, Friedrich K (2007) Tribological properties of epoxy nanocomposites: III. Charact Transf Films Wear 262(5–6):699–706
26. Wei H, Xia J, Zhou W, Zhou L, Hussain G, Li Q, Ostrikov KK (2020) Adhesion and cohesion of epoxy-based industrial composite coatings. Compos Part B 193:108035
27. Perret B, Schartel B, Stöß K, Ciesielski M, Diederichs J, Döring M, Krämer J, Altstädt V (2011) Novel DOPO-based flame retardants in high-performance carbon fibre epoxy composites for aviation. Eur Polym J 47(5):1081–1089
28. Rong H, Guo Y, Chen Z, Chen T, Yu Y, He C, Zhou J, Liu X, Zhang Q, Bu Y, Jiang J (2023) Synergistic effect of phenyl phosphoric acid derivatives and DOPO on multifunctional epoxy resin: fire safety, mechanical properties, transparency and hydrophobicity. Polym Degrad Stab 216:110471
29. Bifulco A, Varganici CD, Rosu L, Mustata F, Rosu D, Gaan S (2022) Recent advances in flame retardant epoxy systems containing non-reactive DOPO based phosphorus additives. Polym Degrad Stab 200:109962
30. Chai M, Liu H, Wu Y, Xue K, Zhang P, Liu L, Huang Y (2024) Flame retardant, toughened, reinforced, transparent and long shelf-life one-component epoxy resins via a P/N/Si-containing hyperbranched polymer. Chem Eng J 152785
31. Meenakshi KS, Sudhan EP, Kumar SA (2012) Development and characterization of new phosphorus based flame retardant tetraglycidyl epoxy nanocomposites for aerospace application. Bull Mater Sci 35:129–136
32. Gan H, Seraji SM, Senanayake RB, Swan SR, Zhang J, Varley RJ (2024) A DOPO modified cyclosiloxane reactive diluent as an effective flame retardant for a high-performance epoxy network. Polymer 309:127427
33. Huo S, Song P, Yu B, Ran S, Chevali VS, Liu L, Fang Z, Wang H (2021) Phosphorus-containing flame retardant epoxy thermosets: recent advances and future perspectives. Prog Polym Sci 1(114):101366
34. Uddin F (2016) Flame-retardant fibrous materials in an aircraft. J Ind Text 45(5):1128–1169
35. Pinni M, Scatteia L. Advanced materials in space architecture: new concepts for space tourism in the coming decades. In: 57th International astronautical congress, pp 3–4

36. Ponnamma D, Cabibihan JJ, Rajan M, Pethaiah SS, Deshmukh K, Gogoi JP, Pasha SK, Ahamed MB, Krishnegowda J, Chandrashekar BN, Polu AR (2019) Synthesis, optimization and applications of ZnO/polymer nanocomposites. Mater Sci Eng C 98:1210–1240
37. Bhakay A, Rahman M, Dave RN, Bilgili E (2018) Bioavailability enhancement of poorly water-soluble drugs via nanocomposites: formulation-processing aspects and challenges. Pharmaceutics 10(3):86
38. Bhat A, Budholiya S, Aravind Raj S, Sultan MTH, Hui D, Md Shah AU, Safri SNA (2021) Review on nanocomposites based on aerospace applications. Nanotechnol Rev 10(1):237–253
39. Prabhakaran R (2015) Nano-composites for space applications. J Aerosp Sci Technol 67(1):1–17
40. Mathews LD, Capricho JC, Peerzada M, Salim NV, Parameswaranpillai J, Hameed N (2022) Recent progress and multifunctional applications of fire-retardant epoxy resins. Mater Today Commun 33:104702
41. Gonçalves FA, Santos M, Cernadas T, Ferreira P, Alves P (2022) Advances in the development of biobased epoxy resins: insight into more sustainable materials and future applications. Int Mater Rev 67(2):119–149

Chapter 5
Nanocomposites Used in the Parts of Aeronautical Engineering in Radio Frequency Identification (RFID) Tag

K. B. Bommegowda and Roopa B. Hegde

Abstract As the aerospace industry continues to demand materials that offer superior performance with reduced weight and increased functionality, nanocomposites will play a key role in driving next-generation RFID technologies. Upcoming research should emphasise on optimising the synthesis processes, improving material compatibility, and exploring novel nanomaterial combinations to further enhance the performance of RFID tags tailored for aeronautical applications. In essence, the synergy between nanocomposite technology and aeronautical RFID systems holds transformative potential for improving operational efficiency, safety and asset management within the aerospace sector.

Keywords Nanocomposites · Aeronautical engineering · RFID tags · Aerospace technology · Thermal stability · Electromagnetic shielding · Aeronautical environments · Data transmission efficiency · Aerospace operations · Miniaturised · Asset tracking · Health monitoring · Supply chain management

5.1 Introduction to Radio Frequency Identification Tag

Radio frequency identification (RFID) technology enables wireless identification and data exchange using radio waves and has evolved into a versatile tool across multiple domains, including logistics, healthcare and manufacturing [1]. RFID systems are built around three fundamental components: tags, readers (or interrogators), and backend data processing infrastructures for handling and interpreting the acquired information. Tags are typically categorised into passive, active and semi-passive

K. B. Bommegowda · R. B. Hegde (✉)
Nitte (Deemed to be University), NMAM Institute of Technology (NMAMIT), Department of Electronics and Communication Engineering, Nitte 574110, Karnataka, India
e-mail: roopabhegde@nitte.edu.in

K. B. Bommegowda
e-mail: bgowda_kbl@nitte.edu.in

S. K. Panda (ed.), *Nanocomposites for Defense, Aeronautical and Aerospace Applications*, Advanced Structured Materials 248,
https://doi.org/10.1007/978-3-032-11568-3_5

types, each defined by their power source and range capabilities. Communication principles in RFID systems rely on electromagnetic coupling and modulated backscatter, operating across regulated frequency bands from low frequency (LF) to microwave ranges. RFID systems follow standardised protocols such as Electronic Product Code (EPC) global Class 1 Generation 2 (EPC Gen2) and the ISO/IEC 18,000 series, to facilitate seamless global interoperability and compatibility across diverse applications.

Application-focused studies with RFID in supply chains [2–4], healthcare [5], food safety [6] and construction [7] have demonstrated their potential to improve efficiency. These studies frequently trace the benefits of RFID systems, including real-time object identification without line-of-sight, enhanced tracking accuracy and operational efficiency, while also noting implementation challenges such as interference, cost barriers and privacy risks [8–10]. Recent contributions emphasise the development of emerging extensions to RFID systems, such as sensor-integrated tags and chipless technologies [11]. Together, these studies form a well-rounded corpus, introducing the architecture, functionality and practical relevance of RFID technology. They lay the foundation to understand operational principles of RFID systems, the contexts in which they are deployed, and the opportunities and constraints that shape their evolution.

5.1.1 Components of RFID Systems:

RFID components work together to enable automated identification and data capture. This section provides details of RFID tags, readers and backend systems that form the foundation of RFID operation and communication.

i. Tags (Transponders): These are small devices comprising a microchip to store and process information, and an antenna to receive and transmit the signals. Tags can be categorised based on their power source.

Passive RFID tags operate without a built-in power supply. Hence, they function by drawing electromagnetic energy from the reader to activate the tag's circuitry and enable backscatter communication. They are highly cost-effective and exhibit extended operational lifespans due to the absence of an internal power source [1]. However, their read range is limited due to the absence of a built-in power supply. Active RFID tags incorporate an integrated internal power source (battery) to enable autonomous signal transmission. These tags can transmit signals over longer distances and offer greater storage capacity. However, their higher cost and finite operational lifespan, dictated by battery limitations, restrict their long-term deployment [2, 12]. Tags comprise antennas and chips for storage [13].

ii. Readers (Interrogators): These devices are integrated with antennas to transmit RF signals for detecting, interrogating and establishing communication with RFID tags. Depending on the application, readers can be either fixed or mobile [14, 15].

iii. Middleware (Data Processing Systems): These are software systems that process the collected data by readers, facilitating data storage, analysis and integration with other applications [2, 16]. Recent developments in RFID technology include the growth of miniaturised tags. For instance, Hitachi has developed ultra-miniaturised RFID chips, referred to as 'RFID powder', with dimensions as small as 0.05 mm × 0.05 mm, capable of storing 38-digit identification numbers within an integrated 128-bit read-only memory (ROM). These dust-sized chips present challenges in antenna attachment, limiting their read range to only millimetres.

5.1.2 Principle of RFID Technology

RFID technology operates on the principle of wireless communication between a reader and a tag using radio frequency signals. This section briefs the communication principles.

i. **Communication**: Primarily operates through electromagnetic coupling or modulated backscatter communication [11, 12, 14]. Passive RFID systems dominate due to simplicity and cost-effectiveness [12].
ii. **Frequency Bands**: RFID systems use LF, HF, UHF and microwave frequencies; each has unique range and interference properties [2, 9].
iii. **Protocols and Standards**: EPC Gen2 and ISO/IEC protocols ensure interoperability, although global standardisation remains a challenge [15].

5.1.3 Applications

RFID technology has found various applications across multiple domains. This is due to its ability to enable efficient, contactless data capture and tracking. Major domains utilising RFID include supply chains, healthcare and emerging IoT-driven environments.

i. **Supply Chain Management**: In the domain of supply chain management, one of the most significant advantages lies in the ability to accurately track inventory and access real-time data across the entire supply chain network [4].
ii. **Healthcare**: In the healthcare sector, patient tracking and medication management represent two of the most prominent and impactful applications of modern healthcare information systems and sensor-based technologies [5, 6].
iii. **Emerging Applications**: Emerging applications in the field of RFID technology are rapidly evolving, driven by advancements in IoT integration, sensor-enabled RFID tags, and the development of chipless RFID systems [11].

Table 5.1 Frequency bands and ranges of RFID and their applications

Name of the band	Frequency band	Range	Applications
Low frequency	125–134 kHz	10 cm	Access control, animal tracking
High frequency	13.56 MHz	1 m	Library books, contactless payments
Ultra-high frequency	860 to 960 MHz	10–100 m	Logistics, retail, baggage tracking
Microwave	2.45 and 5.8 GHz	> 100 m	Aerospace, military, high-speed tolling

5.1.4 Challenges and Innovations

RFID implementation faces several challenges, such as interference, limited read range in passive tags, privacy concerns and high deployment costs in large-scale systems.

i. **Barriers to Adoption**: Deployment costs, material limitations, interference, privacy and standardisation issues hinder adoption [3, 8, 10].
ii. **Advancements**: Emerging designs include chipless RFID, biodegradable tags and IoT-enabled systems [11, 17]. Improvements in antenna optimisation and energy harvesting are critical [13].

Another notable development is the Terahertz Frequency Identification (TFID) tag, demonstrated by MIT researchers in early 2020. These tags are barely 1 square millimetre in size, functioning like larger RFID tags but at a reduced cost, potentially allowing manufacturers to tag any product and track logistics information more efficiently. These innovations highlight the ongoing evolution of RFID technology, leading to more efficient and versatile applications across various industries. The different frequency bands and their applications are tabulated in Table 5.1.

5.1.5 Classification of RFID Tags

(a) **Passive RFID Tags**

Passive RFID tags operate without an internal power source (battery) [18]. Instead, these tags harvest power from an RFID reader's electromagnetic (EM) energy to activate their circuitry and enable data transmission via backscatter communication. When an RFID reader emits an interrogation signal, its antenna converts the electrical signal into radio frequency (RF) energy. This RF energy is then harvested by the passive RFID tag's antenna, which supplies the necessary power to activate the embedded microchip. Once energised, the RFID chip processes the signal and transmits a response containing the stored data back to the reader. Due to their low-power requirements, passive RFID tags efficiently function within the reader's operational range, enabling reliable data transmission. The passive RFID tags remain independent of power sources and hence are unaffected by the operational limitations imposed by finite battery life.

(i) **Passive RFID Frequencies**

As a subset of EM waves, all radio waves propagate within defined frequency ranges, which determine their propagation characteristics, interaction with materials and suitability for various communication applications. For instance, FM radio signals are transmitted within the 88–108 MHz band, allowing receivers to tune into specific frequencies to access distinct radio stations [19]. Similarly, RFID systems operate within designated frequency bands. However, unlike traditional radio receivers, RFID readers rapidly cycle through multiple frequencies within their allocated spectrum. This process is known as frequency hopping or spread spectrum technology, which enables RFID readers to detect and decode tag responses across all available frequencies, improving signal reliability and minimising interference. Passive RFID technologies utilise three main frequency ranges for communication and operation: low frequency (LF), high frequency (HF) and ultra-high frequency (UHF), each offering distinct data rates and application suitability.

(a) **Low Frequency (LF) RFID:**

The LF RFID operates within the 30–300 kHz range, with most of the LF tags functioning at either 125 kHz or 134 kHz. Due to their longer wavelength and low energy levels, LF radio signal transmission is limited to short distances, typically only a few inches. This restricted range minimises the risk of signal interference, allowing only one or two frequency bands to be used within a given environment without significant cross-system disruption. LF RFID is commonly utilised in applications requiring close-proximity scanning, namely animal identification, access control and industrial automation.

(b) **High Frequency (HF) RFID**:

The HF RFID operates between the bands 3–30 MHz, and most of the HF tags utilise 13.56 MHz band. Comparable to low-frequency RFID, HF RFID is designed for short-range communication, typically transmitting data over distances ranging from a few inches to approximately one foot. Due to this limited range, a single frequency band is sufficient for reliable operation without significant interference. A widely recognised application of HF RFID is Near-Field Communication (NFC), a technology that enables secure and efficient data exchange, particularly in contactless payment systems, access control and smart cards.

(c) **Ultra-High Frequency (UHF) RFID**:

UHF RFID technology functions within the 300 MHz to 3 GHz frequency band, offering extended read ranges and faster data transmission rates compared to LF and HF systems. Most UHF tags transmit in the primary 860–960 MHz band, while a secondary 433 MHz band is also used, though less frequently. The 860–960 MHz band is a globally recognised spectrum for UHF RFID applications [1]. However, it is further subdivided into region-specific frequency allocations to comply with regulatory standards in different countries. UHF RFID offers significantly longer read ranges compared to LF and HF RFID, making it ideal for applications, namely asset tracking, supply chain management and inventory control. However, these signals are more susceptible to interference from metal surfaces and liquids, necessitating optimised tag and reader placement for reliable performance.

Types of Passive RFID Tags

The tags are commonly manufactured in two basic forms and they are as follows.

- **Inlays/Labels**—Thin, flexible tags consisting of an antenna and RFID chip embedded in a substrate, often used for applications like inventory tracking and retail.
- **Hard Tags**—RFID tags enclosed in a durable protective casing, designed for harsh environments, asset tracking and industrial applications.

The choice between inlays and hard tags depends on the specific use, environmental conditions and durability. Different types of passive RFID tags available in the market are as shown in Fig. 5.1.

As active RFID systems are employed less frequently than their passive counterparts, the mainstream of contemporary RFID applications is predominantly powered by passive RFID tags and compatible readers. The various applications of passive RFID, along with their corresponding operating frequency ranges categorised by usage popularity, are provided in Table 5.2.

(a) **Active RFID Tags**

Active RFID systems form a crucial subset of wireless identification technologies used in real-time tracking, automation and asset monitoring. In contrast to passive, active RFID tags are furnished with an internal power source (battery), which enables them to autonomously transmit signals without relying on external interrogation. This independent power capability allows active RFID tags to achieve significantly longer communication ranges, typically spanning from several tens to hundreds of meters, making them suitable for industrial, logistic and sensor-integrated applications.

(i) **System Components and Architectures**

Tags: Embedded with microcontrollers, memory, RF transceivers and a power supply. Some also include integrated sensors for environmental monitoring.

Readers: Devices that interrogate tags and receive transmitted data.

Middleware/Protocols: Software stacks and communication algorithms that orchestrate tag-reader interaction, often conforming to specific standards (e.g., ISO/IEC 18,000–7 for 433 MHz systems) [5, 14, 20].

Two primary operation types of the tags: Beaconing Tags—Periodically transmit data on a preset interval, useful in Real-Time Location Systems (RTLS). Transponder Tags—Wait in low-power mode until woken by a reader signal, conserving energy.

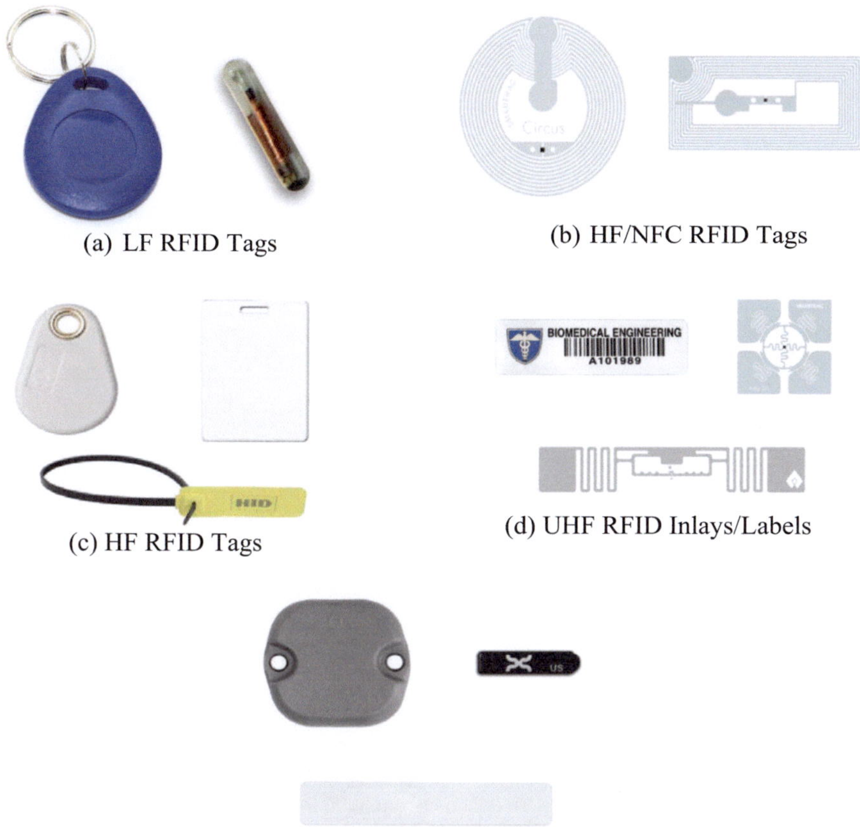

(a) LF RFID Tags

(b) HF/NFC RFID Tags

(c) HF RFID Tags

(d) UHF RFID Inlays/Labels

(e) UHF RFID Hard Tags

Fig. 5.1 Types of passive RFID tags [18]

(ii) **Frequency Bands, Standards and Modulation**

Common operating frequency bands include 433, 868/915 and 2.45 GHz, with 433 MHz appearing frequently due to its compatibility with ISO/IEC 18,000–7 [5, 14, 21]. This standard defines modulation schemes, data rates and anti-collision protocols for active RFID. Spread spectrum techniques (e.g. frequency hopping) are employed in some designs to manage interference and multipath effects in dense or reflective environments [9]. Data encoding and memory access protocols are covered under ISO/IEC 15,961/15962 [20].

Table 5.2 Applications of passive RFID tags [18]

Application	Frequency band	Application description
Most popular		
Access control	HF/NFC	To regulate access to buildings, parking facilities and residential complexes
Animal identification	LF	Tags for the identification and tracking of pets and livestock
Stock management	UHF	To monitor and manage inventory levels in retail environments
Tolling	UHF	Tags to identify toll passes in vehicles
Semi-popular		
Race timing	UHF	To detect and record the precise moment racers cross the start and finish lines
Library Inventory Tracking	HF/NFC (UHF)	To manage library inventory and automate the check-in and checkout of books and other media
Authentication	HF/NFC	Tags for authenticating products such as footwear, vaccines and other critical items to ensure genuineness and prevent counterfeiting
Asset tracking	UHF	Tags are used for tracking the movement and location of assets within facilities to enhance inventory control and operational efficiency
Upcoming		
Work-in-process	UHF	To monitor and track specific items throughout various stages of the manufacturing process
Logistics	UHF	To document and monitor an item throughout all stages of transportation
Food industry	HF/NFC (UHF)	To manage inventory levels, facilitate the purchase of items, and implement theft prevention measures
Pharmaceuticals	HF/NFC (UHF)	Utilising RFID tags to manage inventory, authenticate pharmaceuticals, and track medicines throughout the entire supply chain ensures enhanced accuracy, security and traceability in healthcare logistics

(iii) **Energy and Power Management**

Battery life is a critical limitation of active RFID tags. Several studies explored hardware and software techniques to reduce power usage during both data transmission and idle periods. Low-power protocols reduce energy use through sleep modes and reduced transmission duty cycles [3, 10, 22]. Overhearing mitigation strategies address inefficient energy use when tags remain in an active listening state but are not the intended recipient of transmissions [10]. Dynamic back-off algorithms and CSMA/CA tuning help reduce energy waste, delay and collisions in multi-tag environments [4, 16].

Energy harvesting is proposed as a solution to battery dependence, namely, a MEMS-based energy scavenging technique for a batteryless active RFID tag [17] and the use of energy harvesting supplements (e.g. RF or solar) in support of UHF active tags [15].

(b) **Semi-passive RFID Tags**

These semi-active or battery-assisted passive (BAP) tags represent an advanced implementation of RFID technology [23]. These tags' function is based on fundamental principles, as passive RFID tags, utilising radio frequency communication for identification and data exchange; however, they incorporate an integrated battery, enhancing their functionality by enabling autonomous signal transmission and extended operational range. The battery primarily extends the communication range, increases data storage capacity, and, in specific applications, facilitates sensor integration for real-time data acquisition. Consequently, semi-passive RFID tags are particularly advantageous in scenarios requiring extended read distances and supplementary data collection capabilities.

Semi-passive RFID tags typically operate using the same Gen 2 protocol as passive tags, ensuring full interoperability with existing passive Gen 2 RFID readers. This protocol compatibility enables seamless integration of semi-passive RFID tags into established RFID infrastructures without necessitating substantial system modifications or upgrades [24].

The BAP tags offer an optimal balance between the operational characteristics of passive and active RFID systems, combining the extended read range and enhanced functionality associated with active tags with the battery-assisted, reader-initiated communication model of passive tags by enhancing read range and functionality while maintaining energy efficiency. Their applications in logistics, healthcare and environmental monitoring demonstrate their versatility in modern industries.

(i) **Working Principle**

The BAP tags contain an inbuilt battery that powers the tag's internal electrical system but does not actively transmit a signal. Instead, they rely on backscatter communication, similar to passive RFID tags, where the tag controls and reflects the reader's signal to transmit data. The battery assists in powering the microcontroller and sensors (if present), enabling additional functionalities such as data logging and environmental sensing [25].

(ii) **Advantages**

Extended Read Range: In comparison with passive, semi-passive tags achieve longer read ranges as a result of their battery-assisted operation, which powers the tag's internal electric circuit while still relying on backscatter communication initiated by the reader.

Enhanced Functionality: These tags can support additional features such as temperature sensing, real-time clock functionality and data storage [26].

Lower Power Consumption than Active Tags: Unlike fully active RFID tags, semi-passive tags do not continuously emit signals, thereby consuming less power [27].

(iii) **Applications**

Cold Chain Monitoring: Used in the pharmaceutical and food industries to track temperature-sensitive products.

Asset Tracking: Useful in logistics and inventory management.

Environmental Monitoring: Deployed in smart agriculture for tracking humidity, temperature and soil conditions [28].

Healthcare: Used for patient tracking and medical equipment monitoring.

(iv) **Challenges and Limitations**

Limited Battery Life: Although battery consumption is lower than active RFID tags, the need for a battery limits the tag's operational lifespan.

Higher Cost: Semi-passive RFID tags are typically more costly than passive tags, owing to the addition of an integrated battery and additional electronic circuitry required to support their enhanced operational capabilities.

Reader Dependency: Requires an RFID reader for data retrieval, similar to passive RFID tags [29].

5.2 Working Procedure of Radio Frequency Identification (RFID) Tag

An RFID is a wireless communication technology that utilises EM fields to identify, track and manage objects without requiring line-of-sight (LoS). A typical RFID system comprises three primary components: RFID tags, readers (or interrogators) and a backend data processing infrastructure.

5.2.1 Structure of RFID Tag

An RFID tag consists of three primary components: an antenna, a microchip (IC) and a substrate (base material). The antenna enables bidirectional communication by transmitting and receiving RF signals between the RFID tag and the reader. Its design affects the tag's operating frequency, range and power efficiency. The microchip stores data, processes signals and manages power in passive tags. It also handles encryption and security functions if needed. The substrate serves as the structural foundation, holding the microchip and antenna in place. It is typically made of materials like plastic, paper, or ceramic, reliant on the application and durability requirements. The different RFID tag components are depicted in Fig. 5.2.

RFID tags are encapsulated in protective materials to safeguard their components from environmental factors and mechanical damage. This protective layer enhances the tag's longevity and durability, enabling its use in a diverse range of applications.

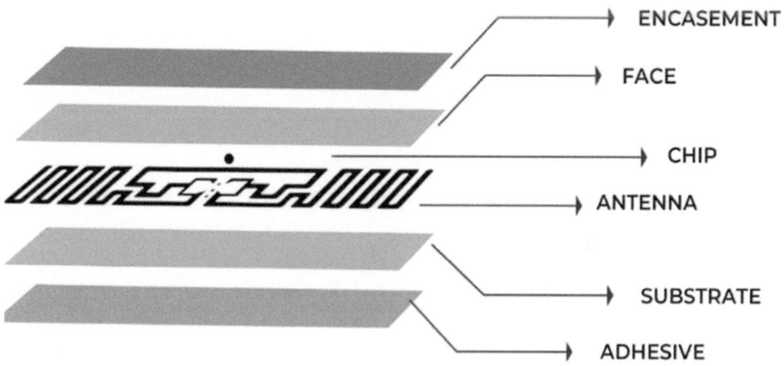

Fig. 5.2 RFID tag components [30]

The modulation and demodulation circuitry encodes and decodes data during communication between the tag and reader. It modulates data onto the RF carrier wave in response to the reader's signal and demodulates incoming signals to retrieve the transmitted information. RFID tags are equipped with built-in memory, extending their functionality beyond just a unique identifier. This memory is both readable and writable during interactions with an RFID reader, enabling dynamic and versatile use of the tag across an extensive range of applications. RFID chips are equipped with four types of memory, namely reserved, EPC, TID and user memory.

The reserved memory bank within an RFID tag is designated to store both the kill password, which permanently disables the tag when activated, and the access password, which restricts unauthorised reading or writing of the tag's data (each 32 bits in length). By default, both passwords are set to '0' to disable them effectively. The kill password permanently deactivates the tag, rendering it inoperable. While this function is rarely used, it proves useful in certain sectors, such as retail, where the RFID tag is often disabled at checkout.

The EPC memory bank is designed to store the EPC, which follows the standard for uniquely identifying retail products. Additionally, the EPC memory can be repurposed for storing personal codes, service numbers, inventory numbers, serial numbers and other customisable data.

The TID memory is dedicated solely to store the unique ID number of the tag, which is assigned by the manufacturer during chip production. Typically, this memory is non-editable. The initial segment of the TID identifies the distinct set of codes of manufacturers. As a result, the TID is generally globally unique. For applications requiring a unique identifier for objects or individuals, using the TID offers a straightforward solution, bypassing the need for EPC programming.

If additional memory beyond the EPC section is required, some chips feature extended user memory, allowing for the storage of more information. However, there is no standardised limit on the number of memory bits that can be written to the user memory of each tag.

5.2.2 The Working Process of RFID Tag

The workflow diagram of RFID tags is shown in Fig. 5.3. The working process of an RFID tag begins when an RFID reader emits an electromagnetic field that activates nearby tags.

Passive tags rely on this energy to operate, while active and semi-passive tags use an internal battery for power. Once activated, the tag's antenna captures the reader's radio frequency signals, and the microchip (IC) processes the signal, encoding stored data such as a unique identification number (UID). The tag subsequently modulates the incident radio waves and reflects the modulated signal to the reader through a process known as backscatter communication. The RFID reader receives this signal, demodulates the data and transmits it to a backend system or computer for further processing. Security checks, such as encryption or authentication, may be performed to verify the tag's authenticity. Upon processing, the retrieved data can be

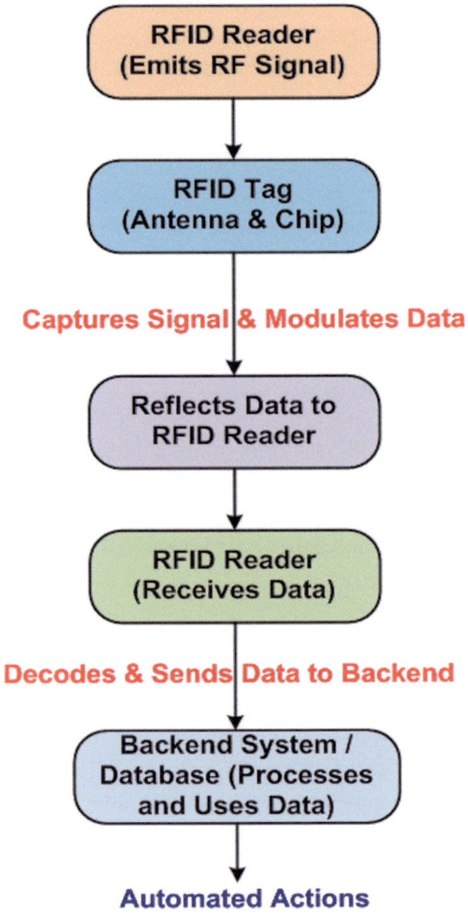

Fig. 5.3 Flow diagram of the RFID tag working

utilised across various applications, including access control, inventory management, asset tracking and supply chain monitoring, thereby enabling real-time visibility and operational efficiency within these systems.

The communication range and operational efficiency of an RFID system are influenced by several factors, including the operating frequency (LF, HF, UHF), antenna design, tag type, reader power output, environmental conditions, and the presence of interfering materials, namely metals and liquids.

5.2.3 RFID Reader

An RFID Reader (Interrogator) is an essential portion of any RFID system. It is responsible for communicating with RFID tags to retrieve or write data. It is a hardware device that releases radio waves via an antenna to detect and interact with RFID tags within its range. It 'interrogates' the tag by sending out a signal, which activates the tag, causing it to respond with its stored data.

An RFID reader emits radio frequency (RF) signals through its antenna to connect with nearby tags. When a tag enters the reader's RF field, it gets activated. In case of passive tags, the tag harvests energy from the reader's signal to power its internal circuitry, while active tags use their built-in battery. Once activated, the tag returns its stored data, typically a unique identification number, to the reader. This return signal is captured by the reader's antenna and processed by its internal receiver module. The received data is then decoded and sent to a connected computer or database through a communication interface, namely USB, Ethernet, Wi-Fi or Bluetooth. The entire process happens wirelessly and can read multiple tags simultaneously, making RFID systems highly efficient for inventory management, access control and asset tracking applications.

(a) **Types of RFID Readers**

i. Fixed Reader: Installed in permanent locations like entry gates, warehouses.
ii. Handheld Reader: Portable, can move around to scan items individually.
iii. Integrated Reader: Combines antenna and reader in a single device.

5.3 Nanocomposites Used in the Parts of Aeronautical Engineering RFID Tags

The aerospace industry continually seeks materials that offer lightweight characteristics, high strength, and multifunctionality to improve performance and efficiency [31]. Nanocomposites are materials that incorporate nanoparticles into a matrix (such as ceramics, polymers or metals) to increase their mechanical, thermal and electrical properties. In aeronautical engineering, nanocomposites performance a vital role in improving performance, reducing weight and enhancing durability. One of their emerging applications is in RFID tags, which are used for tracking and identification in aerospace industries. In the aerospace sector, RFID is used for:

- Tracking aircraft components and tools.
- Reducing maintenance time by providing real-time data on part conditions.
- Ensuring compliance with safety and regulatory standards.
- Enhancing inventory management in aviation manufacturing.

However, traditional RFID tags face challenges in extreme aeronautical environments, such as high temperatures, strong vibrations and electromagnetic interference. Nanocomposites enhance RFID tag performance in the following ways:

- ***Lightweight Design***: Reduces the overall weight of aircraft parts.
- ***High Strength and Durability***: It withstands extreme aeronautical conditions.
- ***Improved Conductivity***: Enhances signal transmission and reception.
- ***Thermal and Chemical Stability***: Protects against heat and corrosive fuels.

Nanomaterials, distinguished by their high specific surface area and unique surface morphology conducive to the integration of flexible components, significantly progress the operational presentation and efficiency of RFID sensor systems [32–35]. Strategic material selection and the investigation of varied structural configurations play a critical role in enhancing sensor reliability, selectivity and dynamic response characteristics, including rapid response and recovery to external stimuli [36–40]. The incorporation of diverse materials in the production of RFID sensor components facilitates the seamless incorporation of wireless communication functionalities with advanced sensing capabilities [41–45].

5.3.1 Nanomaterials for RFID-Based Sensing System

RFID-based sensing devices enable wireless detection systems capable of accurately monitoring a diverse range of external stimuli, thereby expanding the functional scope of conventional RFID technology beyond identification and tracking. These systems leverage a spectrum of nanomaterials to enhance sensing performance and broaden their functional capabilities. ***Polymers*** are widely employed as substrate materials in RFID sensors owing to their mechanical flexibility, cost-effectiveness and ease of fabrication. Among them, conducting polymers attract significant interest as sensing elements due to their ability to exhibit selective responses towards specific analytes. Additionally, ***carbon-based nanomaterials*** and ***metal nanoparticles*** are extensively utilised as conductive inks, patterned onto RFID components such as sensing interfaces and antennas to enhance electrical performance [46]. Ongoing research continues to explore a diverse range of materials with tailored compositions and nanostructures to advance the functional capabilities and sensitivity of RFID-based sensing devices.

Polymers are preferred as primary substrate materials for sensor tags due to their flexibility, processability and cost-efficiency [47]. Commonly used polymer substrates include polyethene terephthalate, polyethene naphthalate, polyethersulfone, polyimide, polycarbonate and polydimethylsiloxane [48]. The polyethene terephthalate-based RFID sensor demonstrated excellent mechanical stability under bending conditions and maintained structural and functional integrity when exposed to various environmental stresses [49].

The dielectric constant of the substrate plays a critical role in determining the sensor's sensitivity, as it directly influences the electromagnetic response mechanism of the device under varying environmental conditions such as temperature, humidity and mechanical stress, which can alter the dielectric properties and thereby affect signal strength and accuracy [50]. Additionally, the dielectric properties of the substrate significantly influence critical performance parameters of an RFID system, including overall readout efficiency, impedance matching and antenna bandwidth, thereby affecting the system's communication reliability and sensing accuracy [51]. Hence, careful selection of polymers is essential during design and fabrication.

Conducting polymers exhibit a range of tuneable properties attributed to the conjugated construction of their polymer chains. Notably, an enlarged density of charge carriers within these chains enhances electrical conductivity while enabling high sensitivity toward specific analytes. This is achieved through the strategic selection of monomer units and the incorporation of various dopants, which modulate the electronic structure and interaction sites of the polymer matrix. Commonly employed conducting polymers in RFID sensor systems include polyaniline, polypyrrole, polystyrene sulfonate and polythiophene [52]. The intrinsically high electrical conductivity of conducting polymers renders them suitable for antenna fabrication, for sensing tags, contributing to efficient signal transmission and enhanced device performance [53]. Nanostructured forms of conducting polymers improve the effective exposure of target analytes by providing an increased number of interaction sites, attributed to their high surface-area-to-volume ratio [54].

Metal nanostructures are incorporated as essential components in sensing tags owing to their high specific surface area, admirable electrical conductivity, ease of fabrication and superior mechanical robustness for enhanced sensing performance and device reliability [55]. In RFID sensor systems, metal nanostructures are commonly employed in the fabrication of conductive patterns or antennas, serving as potential alternatives to conventional integrated circuit (IC) chips [56]. The electrical conductivity of metal nanoparticle inks plays a pivotal role in determining the performance of conductive patterns and antennas, directly influencing parameters such as signal transmission efficiency, antenna gain and overall system reliability. Increased electrical conductivity shows enhanced electron mobility within the material, subsequent in lower electrical resistance and enabling the efficient and rapid transmission of signals throughout the circuit [57].

Carbon-based materials are gaining fast attention due to their low-density with high specific strength, offering a combination of desirable properties, including lightweight nature, superior mechanical strength, excellent corrosion resistance and outstanding thermal stability. These characteristics make carbon-based materials highly suitable for the production of components in sensor systems, namely sensor antennas and sensing interfaces, where durability, sensitivity and environmental stability are critical [58].

5.3.2 Applications of Nanocomposite-Based RFID in Aeronautical Engineering

Nanocomposite-based RFID systems incorporate advanced nanomaterials into RFID tags and sensors, enhancing their mechanical, thermal and electromagnetic performance. In aeronautical engineering, these enhanced systems enable a range of high-value applications, as illustrated in Fig. 5.4. The most common types of nanocomposites used in RFID tags and their properties, and advantages are provided in Table 5.3.

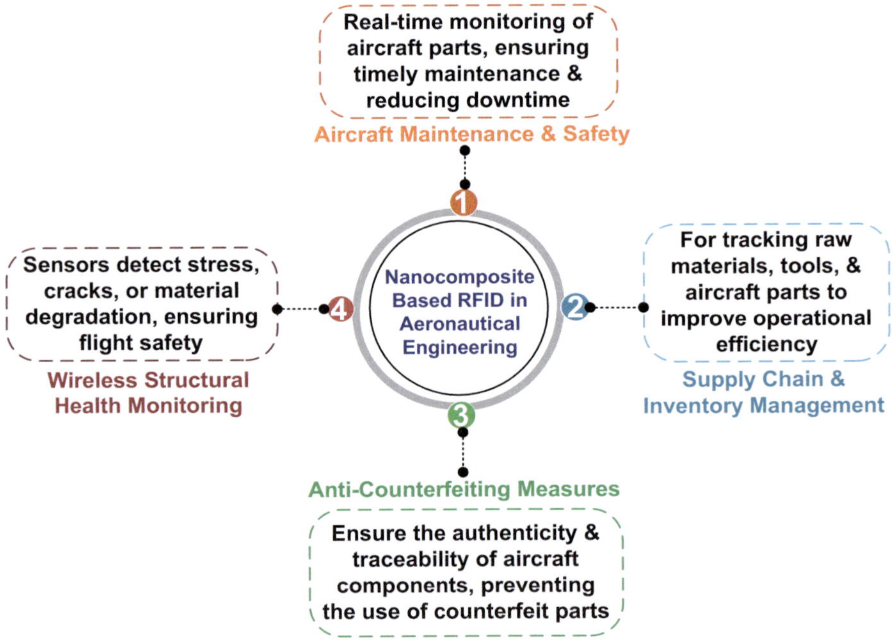

Fig. 5.4 Applications of nanocomposite RFID in aeronautics

Table 5.3 Types of nanocomposites used in RFID tags

Type of the composites	Applications	Properties	Advantages
Carbon nanotube nanocomposites	Used as an antenna material in RFID tags	High electrical conductivity, lightweight, flexible	Enhances signal transmission efficiency
Graphene-based nanocomposites	Used in RFID tag substrates for flexible and stretchable electronics	High strength, thermal resistance and electrical conductivity	Improves durability and reliability under aeronautical conditions
Polymer nanocomposites	Used in RFID casings and encapsulation	Lightweight, corrosion-resistant, high-impact strength	Protects RFID circuits from external damage
Metal nanoparticle-embedded composites	Used to enhance RFID sensor sensitivity and EMI shielding	High reflectivity and conductivity	Reduces interference and improves signal clarity

5.3.3 Enhancing RFID Performance Through Nanocomposite Integration

RFID technology has established itself as a pivotal tool for wireless object identification and tracking in the logistics, healthcare, aerospace and retail industries. However, conventional systems often encounter performance limitations, including reduced read ranges, signal attenuation, poor environmental resilience and challenges in antenna miniaturisation for high-frequency applications [59]. Recent advancements in nanocomposite materials, namely hybrid structures composed of a matrix (polymer) reinforced with fillers (nanoscale), namely graphene and carbon nanotubes, and metallic nanoparticles, have revealed great potential in addressing these shortcomings. When integrated into RFID tags, particularly in the antenna and substrate components, nanocomposites provide several tangible benefits:

(a) ***Antenna Conductivity and Efficiency***

The antenna is a critical component of an RFID tag, responsible for sending and receiving signals. Incorporating nanomaterials such as graphene and carbon nanotubes into antenna structures improves electrical conductivity, resulting in longer read ranges and higher signal sensitivity. For instance, graphene-based conductive inks enable inkjet-printed antennas that are flexible, lightweight and more efficient than traditional metal-based antennas [60]. Nanomaterials, namely graphene and silver nanowires, substantially reduce electrical resistance, thereby improving antenna efficiency and enabling thinner, lightweight designs [61]. High-gain antennas enhance power conversion efficiency and enable increased power acquisition, making them the preferred choice in many applications. The performance of an antenna can be characterised by parameters such as gain, operating frequency band, radiation pattern, polarisation, physical dimensions and intended field of application [62]. Consequently, the development of innovative antenna designs aimed at enhancing the performance of sensor devices across diverse application scenarios has emerged as a prominent focus in current research. The different RFID antennas and their technical specifications are summarised in Table 5.4.

(b) ***Structural Integration and Embedding***

Nanocomposites allow for the embedding of RFID tags directly into aircraft composite structures such as wings, fuselage or turbine components. This not only protects the RFID tags from environmental damage but also enables seamless asset tracking without additional weight or interference. Recent advancements in carbon fibre-reinforced composites have shown successful integration of RFID systems without compromising mechanical integrity [63].

(c) ***EMI Shielding and Signal Integrity***

Table 5.4 Various categories of antennas

Type of antenna	Carrier frequency	Antenna size	Measuring range	References
Metallic surface mountable antennas	UHF	30 × 18 × 3.2 mm^3	1.5 m	[65]
		106 × 44 × 5 mm^3	NA	[66]
		104 × 31 × 7.6 mm^3	14.6 m	[67]
		41.5 × 55 × 3 mm^3	Metal: 6.1 m Dielectrics: 14.1 m	[68]
3D antennas	UHF	3 × 3 × 3 cm^3	NA	[69]
		50 × 50 × 50 mm^3		[70]
		2 × 2 × 1.2 cm^3		[71]
Multi-band antennas	UHF-HF	85 × 54 × 0.8 mm^3	NA	[72]
	UHF-MW	L: 19 cm		[73]
		30 × 30 mm^2		[74]
Omnidirectional and directional antennas	UHF	98.7 × 14.2 mm^2 (866 MHz)	NA	[75]
		88 × 14.2 mm^2 (915 MHz)		
		84 × 14.2 mm^2 (953 MHz)		
		43 × 43 mm^2	9.5 m	[76]
	UHF near-field	R:80 mm	120 mm	[77]

Electromagnetic interference is a significant concern in aircraft due to the high density of electronic systems. Graphene and ferrite nanoparticle-reinforced composites provide excellent EMI shielding, which ensures the RFID tags function reliably even in high-noise environments. These composites maintain signal integrity and reduce the risk of misreads or system failures [64].

(d) ***Improved Mechanical and Thermal Stability***:

Nanocomposite-based substrates exhibit superior flexibility, durability and resistance to mechanical stress, moisture and temperature fluctuations, making them suitable for harsh environments such as aerospace and biomedical applications [78].

(e) ***Tunable Dielectric Properties***:

The dielectric constant of nanocomposites can be tailored through the choice and concentration of nanofillers, optimising antenna impedance matching and enabling miniaturisation without compromising performance [79].

(f) ***Extended Read Range and Higher Sensitivity***:

By improving the electromagnetic properties of RFID tag components, nanocomposites help achieve extended read ranges and heightened sensitivity, even in electromagnetically noisy or metallic environments [79].

5.4 Conclusion

The integration of nanocomposites into aeronautical engineering, specifically in the fabrication and enhancement of RFID tags, represents a significant advancement in aerospace technology. Through this chapter, we have explored how nanocomposites, owing to their exceptional mechanical strength, lightweight nature, thermal stability and electromagnetic shielding properties, offer distinct advantages over conventional materials in demanding aeronautical environments. The application of these advanced materials in RFID tags enhances their durability, operational reliability, and data transmission efficiency, even under extreme conditions of temperature, pressure and vibration typically encountered in aerospace operations. Furthermore, the integration of nanomaterials and metallic nanoparticles has opened new avenues for the development of flexible, miniaturised and multifunctional RFID systems suitable for asset tracking, structural health monitoring and supply chain management in aeronautics. As the aerospace industry continues to demand materials that offer superior performance with reduced weight and increased functionality, nanocomposites will play a key role in driving next-generation RFID technologies. Upcoming research should emphasise on optimising the synthesis processes, improving material compatibility and exploring novel nanomaterial combinations to further enhance the performance of RFID tags tailored for aeronautical applications. In essence, the synergy between nanocomposite technology and aeronautical RFID systems holds transformative potential for improving operational efficiency, safety and asset management within the aerospace sector.

References

1. Finkenzeller K (2010) RFID handbook: fundamentals and applications in contactless smart cards and identification, 3rd ed., Wiley
2. Weinstein R (2005) RFID: a technical overview and its application to the enterprise. IT Professional 7(3):27–33. https://doi.org/10.1109/MITP.2005.69
3. Li S, Visich JK (2006) Radio frequency identification: supply chain impact and implementation challenges. Int J Int Supply Manage 2(4). https://doi.org/10.1504/IJISM.2006.009643
4. Lee YM, Cheng F, Leung (2004) Exploring the impact of RFID on supply chain dynamics. In: Proceedings of the 2004 winter simulation conference. https://doi.org/10.1109/WSC.2004.1371441
5. Ahsan K, Shah H, Kingston P (2010) RFID applications: an introductory and exploratory study. Int J Comput Sci Iss 7(1). https://doi.org/10.48550/arXiv.1002.1179
6. Kumar P, Reinitz HW, Simunovic J, Sandeep KP, Franzon PD (2009) Overview of RFID technology and its applications in the food industry. J Food Sci 74:R101–R106. https://doi.org/10.1111/j.1750-3841.2009.01323.x
7. Sun C, Jiang F, Jiang S (2013) Research on RFID Applications in Construction Industry. J Networks 8:1221–1228. https://doi.org/10.4304/jnw.8.5.1221-1228
8. Veena AP, Rathor KO (2014) RFID technology: a performance differentiator. Library Herald 52. https://doi.org/10.5958/j.0976-2469.52.1.006
9. Kaur M, Sandhu MK, Mohan N, Sandhu P (2011) RFID technology principles, advantages, limitations and its applications. Int J Comput Electr Eng 3(1):151–157. https://doi.org/10.7763/IJCEE.2011.V3.306
10. Xiao Y, Senhua Y, Kui W, Ni Q, Janecek C, Nordstad J (2007) Radio frequency identification: technologies, applications, and research issues. Wirel Commun Mob Comput 7:457–472. https://doi.org/10.1002/wcm.365
11. Ezzeddine H, Merakeb Y, Huillery J, Breard A, Duroc Y (2023) Pulsing the passive RF identification technology: the pulsed wave mode. IEEE Microwave Mag 24(10):40–50. https://doi.org/10.1109/MMM.2023.3293613
12. Chawla V, Ha DS (2007) An overview of passive RFID. IEEE Commun Magaz 45(9):11–17. https://doi.org/10.1109/MCOM.2007.4342873
13. Preradovic S, Karmakar NC, Balbin I (2008) RFID transponders. IEEE Microwave Mag 9(5):90–103. https://doi.org/10.1109/MMM.2008.927637
14. Rao K (1999) An overview of backscattered radio frequency identification system (RFID), 1999 Asia Pacific microwave conference. APMC'99. Microwaves Enter the 21st Century. Conference Proceedings (Cat. No.99 TH8473)1999. https://doi.org/10.1109/APMC.1999.833700
15. Saleem H, Zamin M, Khan A, Afzal S (2012) Review of various aspects of radio frequency identification (RFID) technology. IOSR J Comput Eng 8(1):01–06. https://doi.org/10.9790/0661/0810106
16. Chuang M-L, Shaw WH (2005) How RFID will impact supply chain networks. In: Proceedings of IEEE international engineering management conference, pp 231–235. https://doi.org/10.1109/IEMC.2005.1559124
17. Falinski W (2006) An overview of radio frequency identification (RFID) tags technology. Int Soc Opt Eng/Proc SPIE, 634720–634727. https://doi.org/10.1117/12.714563
18. What is Passive RFID, Atlas RFID Store, www.atlasrfidstore.com/rfid-insider/what-is-passiverfid/?srsltid=AfmBOopqI_7ypHm2vC_EjXcCHFwxSGabNcqaLHlgTm6wjOEvMPEfV279. Accessed 29 Mar 2025

19. Ramo S, Whinnery JR, Van Duzer T (1994) Fields and waves in communication electronics, 3rd ed. Wiley
20. Domdouzis K et al (2007) Radio-frequency identification (RFID) applications: a brief introduction. Adv Eng Inf 21(4):350–355. https://doi.org/10.1016/j.aei.2006.09.001
21. Chuang M-L, Shaw W (2007) RFID: integration stages in supply chain management. IEEE Eng Manage Rev 35(2):80–87. https://doi.org/10.1109/EMR.2007.899757
22. Want R (2004) The magic of RFID. Queue 2(7):40–48. https://doi.org/10.1145/1035594.1035619
23. RFID Tags and Labels. RFID4U Store, https://rfid4ustore.com/rfid-tags-labels/rfid. Accessed 2 Apr. 2025
24. Solar H et al (2023) Semi-passive UHF RFID sensor tags: a comprehensive review. IEEE Access 11:135583–135599. https://doi.org/10.1109/ACCESS.2023.3336761
25. Ahson S, Ilyas M (2008) RFID handbook: applications. Security, and Privacy. CRC Press, Technology
26. Dobkin DM (2012) The RF in RFID: passive UHF RFID in practice. Elsevier
27. Banks J, Hanny D, Pachano M, Thompson L (2007) RFID Applied. Wiley
28. Liu H (2013) Advances in RFID technology and applications. IntechOpen
29. https://rfid4ustore.in/blog/the-anatomy-of-rfid-tags-components-and-functionality/ (Date of Access: 08 July 2025)
30. Ashby MF (2010) Materials selection in mechanical design, 4th ed. Butterworth-Heinemann
31. Yao S, Ren P, Song R, Liu Y, Huang Q, Dong J, O'Connor BT, Zhu Y (2020) Nanomaterial-enabled flexible and stretchable sensing systems: Processing, integration, and applications. Adv Mater 32:1902343. https://doi.org/10.1002/adma.1902343
32. Li W, Zhan X, Song X, Si S, Chen R, Liu J, Wang Z, He J, Xiao X (2019) A review of recent applications of ion beam techniques on nanomaterial surface modification: design of nanostructures and energy harvesting. Small 31:1901820. https://doi.org/10.1002/smll.201901820
33. Moinudeen GK, Ahmad F, Kumar D, Al-Douri Y, Ahmad S (2017) IoT applications in future foreseen guided by engineered nanomaterials and printed intelligence technologies: a technology review. IEEE Internet Things J 6:106–148. https://doi.org/10.1109/JIOT.2017.2783635
34. Choy KL, Ho GTS, Lee CKH (2014) An RFID-based storage assignment system for enhancing the efficiency of order picking. J Intell Manuf 28:111–129. https://doi.org/10.1007/s10845-014-0883-4
35. Subramanian V, Frechet JMM, Chang PC, Huang DC, Lee JB, Molesa SE, Redinger DR, Volkman SK (2005) Progress toward development of all-printed RFID tags: materials, processes, and devices. Proc IEEE 93:1330–1338. https://doi.org/10.1109/JPROC.2005.852872
36. Saarinen K, Bjorninen T, Ukkonen L, Frisk L (2014) Reliability analysis of RFID tags in changing humid environment. IEEE Trans Compon Packag Manuf Technol 4:77–85. https://doi.org/10.1109/TCPMT.2013.2297303
37. Potyrailo RA, Nagraj N, Tang Z, Mondello FJ, Surman C, Morris W (2012) Battery-free radio frequency identification (RFID) sensors for food quality and safety. J Agric Food Chem 60:8535–8543. https://doi.org/10.1021/jf301104b
38. Potyrailo RA, Morris WG, Sivavec T, Tomlinson HW, Klensmeden S, Lindh K (2008) RFID sensors based on ubiquitous passive 13.56-MHz RFID tags and complex impedance detection. Wireless Commun. Mobile Comput. 9:1318–1330. https://doi.org/10.1002/wcm.706
39. Gee KM, Andandarajah P, Colline D (2020) Current progress towards the integration of thermocouple and chipless RFID technologies and the sensing of a dynamic stimulus. Micromachines 11(10):1019. https://doi.org/10.3390/mi11101019
40. Baumbauer CL, Anderson MG, Ting J, Sreekumar A, Rabaey JM, Arias AC, Thielens A (2020) Printed, flexible, compact UHF-RFID sensor tags enabled by hybrid electronics. Sci Rep 10:16543. https://doi.org/10.1038/s41598-020-73435-5
41. Yang W, Cheng X, Guo Z, Sun Q, Wang J, Wang C (2023) Design, fabrication and applications of flexible RFID antennas based on printed electronic materials and technologies. J Mater Chem C 11:406–425. https://doi.org/10.1039/D2TC05056B

42. Sun X, Zhan C, Li H, Yu H, Zhang J, Qiu H, Liang J, Wu J, Su M, Shi Y et al (2022) Wearable near-field communication sensors for healthcare: materials, fabrication and application. Micromachines 13(5):784. https://doi.org/10.3390/mi13050784
43. Sanchez-Romaguera V, Wünscher S, Turki BM, Abbel R, Barbosa S, Tate DJ, Oyeka D, Batchelor JC, Parker EA, Schubert US et al (2015) Inkjet printed paper-based frequency selective surfaces and skin mounted RFID tags: the interrelation between silver nanoparticle ink, paper substrate and low temperature sintering technique. J Mater Chem C 3:2132–2140. https://doi.org/10.1039/C4TC02188D
44. Sarycheva A, Polemi A, Liu Y, Dandekar K, Anasori B, Gogotsi Y (2018) 2D titanium carbide (MXene) for wireless communication. Sci Adv 4,:eaau0920. https://doi.org/10.1126/sciadv.aau0920
45. Dhawan SK, Dey DC, Kumar R (2020) Nanomaterials-based conductive inks for printed electronics: a review. Mater Today: Proc 28:1725–1730. https://doi.org/10.1016/j.matpr.2020.06.004
46. Yang L, Rida A, Vyas R, Tentzeris MM (2007) RFID tag and RF structures on a paper substrate using inkjet-printing technology. IEEE Trans Microw Theory Tech 55(12):2894–2901. https://doi.org/10.1109/TMTT.2007.909886
47. Yakimets I, MacKerron D, Giesen P, Kilmartin KJ, Goorhuis M, Meinders E, MacDonald WA (2010) Polymer substrates for flexible electronics: achievements and challenges. Adv Mater Res 93–94:5–8. https://doi.org/10.4028/www.scientific.net/AMR.93-94.5
48. Zhang L, Tan Q, Kou H, Wu D, Zhang W, Xiong J (2019) Highly sensitive NH3 wireless sensor based on Ag-RGO composite operated at room temperature. Sci Rep 9:9942. https://doi.org/10.1038/s41598-019-46374-5
49. Sharma AK, Singh AK, Akhtar MJ (2014) Dielectric characterisation of substrate materials for RFID applications and their effect on tag performance. IEEE Trans Antennas Propag 62(1):230–238. https://doi.org/10.1109/TAP.2013.2287285
50. Zaid J, Farahani M, Denidni TA (2017) Magneto-dielectric substrate-based microstrip antenna for RFID applications. IET Microw. Antennas Propag. 11:1389–1392. https://doi.org/10.1049/iet-map.2016.0799
51. Yoon H, Jang J (2009) Conducting-polymer nanomaterials for high-performance sensor applications: issues and challenges. Adv Funct Mater 19:1567–1576. https://doi.org/10.1002/adfm.200801324
52. Guerchouche K, Herth E, Calvet LE, Roland N, Loyez C (2017) Conductive polymer-based antenna for wireless green sensors applications. Microelectron Eng 182:46–52. https://doi.org/10.1016/j.mee.2017.07.004
53. Jun J, Oh J, Shin DH, Kim SG, Lee JS, Kim W, Jang J (2016) Wireless, room temperature volatile organic compound sensor based on polypyrrole nanoparticle immobilized ultrahigh frequency radio frequency identification tag. ACS Appl Mater Interfaces 8:33139–33147. https://doi.org/10.1021/acsami.6b11829
54. Murray RW (2008) Nanoelectrochemistry: metal nanoparticles, nanoelectrodes, and nanopores. Chem Rev 108:2688–2720. https://doi.org/10.1021/cr0783776
55. Raut NC, Al-Shamery K (2018) Inkjet printing metals on flexible materials for plastic and paper electronics. J Mater Chem C 6:1618. https://doi.org/10.1039/C7TC05147D
56. Sahu A, Aaen PH, Lewandowski A, Shkunov M, Rigas G, Blanchard PT, Wallis TM, Devanbhaktuni VK (2017) Robust microwave characterisation of inkjet-printed coplanar waveguides on flexible substrates. IEEE Trans Instrum Meas 66:3271–3279. https://doi.org/10.1109/TIM.2017.2679704
57. Titirici MM, White RJ, Brun N, Budarin VL, Su DS, Del Monte F, Clark JH, MacLachlan MJ (2015) Sustainable carbon materials. Chem Soc Rev 44:250. https://doi.org/10.1039/C4CS00223J
58. Finkenzeller K (2010) RFID handbook: fundamentals and applications in contactless smart cards and identification, 3rd ed. Wiley
59. Roh S, Nguyen TD, Lee JS (2024) Applications of nanomaterials in RFID wireless sensor components. Appl Sci 14(3):1216

60. Mousavi SM et al (2021) Graphene-based materials for advanced RFID applications: a review. Carbon 182:837–857
61. Cui L, Zhang Z, Gao N, Meng Z, Li Z (2019) Radio frequency identification and sensing techniques and their applications—A review of the state-of-the-art. Sensors 19(18):4012. https://doi.org/10.3390/s19184012
62. Hoa S (2022)Advanced carbon-fibre composite materials for RFID tag antenna applications. Appl Comput Electromagnetics Soc J
63. Kargar F et al (2018) Graphene composites with dual functionality: electromagnetic shielding and thermal management. ArXiv preprint arXiv:1808.03401. https://doi.org/10.48550/arXiv.1808.03401
64. Chen SL (2009) A miniature RFID tag antenna design for metallic objects application. IEEE Antennas Wireless. Propagat Lett 8:1043–1045
65. Xu L, Tian L, Hu B (2007) A novel broadband UHF RFID tag antenna mountable on metallic surface. In: Proceedings 2007 international conference wireless communication, network mobile computer. Shanghai, China, pp 2128–2131
66. Hamani A, Yagoub MCE, Vuong TP, Touhami R (2016) A novel broadband antenna design for UHF RFID tags on metallic surface environments. IEEE Antennas Wirel Propag Lett 16:91–94
67. Li H, Zhu J, Yu Y (2017) Compact single-layer RFID tag antenna tolerant to background materials. IEEE Access 5:21070–21079
68. Kruesi CM, Vyas RJ, Tentzeris MM (2009) Design and development of a novel 3-D cubic antenna for wireless sensor networks (WSNs) and RFID applications. IEEE Trans Antennas Propag 57(10):3293–3299
69. Jin X, Dang X, Yang F (2014) Compact 3-D cubic loop antennas with omnidirectional patterns. In Proceeding 2014 IEEE antennas propagation society international symposium (APSURSI), Memphis, TN, USA, Jul. 6–11, 2014, pp 380–381
70. Galehdar A, Thiel DV, O'Keefe SG (2009) Design methods for 3D RFID antennas located on a conducting ground plane. IEEE Trans Antennas Propag 57(10):339–346
71. Ma ZL, Jiang LJ, Xi J, Ye TT (2012) A single-layer compact HF-UHF dual-band RFID tag antenna. IEEE Antennas Wirel Propag Lett 11:1257–1260
72. Keyrouz S, Visser HJ, Tijhuis AG (2013) Multi-band simultaneous radio frequency energy harvesting. In: Proceedings 2013 7th european conference on antennas and propagation (EuCAP), Gothenburg, Sweden, Apr. 8–12, 2013, pp 3058–3061
73. Li HH, Mou XQ, Ji Z, Yu H, Li Y, Jiang L (2011) Miniature RFID tri-band CPW-fed antenna optimised using ISPO algorithm. Electron Lett 47(3):161–162. https://doi.org/10.1049/el.2010.2750
74. Tang ZJ, He YG, Wang Y (2011) Broadband UHF RFID tag antenna with quasi-isotropic radiation performance. AEU-Int J Electron C 65(9):859–863. https://doi.org/10.1016/j.aeue.2011.01.012
75. Liu Q, Yu Y, He S (2013) Capacitively loaded, inductively coupled fed loop antenna with an omnidirectional radiation pattern for UHF RFID tags. IEEE Antennas Wirel Propag Lett 12:1161–1164. https://doi.org/10.1109/LAWP.2013.2262229
76. Zeng Y, Chen ZN, Qing X, Jin JM (2018) A directional, closely spaced zero-phase-shift-line loop array for UHF near-field RFID reader antennas. IEEE Trans Antennas Propag 66(10):5639–5642. https://doi.org/10.1109/TAP.2018.2858309
77. Gupta RK, Kennel E (2010) Polymer nanocomposites handbook. CRC Press
78. Brinker CJ et al (2012) Nanocomposite dielectric materials for miniaturised RFID tags. Adv Funct Mater 22(13):2769–2776. https://doi.org/10.1002/adfm.201102566
79. Souri H, Banerjee H, Jusufi A, Radacsi N, Stokes A, Park I, Sitti M, Amjadi M (2020) Wearable and stretchable strain sensors: materials, sensing mechanisms, and applications. Adv Intell Syst 2. https://doi.org/10.1002/aisy.202000039

Chapter 6
Nanocomposites for Defense, Aeronautical, and Aerospace Applications: Challenges

Wejden Gazehi, Rania Loukil, and Mongi Besbes

Abstract Challenges in using nanocomposites for defense and space include cost of production, scaling up manufacturing processes, ensuring material stability and quality control, managing health and environmental risks, and the need for advanced characterization techniques to understand their complex properties. Despite their potential for being lightweight, improved strength, improved thermal properties, and advanced functionalities such as stealth and sensor integration, integrating these advanced materials into existing systems also poses obstacles.

Keywords Durability · Efficiency across defense · Aeronautical · Aerospace sectors · Propelled materials · Nanocomposites · Materials engineered · Nanoscale · Exhibit superior · Conventional counterparts · Revolutionize critical

6.1 Introduction

The relentless pursuit of enhanced performance, durability, and efficiency across defense, aeronautical, and aerospace sectors has propelled materials science to the forefront of innovation [1]. Among the most promising avenues of exploration lies the study of nanocomposites—materials engineered at the nanoscale to exhibit superior properties compared to their conventional counterparts. The integration of these advanced materials holds the potential to revolutionize critical applications, from next-generation military armaments to lightweight, fuel-efficient aircraft and robust spacecraft [2]. However, the path to widespread implementation is fraught with

W. Gazehi · R. Loukil · M. Besbes (✉)
University of Carthage, Sidi Bou Said, Av. de la République, 1054, Carthage, Tunis, Tunisia
e-mail: mongi.besbes@ucar.tn

W. Gazehi
e-mail: wejden.gazehi@istic.ucar.tn

R. Loukil
e-mail: rania.loukil@istic.ucar.tn

S. K. Panda (ed.), *Nanocomposites for Defense, Aeronautical and Aerospace Applications*, Advanced Structured Materials 248,
https://doi.org/10.1007/978-3-032-11568-3_6

challenges that demand meticulous research and innovative solutions. Nanocomposites, by definition, are multiphase solid materials where at least one dimension of one of the phases is in the nanometer scale (1–100 nm). This nanoscale manipulation imparts unique characteristics, such as increased strength, enhanced thermal stability, improved electrical conductivity, and superior wear resistance. In defense, this translates to lighter yet more resilient armor, advanced sensor systems, and high-performance structural components. In aeronautics, nanocomposites offer the prospect of reduced aircraft weight, leading to significant fuel savings and decreased emissions, alongside improved structural integrity and resistance to environmental degradation [2]. In aerospace, where materials face extreme conditions, nanocomposites can provide the necessary protection against radiation, thermal stress, and mechanical fatigue, enabling longer and more ambitious space missions. The allure of nanocomposites stems from their ability to overcome the limitations of traditional materials [3]. For instance, the incorporation of carbon nanotubes or graphene into polymer matrices can dramatically enhance their mechanical strength and stiffness, while maintaining a low density. Similarly, the addition of ceramic nanoparticles can improve the thermal stability and wear resistance of metallic alloys [4]. However, realizing the full potential of these materials requires addressing a multitude of challenges that span from synthesis and processing to characterization and long-term performance (Fig. 6.1).

One of the primary challenges lies in achieving uniform dispersion of nanoparticles within the matrix material [5]. Agglomeration of nanoparticles can lead to stress concentration points, compromising the overall mechanical properties of the

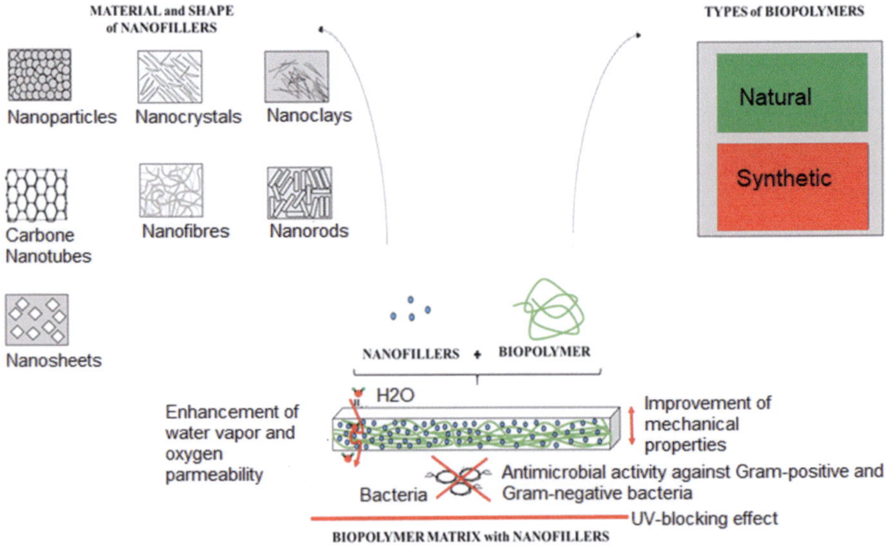

Fig. 6.1 Improvement the mechanical property of the material

nanocomposite [6]. Effective dispersion requires careful control of synthesis parameters, surface modification of nanoparticles, and optimized processing techniques. Furthermore, the interfacial bonding between nanoparticles and the matrix material plays a crucial role in determining the load transfer efficiency and overall performance of the nanocomposite [7]. Achieving strong interfacial bonding often necessitates the use of coupling agents or surface treatments, which add complexity to the manufacturing process. Scalability and cost-effectiveness are also significant hurdles. Many nanocomposite synthesis and processing techniques are still in their infancy, making it difficult to produce these materials in large quantities and at a reasonable cost. Developing scalable and cost-effective manufacturing processes is essential for the widespread adoption of nanocomposites in defense, aeronautical, and aerospace applications. Additionally, the long-term performance of nanocomposites in harsh environments is a critical concern. These materials must maintain their properties under extreme temperatures, high stresses, and prolonged exposure to radiation and corrosive substances. Understanding the degradation mechanisms and developing strategies to mitigate them are crucial for ensuring the reliability and durability of nanocomposite-based components.

Characterization of nanocomposites is another area that requires advanced techniques. Due to the nanoscale dimensions of the reinforcing phases, traditional characterization methods may not be sufficient to fully understand the structure–property relationships. Techniques such as transmission electron microscopy (TEM), atomic force microscopy (AFM), and nanoindentation are essential for probing the nanoscale features and mechanical properties of nanocomposites. Furthermore, computational modeling and simulation play a vital role in predicting the behavior of nanocomposites under various loading conditions and environmental exposures. In the defense sector, the development of advanced armor systems is a key area of focus. Nanocomposites offer the potential to create lightweight yet highly resistant armor that can protect soldiers and vehicles from ballistic threats. The incorporation of nanoparticles, such as silicon carbide or boron nitride, into polymer or metal matrices can significantly enhance the ballistic performance of armor materials. Additionally, nanocomposites are being explored for use in advanced sensor systems, where their enhanced electrical and optical properties can enable the development of highly sensitive and durable sensors.

In aeronautics, the drive for fuel efficiency and reduced emissions has led to a growing interest in lightweight materials. Nanocomposites offer the potential to create aircraft structures that are lighter and stronger than traditional materials, leading to significant fuel savings and reduced environmental impact. The use of carbon nanotube-reinforced polymer composites in aircraft wings and fuselages is one example of this trend. Furthermore, nanocomposites are being explored for use in aircraft engines, where their enhanced thermal stability can enable higher operating temperatures and improved efficiency. In aerospace, the harsh environment of space poses unique challenges for materials. Nanocomposites offer the potential to create spacecraft structures that are resistant to radiation, thermal stress, and mechanical fatigue. The use of ceramic matrix composites in thermal protection systems and radiation shielding is one example of this application. Additionally, nanocomposites are being explored for use in advanced propulsion systems, where their enhanced thermal and mechanical properties can enable the development of more efficient and reliable engines.

Figure 6.2 depicts a futuristic aircraft in flight, showcasing various innovative applications of nanotechnology in the aerospace sector. On the left side of the aircraft, labels highlight smart materials and coatings, such as self-healing materials, specialized paints, hydrophobic surfaces, and nano-structured coatings—all designed to enhance the aircraft's durability and protection. In the lower left corner, the green text emphasizes advancements in energy storage and production, including hydrogen storage in carbon nanotubes, the use of distributed energy sources such as fuel cells, miniature energy storage capsules, and nanoscale photovoltaic cells, all contributing to increased energy autonomy. At the center of the aircraft, technologies such as self-monitoring systems and morphing structures are mentioned, underscoring the aircraft's ability to adapt and reconfigure in real time to optimize performance. On the right side, communication and sensor applications are presented, including the enhancement of electrical conductivity through nanocomposites, the integration of antennas within nanocavities, and the use of wireless sensors to monitor airflow laminarity features that aim to reduce drag and improve efficiency. Finally, the lower fuselage highlights self-cleaning and anti-icing surfaces, which are crucial for ensuring safety and minimizing maintenance requirements. Overall, the image illustrates a concept of a smart, sustainable aircraft fully enhanced by the advancements of nanotechnology.

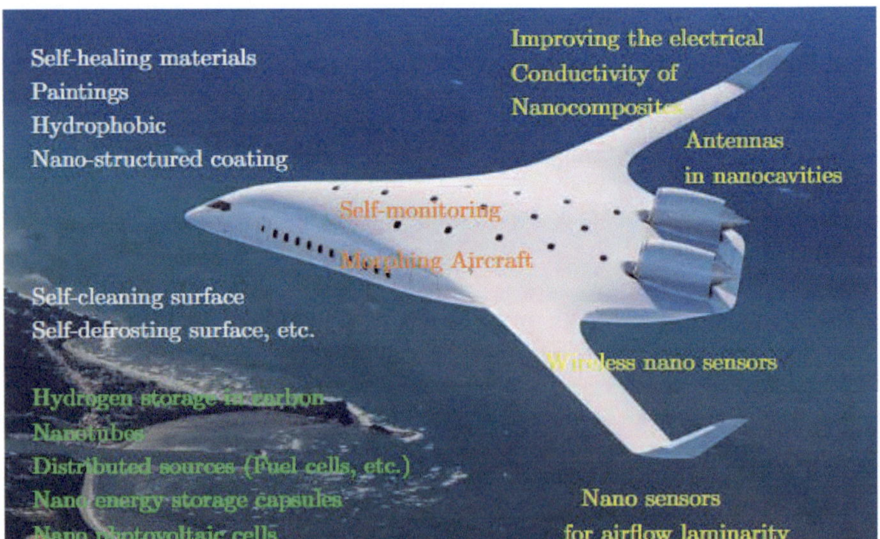

Fig. 6.2 Nanotechnology-enabled innovations in aerospace engineering

In conclusion, nanocomposites hold immense promise for revolutionizing defense, aeronautical, and aerospace applications. However, realizing their full potential requires addressing a multitude of challenges related to synthesis, processing, characterization, and long-term performance. Continued research and development efforts are essential to overcome these challenges and pave the way for the widespread adoption of nanocomposites in these critical sectors.

6.2 Nanocomposites for Defense, Applications

6.2.1 Generality

Nanocomposites are materials composed of a matrix (such as a polymer, metal, or ceramic) combined with nanoscale reinforcements (typically 1–100 nm in size) [8, 9], like carbon nanotubes, graphene, nanoclays, or metal nanoparticles. These reinforcements dramatically enhance the mechanical, thermal, electrical, and barrier properties of the base material. Nanocomposites are particularly valuable in defense applications because they address multiple critical performance requirements simultaneously [10]. Traditional materials often force a trade-off between strength and weight, but nanocomposites thanks to their nanoscale reinforcements like carbon nanotubes, nanoclays, or metal oxides can deliver exceptionally high strength-to-weight ratios, making them ideal for lightweight armor, aerospace structures, and unmanned vehicles [11]. Their enhanced thermal stability enables them to withstand the extreme temperatures encountered in propulsion systems, engine components, and high-speed projectiles. Moreover, nanocomposites exhibit superior resistance to wear, corrosion, and impact, which extends the service life and reliability of equipment used in harsh environments, such as naval vessels or desert operations [12]. Perhaps most importantly, these materials can be engineered for multifunctional capabilities, such as combining mechanical strength with electromagnetic interference (EMI) shielding, stealth (radar-absorbing) properties, or self-healing functions. This makes nanocomposites not just materials, but strategic enablers of next-generation defense technologies that demand durability, adaptability, and performance in one integrated package [13].

6.2.2 Applications

A more detailed breakdown of nanocomposite applications in defense, with examples of specific nanocomposite materials used in each category.

6.2.2.1 Armor and Ballistics

To protect soldiers and vehicles while reducing weight for greater mobility for Bulletproof vests, helmets, vehicle, and aircraft armor. For example, we cite Kevlar reinforced with carbon nanotubes (CNTs). In fact, CNTs improve tensile strength and energy absorption. Its results are in lighter, more flexible armor with equal or better protection than traditional systems. Besides Polymer nanocomposites with boron carbide nanoparticles which are used in ceramic armor for high hardness and low density [14].

6.2.2.2 Stealth Technology (Radar Absorbing Materials—RAM)

Stealth Technology: Nanocomposite-Based Radar Absorbing Materials (RAM)

Nanocomposites play a critical role in reducing the radar cross-section (RCS) of modern defense systems, including aircraft, drones, and naval vessels. These materials attenuate incident electromagnetic waves, thereby decreasing the vehicles' radar signature [15].

1. **Epoxy Nanocomposites Based on $Fe_3O_4$and Graphene**

Epoxy composites reinforced with magnetite (Fe_3O_4) nanoparticles or graphene exhibit outstanding electromagnetic absorption properties. For instance, epoxy composites containing 8 wt% graphene have demonstrated an electromagnetic shielding effectiveness of 70 dB in the EHF band (220–325 GHz), primarily through absorption, with transmission below 1% at 300 GHz [16]. The absorption efficiency depends on the dielectric and magnetic properties of the material, characterized by complex permittivity ($\varepsilon = \varepsilon' - j\varepsilon''$) and complex permeability ($\mu = \mu' - j\mu''$) [17]. Optimal absorption impedance is achieved when the impedance of the material (Z) matches that of free space ($Z_0 \approx 377\ \Omega$), thereby minimizing the reflection of incident waves (Fig. 6.3).

This structure represents a dense, extended framework composed of iron (Fe) and oxygen (O) atoms, typical of hydrated iron oxides such as ferrihydrite (Fig. 6.4).

- **Iron (Fe)**: Displayed in **yellow-orange**, iron atoms form a network of interconnected metal centers, often coordinated octahedrally by oxygen atoms.
- **Oxygen (O)**: Shown in **red**, oxygen atoms bridge between iron centers, forming Fe–O–Fe linkages. These may involve oxide (O^{2-}) or hydroxide (OH^-) species, depending on hydration.

2. **Flexible Conductive Nanocomposites: Polyaniline and Carbon Nanotubes**

Composites based on polyaniline (PANi) and multiwalled carbon nanotubes (MWCNTs) offer both flexibility and effective electromagnetic absorption, making them suitable for curved surfaces and moving components. Multilayer structures reinforced with PANi-modified MWCNTs have demonstrated a reflection loss (RL)

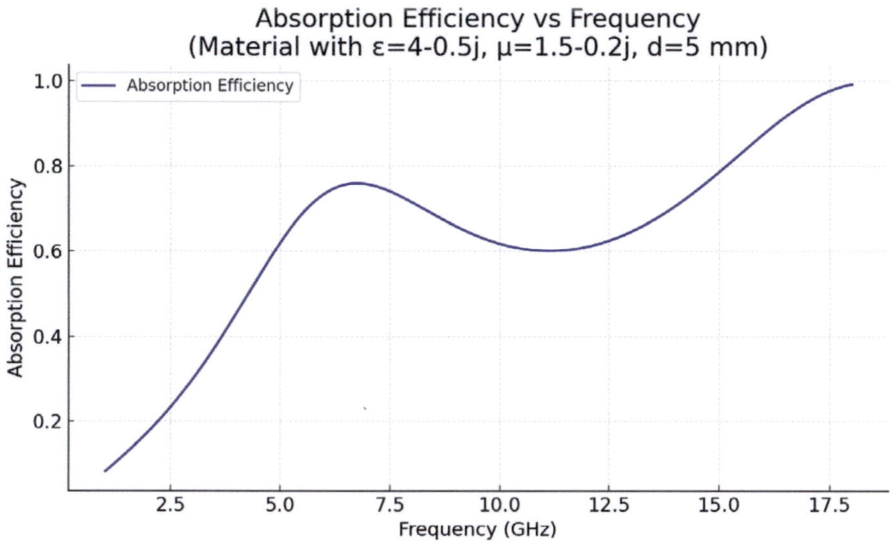

Fig. 6.3 Variation of the absorption efficiency according to frequency

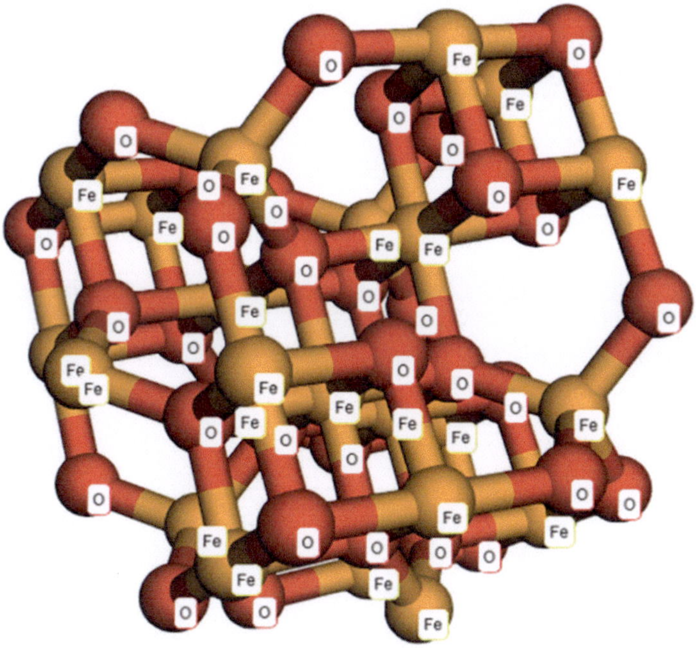

Fig. 6.4 Ferrihydrite-like iron-oxide network

of – 24.53 dB at 10 GHz, with absorption exceeding 5 dB across the entire X-band (8–12 GHz) [18].

The reflection loss (RL) can be calculated from the reflection coefficient (Γ) as:

$$\mathrm{RL(dB)} = -20\log_{10}|\Gamma|$$

where Γ depends on the electromagnetic properties and thickness of the material.

3. **PANi–Graphene-Based Absorbing Structures with Graded Conductivity**

An innovative approach involves non-resonant structures composed of PANi–graphene composites with graded conductivity following the golden ratio (Fig. 6.5). These structures, embedded in a Plexiglas substrate, have demonstrated a radar cross-section (RCS) reduction of 11.42 dB in the X-band (7.8–14 GHz) [19].

The conductivity and thickness of each layer are defined as:

$$\sigma_i = \sigma_0 \times \phi^{(i-1)}$$
$$d_i = d_0 \times \phi^{(i-1)}$$

where,

σi conductivity of layer i
di thickness of layer i

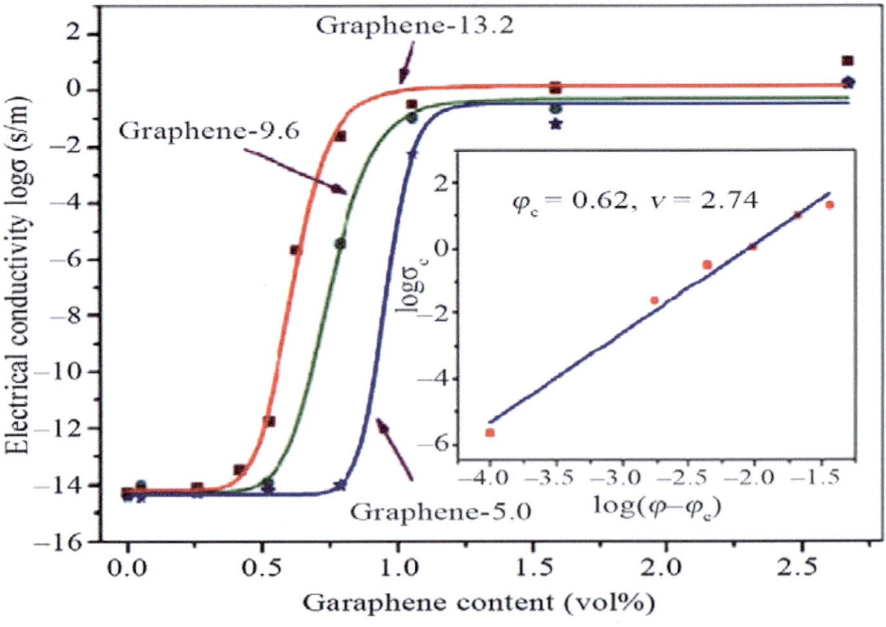

Fig. 6.5 Electrical conductivity variation depending on graphene content [20]

σ_0, d_0 initial conductivity and thickness
ϕ golden ratio (≈ 1.618)

This configuration enables a gradual impedance transition, enhancing absorption over a broad frequency range.

4. **Reflection Loss (RL) Curves of Fe_3O_4/Graphene Composites**

A study published in *Scientific Reports* presents the absorption performance of composites based on Fe_3O_4 and graphene. The reflection loss (RL) curves reveal a minimum RL of approximately – 45 dB at 11 GHz for a composite containing 5 wt% graphene, with an effective absorption bandwidth exceeding 6 GHz (Fig. 6.6).

6.2.2.3 Structural Components

To enhance the mechanical performance and extend the operational lifespan of military hardware, nanocomposites are increasingly used in structural components such as aircraft fuselages, unmanned aerial vehicles (UAVs), missiles, and armored vehicle panels. One prominent example is the use of carbon fiber reinforced polymers (CFRPs) enhanced with nanoclay or graphene nanoplatelets, which significantly improve fatigue resistance, stiffness, and damage tolerance, all while keeping the structure lightweight [22]. These enhancements are crucial for enduring the high mechanical stress and dynamic loads encountered in military operations. Another widely used material is the aluminum matrix nanocomposite, particularly aluminum reinforced with silicon carbide (SiC) or aluminum oxide (Al_2O_3) nanoparticles [23]. These composites are employed in aircraft structural parts and vehicle frames, offering superior strength-to-weight ratios and excellent wear and corrosion resistance. Overall, these nanocomposites contribute to more durable, efficient, and resilient military platforms [24].

Figure 6.7 depicts a **chain-like molecular structure of silicon carbide (Si–C)**:

- **Silicon (Si)** atoms are shown in **yellow**.
- **Carbon (C)** atoms are shown in **gray**.
- Each silicon atom is covalently bonded to a carbon atom.
- The **bond lengths** (1.88 Å for Si–C and 1.89 Å for C–Si) and **bond angles** (~ 109.3° to ~ 109.6°) are labeled, indicating **tetrahedral geometry**, typical of **sp^3hybridization**.
- The structure forms a **repeating polymeric chain** with alternating silicon and carbon atoms, characteristic of **polycarbosilanes** or **semiconducting materials** such as silicon carbide.

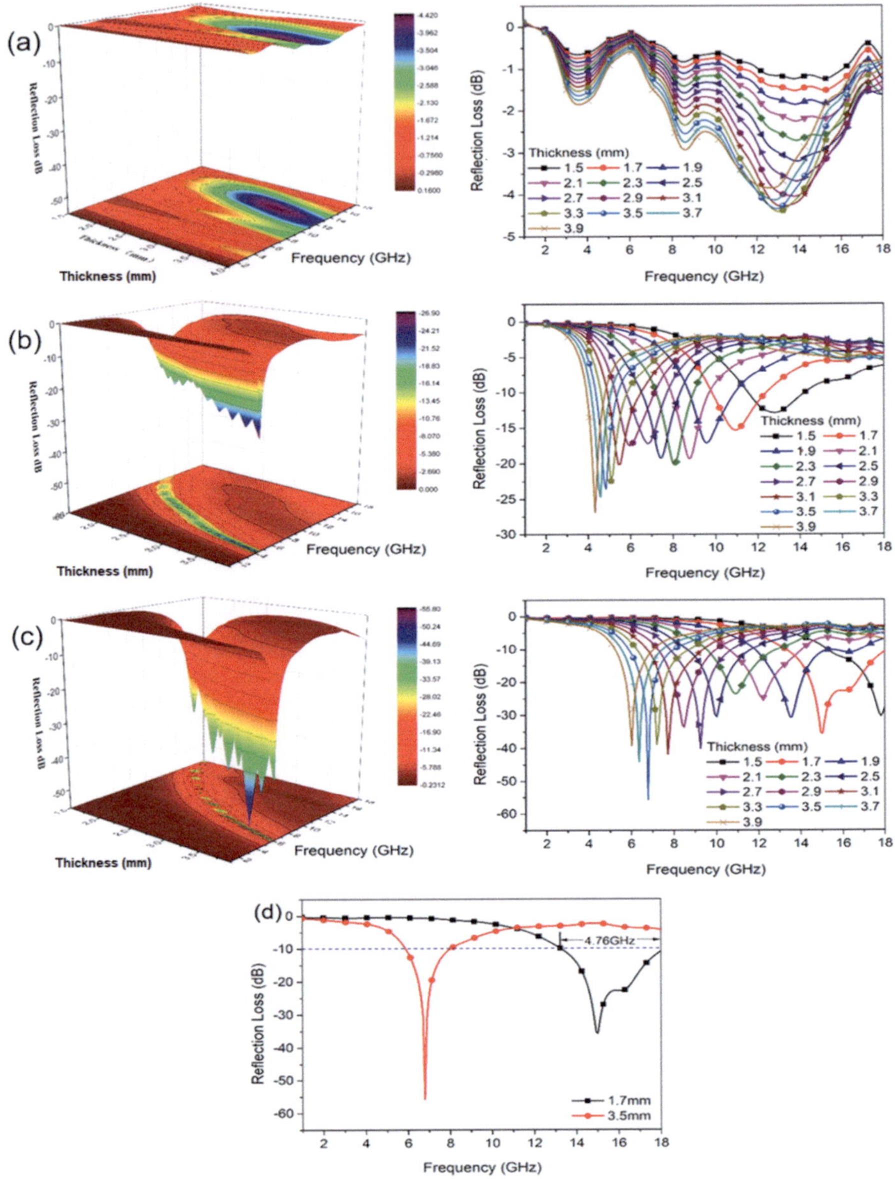

Fig. 6.6 Reflection loss (RL) curves of Fe_3O_4/graphene composites [21]

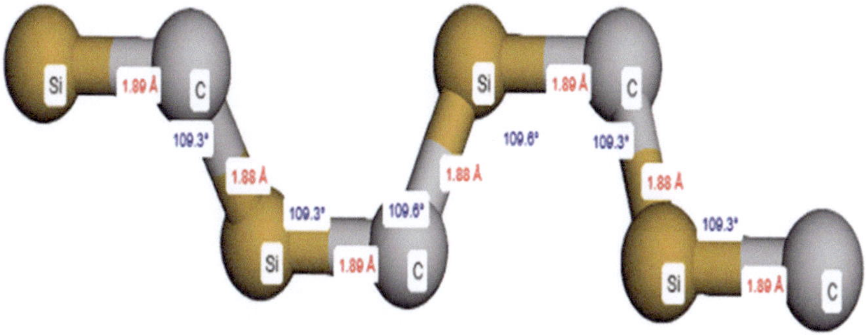

Fig. 6.7 Molecular structure of silicon carbide (Si–C)

6.2.2.4 Sensors and Military Electronics: The Role of Nanocomposites

Nanocomposites play a pivotal role in the development of modern military electronic systems, enabling the creation of lighter, more durable, and multifunctional components. These advanced materials are utilized in a variety of applications, including wearable sensors [25], drone control systems, and electronic warfare equipment, where flexibility [26], miniaturization, and resilience are critical.

Flexible Circuits Based on Silver Nanoparticles

Polymers reinforced with silver nanoparticles (AgNPs) enable the fabrication of flexible and stretchable circuits, ideally suited for integration into wearable electronics used by soldiers for real-time monitoring of health and environmental conditions. For instance, a study demonstrated that incorporating AgNPs into electrospun polystyrene-butadiene-styrene (SBS) fibers results in a sandwich-structured piezoresistive nanocomposite fabric (SSPWN), which exhibits excellent sensitivity to mechanical deformation [27].

There are four silver (Ag) atoms labeled on each vertex. This type of model is often used in chemistry to represent metal clusters or nanostructures, possibly illustrating the Ag_4 (silver tetramer) cluster in a simplified or stylized format (Fig. 6.8).

Sensor Electrical Response

The Electrical Response of These Sensors Can Be Modeled by the Relative Resistance Change ($\Delta R/R_0$) as a Function of Strain (ε):

$$\Delta R/R_0 = k \cdot \varepsilon$$

where k is the gauge factor, characterizing the sensitivity of the sensor.

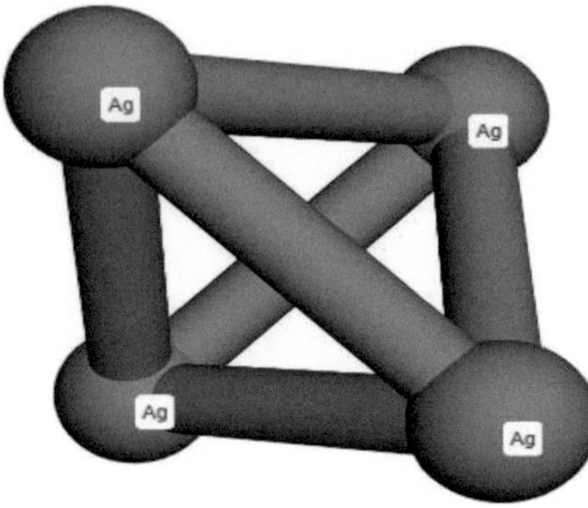

Fig. 6.8 Silver tetra-atomic cluster (Ag_4)

Graphene Oxide-Based Nanocomposites for Chemical and Biological Sensing

Graphene oxide (GO)-based nanocomposites exhibit exceptional electrical conductivity and high surface sensitivity, making them highly effective in chemical and biological sensing devices. These sensors can accurately detect hazardous substances, contributing to enhanced situational awareness and operational safety on the battlefield. For instance, electrochemical sensors fabricated from graphene-based nanomaterials have demonstrated high efficiency in detecting various chemical and biological agents [28].

The sensitivity of these sensors can be expressed by the current change (ΔI) as a function of analyte concentration (C):

$$\Delta I = S \cdot C$$

where S is the sensitivity of the sensor.

Applications in Drones and Electronic Warfare Systems

Nanocomposites are also employed in drone control systems and electronic warfare equipment. For example, the Black Hornet 4 nano-drone, designed to provide discreet situational awareness for small combat units, integrates advanced sensors for reconnaissance and target identification [29].

6.2.2.5 Propellants and Explosives

Nanocomposites are transforming the field of energetic materials in defense by enhancing control, energy density, and safety in systems such as rocket propellants, demolition explosives, and missile warheads.

Solid Propellants Based on Nano-aluminum

The incorporation of aluminum nanoparticles (nano-Al) combined with ammonium perchlorate (AP) in solid propellants significantly increases the burn rate and energy density of the fuel [30].

The combustion of aluminum is represented by the following chemical reaction:

$$4\mathrm{Al} + 3\mathrm{O}_2 \rightarrow 2\mathrm{Al}_2\mathrm{O}_3$$

Figure 6.9 shows a **three-dimensional molecular structure** composed primarily of **aluminum (Al)** and **oxygen (O)** atoms. Here is a detailed description:

- The **aluminum atoms** are shown in **gray** and labeled "Al."
- The **oxygen atoms** are shown in **red** and labeled "O."
- The structure likely represents a **polymeric aluminum-oxygen compound**, typical of **aluminosilicates** or **aluminum hydroxide clusters** (such as **alumoxanes** or **boehmite-like structures**).
- The **bonds between atoms** form a complex, branched **3D network**.

For AP-based composite propellants, the empirical burn rate law is expressed as:

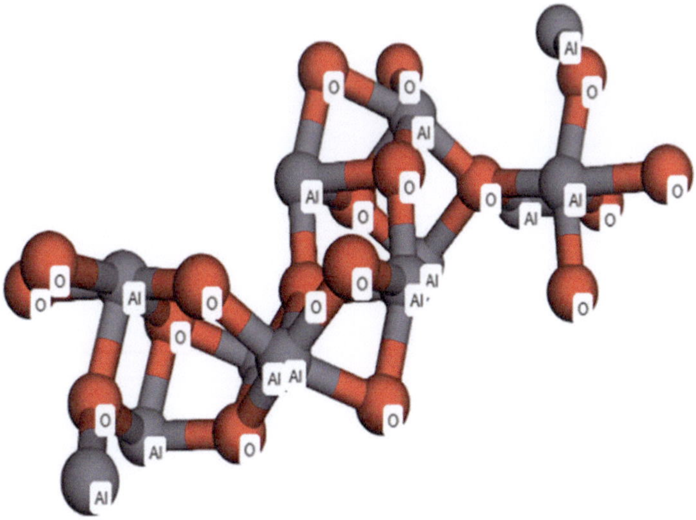

Fig. 6.9 Nano-Al_2O_3 (alumine nanostructured)

$$r = aP_c^n$$

- r the burn rate is conventionally denoted by a lowercase r (for rate),
- P represents the chamber pressure.
- n is the exponent applied to the pressure.
- a is an empirical coefficient specific to the propellant's composition, and nn is the pressure exponent, which typically falls between 0.3 and 0.5.

Studies have shown that the addition of nano-aluminum in HTPB/AP/Al propellant formulations can increase the combustion rate by up to 80%, while maintaining a nearly constant pressure exponent [31].

Polymer-Bonded Explosives (PBX) with Nano-metallic Additives

Polymer-bonded explosives (PBX) enhanced with nano-metallic additives exhibit more controlled and efficient detonation, while also reducing the material's sensitivity to accidental stimuli such as impact, friction, or heat. For example, Fig. 6.10 shows the incorporation of aluminum nanoparticles in PBX formulations which decreases the sensitivity to impact and friction, while simultaneously increasing the detonation velocity.

Studies have shown that nanoscale energetic materials (nEMs) significantly reduce the impact and friction sensitivity of solid propellants and explosives compared to their microscale counterparts.

These innovations are crucial for improving the precision, performance, and reliability of modern defense munitions.

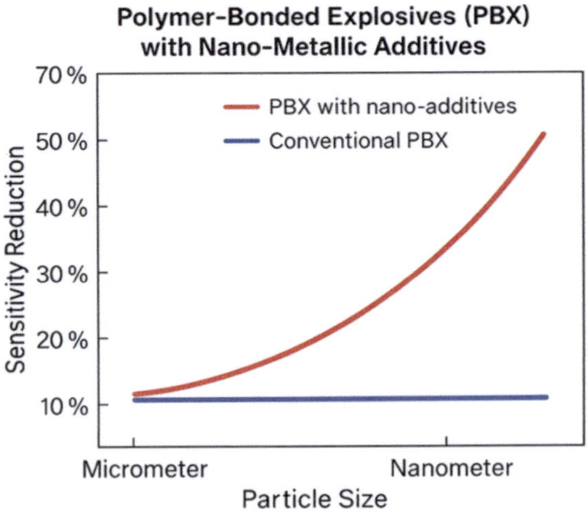

Fig. 6.10 The variation of sensitivity Reduction versus Particle size for PBX

6.2.2.6 Fire Retardant and Protective Coatings

To enhance the safety of military vehicles and equipment operating in fire-prone or high-temperature environments, nanocomposite-based coatings are increasingly employed in systems such as tanks, aircraft interiors, and ship hulls. One effective solution is the use of polymer/clay nanocomposites, which are engineered to significantly delay ignition, reduce the spread of flames, and lower smoke emission during fire incidents. These properties are essential for protecting personnel and sensitive equipment in confined or critical areas. Additionally, paints reinforced with silica or titanium dioxide (TiO_2) nanoparticles are used for thermal shielding, helping to reflect heat and improve resistance to UV radiation. Enhancing the safety of military vehicles and equipment operating in fire-prone or high-temperature environments necessitates advanced materials. Nanocomposite-based coatings have emerged as a pivotal solution, offering improved fire resistance, thermal shielding, and UV protection. Polymer/clay nanocomposites, particularly those incorporating montmorillonite (MMT) clay, have demonstrated significant improvements in fire retardancy. The mechanism involves the formation of a protective char layer during combustion, which acts as a barrier to heat and mass transfer. The effectiveness of these nanocomposites can be quantified using the Limiting Oxygen Index (LOI), which indicates the minimum concentration of oxygen required to sustain combustion. Higher LOI values correspond to better flame retardancy. For instance, a study by the National Institute of Standards and Technology (NIST) observed that polyamide 6/clay nanocomposites exhibited increased LOI values compared to pure polyamide 6, indicating enhanced flame resistance. Nanoparticle-Reinforced Coatings for Thermal and UV Protection. Incorporating nanoparticles such as silica (SiO_2) and titanium dioxide (TiO_2) into coatings enhances their thermal stability and UV resistance. TiO_2 nanoparticles, in particular, are known for their high refractive index and strong UV light absorption capabilities.

Figure 6.11 displays a molecular unit of titanium dioxide, which may represent a local coordination environment found in crystalline phases like anatase, rutile, or brookite.

- **Titanium (Ti)**: Shown in dark **gray**, the titanium atom is centrally located in an octahedral geometry, coordinated by six surrounding oxygen atoms.
- **Oxygen (O)**: Colored in **red**, oxygen atoms form Ti–O bonds at the vertices of the octahedron, contributing to the lattice structure and determining the electronic properties.

TiO_2 is a widely studied semiconductor, notable for its photocatalytic activity, UV absorption, and use in solar cells, pigments, and environmental purification technologies. The UV protection efficiency of these coatings can be evaluated using UV–Vis spectroscopy, measuring the absorbance of UV radiation. Coatings with higher absorbance in the UV region offer better protection. Coatings extend the service life of surfaces exposed to extreme environmental conditions while also contributing to passive fire protection. The multifunctional nature of these nanocomposites makes

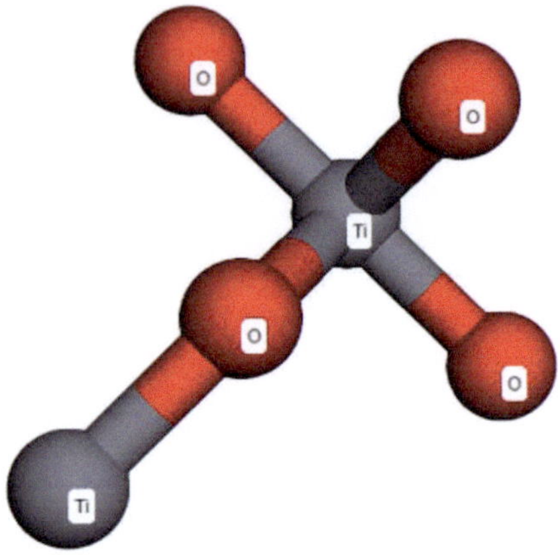

Fig. 6.11 Titanium dioxide unit (TiO_2)

them highly valuable in maintaining safety and structural integrity across a wide range of defense platforms.

6.2.2.7 Environmental and Chemical Protection

In defense settings where personnel and equipment are exposed to harsh environmental or chemical threats, nanocomposites offer critical protection through advanced barrier and antimicrobial properties. This is especially important in the development of NBC (Nuclear, Biological, Chemical) protective suits and sensor housings used in contaminated or high-risk zones (Table 6.1). One effective material is polyurethane films embedded with layered silicate nanofillers, which act as high-performance gas barriers, significantly reducing the permeability of hazardous chemicals and gases. These films enhance the protective function of clothing and enclosures without adding bulk. Additionally, fabrics infused with zinc oxide (ZnO) or silver nanoparticles provide strong antimicrobial and chemical resistance, making them ideal for use in protective suits that guard against biological agents and infectious threats. These nanocomposite materials contribute to improved survivability and operational readiness in chemically or biologically hazardous environments.

6.3 Nanocomposites for Aeronautical Applications

6.3.1 Generality

Nanocomposites are at the forefront of innovation in the aerospace industry, providing transformative advantages in terms of weight reduction, fuel efficiency, mechanical strength, and thermal resistance. In the aeronautical sector, where every gram matters

Table 6.1 Nanocomposite examples in defense area

Defense area	Nanocomposite material example	Key benefit
Armor and ballistics	Kevlar + CNTs, polymer + boron carbide	Lightweight, high-strength protection
Stealth tech	Epoxy + Fe_3O_4, Polyaniline + CNTs	Radar absorption
Structural components	CFRP + nanoclay, Al + SiC	Mechanical durability, low weight
Electronics and sensors	Silver NP polymers, graphene oxide sensors	Flexible, high-performance electronics
Propellants and explosives	Nano-Al + AP, PBX + nano-metals	Enhanced energy, stability
Fire retardant coatings	Polymer + nanoclay, SiO_2 nano paint	Thermal shielding, fire resistance
Environmental protection	Polyurethane + nanoclays, ZnO/Ag-infused fabrics	Chemical, microbial resistance

and structural integrity is critical under extreme operational conditions, nanocomposites enable the design of next-generation aircraft that are both lighter and more resilient. Their ability to combine multiple functionalities in a single material system makes them particularly attractive for use in airframes, propulsion systems, avionics, and environmental control systems [20].

6.3.2 Airframe Structures

One of the primary uses of nanocomposites in aeronautics is in the construction of airframe components such as fuselage panels, wings, and tail assemblies. Carbon fiber reinforced polymers (CFRPs) enhanced with carbon nanotubes (CNTs), graphene nanoplatelets, or nanoclays are employed to improve stiffness, tensile strength, and impact resistance. These materials offer significant weight reductions compared to traditional metal alloys, directly contributing to reduced fuel consumption and increased payload capacity. Moreover, the improved fatigue life and crack propagation resistance of nanocomposite structures result in lower maintenance requirements and longer service intervals for aircraft [29].

6.3.3 Engine Components and Thermal Management

Aircraft engines are exposed to high temperatures, mechanical stress, and corrosive environments. Nanocomposites provide the necessary thermal stability and mechanical integrity to withstand these conditions. Ceramic matrix nanocomposites

(CMNCs), such as silicon carbide (SiC) reinforced with carbon nanotubes or zirconia nanoparticles, are used in turbine blades, nozzles, and combustion chamber linings. These materials maintain structural strength at elevated temperatures, allowing engines to operate more efficiently with improved thermal insulation and reduced cooling demands. Additionally, phase change nanocomposite materials are being explored for use in thermal management systems to regulate temperature in both civilian and military aircraft avionics [32].

6.3.4 Lightning Strike Protection

Aircraft structures made from composite materials are particularly vulnerable to lightning strikes due to their low electrical conductivity. To address this, conductive nanocomposites are integrated into the outer layers of aircraft skins and radomes. For example, carbon nanotube-based nanocomposites or graphene-enhanced coatings create conductive networks that safely dissipate lightning energy across the surface, preventing structural damage. These materials offer a lightweight alternative to metallic meshes traditionally used for lightning protection, thereby maintaining aerodynamic performance without adding significant weight [33].

6.3.5 Avionics and Sensor Integration

Modern aircraft increasingly rely on embedded electronics and sensors for performance monitoring, navigation, and communication. Nanocomposite materials allow for the development of flexible, lightweight, and durable sensors that can be seamlessly integrated into the aircraft structure. Conductive polymer nanocomposites using silver nanoparticles or carbon-based fillers enable the creation of stretchable circuits and antennas that conform to complex geometries. These smart materials can monitor strain, temperature, and damage in real time, supporting predictive maintenance and enhancing operational safety [34].

6.3.6 Noise and Vibration Damping

Aircraft cabin noise and structural vibration are critical concerns for passenger comfort and structural longevity. Nanocomposites play a vital role in acoustic damping and vibration control. For instance, nanoclay or carbon black-filled elastomeric nanocomposites are used in engine mounts, landing gear components, and fuselage linings to absorb mechanical energy and reduce noise transmission. These materials not only improve comfort but also protect sensitive onboard instruments from mechanical disturbances [35].

6.3.7 Fuel System and Barrier Applications

Nanocomposite coatings are used to line fuel tanks and pipelines, offering enhanced barrier properties against fuel permeation and corrosion. For example, epoxy resins embedded with layered silicates or graphene oxide create dense molecular structures that minimize fuel vapor leakage and resist chemical degradation. These systems ensure long-term reliability and safety in aviation fuel systems while complying with stringent environmental regulations [36].

6.3.8 Fire Retardancy and Safety

Safety in aviation also relies on materials that resist ignition and delay fire propagation. Nanocomposites incorporating flame retardant nanofillers such as aluminum hydroxide, magnesium hydroxide, or organically modified montmorillonite are used in interior cabin panels, insulation foams, and cargo linings. These materials offer low smoke generation, reduced flammability, and improved thermal insulation, all of which are critical for meeting fire safety standards in commercial and military aviation (Table 6.2).

Table 6.2 Nanocomposite examples in aeronautical area

Aeronautical area	Nanocomposite material example	Key benefits
Fuselage and wing structures	Epoxy + carbon nanotubes (CNTs)	Lightweight, high mechanical strength, fatigue resistance
Cabin interior components	Thermoplastic + nanoclay	Fire resistance, improved dimensional stability, lightweight
Landing gear components	Aluminum matrix + nano-SiC	Increased hardness, wear resistance, load-bearing capacity
Avionics and EMI shielding	Polyimide + graphene Nanoplatelets	Excellent electrical conductivity, EMI suppression, thermal stability
Thermal insulation panels	Phenolic resin + nano-Al_2O_3	Enhanced thermal resistance, flame retardancy
Flight control surface skins	Carbon fiber + CNT-reinforced epoxy	Improved stiffness-to-weight ratio, vibration damping

6.4 Nanocomposites for Aerospace Applications

6.4.1 Generality

Nanocomposites have emerged as transformative materials in the aerospace sector, enabling advancements in both civil and military aviation. Their unique combination of high mechanical performance, low weight, thermal stability, and multifunctional capabilities makes them ideal candidates for components that must endure extreme operational conditions while maintaining structural integrity and performance.

6.4.2 Structural Components in Aircraft

In aerospace engineering, reducing weight while maintaining or improving structural performance is a fundamental design goal. Nanocomposites achieve this by integrating nanoscale reinforcements, such as carbon nanotubes (CNTs), graphene nanoplatelets, or nanoclays, into polymer or metal matrices. These reinforcements enhance properties such as tensile strength, stiffness, and fatigue resistance. A prime example is carbon fiber reinforced polymers (CFRPs) infused with carbon nanotubes or graphene, which are used in aircraft fuselages, wings, and tail structures. These nanocomposites improve damage tolerance and fatigue life, allowing for longer maintenance intervals and higher operational reliability. Similarly, aluminum matrix nanocomposites reinforced with silicon carbide (SiC) or boron nitride nanoparticles are utilized in high-stress structural parts, offering superior strength-to-weight ratios and enhanced corrosion resistance.

6.4.3 Thermal and Fire Protection

Aircraft often operate in environments where components are exposed to high temperatures, particularly near engines or in hypersonic vehicles. Nanocomposite materials reinforced with ceramic fillers such as alumina (Al_2O_3), zirconia (ZrO_2), or silicon nitride (Si_3N_{44}) are commonly employed in **thermal barrier coatings (TBCs)** to insulate engine parts and mitigate thermal fatigue. In addition, polymer-based nanocomposites incorporating nanoclays or intumescent additives are used as **fire-retardant coatings** inside aircraft cabins, enhancing passenger safety by delaying flame propagation and reducing smoke generation.

1. **Thermal Barrier Coatings (TBCs)**

TBCs are critical for protecting aircraft engine components from extreme temperatures. Ceramic materials like alumina (Al_2O_3), yttria-stabilized zirconia (YSZ), and silicon nitride (Si_3N_4) are widely used due to their low thermal conductivity and high-temperature stability.

Effective Thermal Conductivity Equation

The effective thermal conductivity (k_{eff}) of a nanocomposite can be estimated using the **Maxwell–Garnett equation**:

$$K_{\text{eff}} = k_m \frac{\left(k_p + 2k_m + 2f\left(k_p - k_m\right)\right)}{\left(k_p + 2k_m - f\left(k_p - k_m\right)\right)}$$

where:

k_{eff} The effective thermal conductivity of the nanocomposite
k_m The thermal conductivity of the continuous matrix phase
k_p The thermal conductivity of the dispersed particle (inclusion) phase
f The volume fraction of the dispersed particles

This equation illustrates how incorporating ceramic fillers can decrease the overall thermal conductivity of the composite material, thereby enhancing its insulation properties.

6.4.4 EMI Shielding and Lightning Protection

With the growing reliance on electronic systems in modern aircraft, shielding sensitive electronics from electromagnetic interference (EMI) is crucial. Conductive nanocomposites, such as those containing carbon nanotubes, graphene, or metallic nanoparticles (e.g., silver or copper), are used as EMI shielding layers within avionics enclosures, cable coatings, and fuselage linings. These materials not only reduce electromagnetic noise but also provide weight savings compared to traditional metallic shields. Furthermore, aerospace vehicles are at risk of lightning strikes, especially at high altitudes. Nanocomposites with enhanced electrical conductivity are used in outer fuselage layers to safely dissipate electrical discharges, protecting the underlying structure and systems.

6.4.5 Fuel Efficiency and Aerodynamics

By enabling lightweight designs without compromising strength, nanocomposites contribute directly to improved fuel efficiency. This is particularly important for commercial aircraft, where even marginal reductions in weight translate into significant fuel savings over time. Moreover, nanostructured surface coatings can improve aerodynamic properties by reducing drag through surface texturing or self-cleaning effects based on nanostructured hydrophobic coatings.

Table 6.3 Nanocomposite examples in aerospace area

Aerospace area	Nanocomposite material example	Key benefits
Aircraft structural components	Epoxy + carbon nanotubes (CNTs)	High strength-to-weight ratio, improved fatigue resistance
Thermal protection systems	SiC matrix + nano-ZrO_2	Extreme heat resistance, thermal shock protection
EMI shielding and electronics	PEEK + Graphene nanoplatelets	High electrical conductivity, lightweight, effective EMI shielding
Interior panels and insulation	Thermoplastic + nanoclay	Flame retardancy, reduced gas permeability, lightweight
Turbine blades and engine parts	Aluminum matrix + nano-SiC	Enhanced wear resistance, high temperature stability
Smart surfaces and sensors	Epoxy + graphene oxide	Real-time sensing, self-healing potential, conductivity

6.4.6 *Self-Healing and Smart Materials*

One of the emerging frontiers in aerospace nanocomposites is the development of self-healing materials. These systems incorporate microcapsules or nanocapsules containing healing agents within the matrix. When damage occurs (e.g., a microcrack), the capsules rupture, releasing the healing agent and autonomously repairing the material. This capability can extend the lifespan of aircraft components and reduce maintenance costs.

Additionally, smart nanocomposites equipped with sensing functions (e.g., through embedded piezoresistive CNT networks) can monitor strain, temperature, or damage in real-time. These materials are paving the way for "intelligent" airframes capable of self-diagnosis and adaptive response (Table 6.3).

6.5 Challenges

While nanocomposites offer outstanding advantages such as weight reduction, enhanced strength, thermal stability, and multifunctionality, several challenges continue to limit their widespread adoption in high-performance sectors like defense, aeronautics, and aerospace. These challenges include:

- Manufacturing limitations, such as nanoparticle dispersion, agglomeration, and scalability.
- Material performance concerns, including interfacial compatibility, long-term environmental stability, and reproducibility.

- Economic barriers, notably the high cost of nanofillers and the complexity of integrating them into existing supply chains.
- Health, safety, and environmental risks, due to limited data on the toxicity and end-of-life management of nanomaterials.
- Regulatory and certification difficulties, given the absence of standardized testing, long qualification processes, and evolving modeling tools.
- Integration issues, involving compatibility with legacy systems, challenges in multimaterial bonding, and complexities in repair and maintenance.

Overcoming these obstacles requires continued interdisciplinary research, investment in scalable and safe processing technologies, and the development of international standards and regulatory frameworks. Addressing these challenges will be key to unlocking the full potential of nanocomposites in future aerospace and defense systems.

References

1. Loukil R, Gazehi W, Besbes M (2024) Improving the classification of a nanocomposite using nanoparticles based on a meta-analysis study, recurrent neural network and recurrent neural network Monte-Carlo algorithms. Nanocomposites 10(1):309–337
2. Gazehi W, Loukil R, Besbes M (2025) Classification of a nanocomposite using a combination between recurrent neural network based on transformer and Bayesian network for testing the conductivity property. Expert Syst Appl 270:126518
3. Ruoff RS, Qian D, Liu WK (2003) Mechanical properties of carbon nanotubes: theoretical predictions and experimental measurements. C R Phys 4(9):993–1008
4. Ramachandran K, Boopalan V, Bear JC, Subramani R (2022) Multi-walled carbon nanotubes (MWCNTs)-reinforced ceramic nanocomposites for aerospace applications: a review. J Mater Sci 57(6):3923–3953
5. Ma Z, Jiang R, Jing J, Kang S, Ma L, Zhang K, Li J, Zhang Y, Qin J, Yun S, Zhang G (2024) Lightweight dual-functional segregated nanocomposite foams for integrated infrared stealth and absorption-dominant electromagnetic interference shielding. Nano-Micro Letters 16(1):223
6. Ravindran P, Kumar KM, Rangasamy S, Kumar KRV (2023) Plain Kevlar and a CNT-reinforced Kevlar epoxy polymer composite: comparative study of its mechanical, low velocity and ballistic impact properties. Iran Polym J 32(2):239–249
7. Kim SH, Lee SY, Zhang Y, Park SJ, Gu J (2023) Carbon-based radar absorbing materials toward stealth technologies. Adv Sci 10(32):2303104
8. Barani Z, Kargar F, Godziszewski K, Rehman A, Yashchyshyn Y, Rumyantsev S, Cywiński G, Knap W, Balandin AA (2020) Graphene epoxy-based composites as efficient electromagnetic absorbers in the extremely high-frequency band. ACS Appl Mater Interfaces 12(25):28635-28644
9. Rathod VT, Kumar JS, Jain A (2017) Polymer and ceramic nanocomposites for aerospace applications. Appl Nanosci 7:519–548
10. Indrusiak T, Pereira IM, Pontes K, Pereira EC, Peixoto GG, Migliano AC, Soares BG (2021) Hybrid carbonaceous materials for radar absorbing poly (vinylidene fluoride) composites with multilayered structures. SPE Polymers 2(1):62–73
11. Haritha T, Ramji K, Subrahmanyam C, Krushnamurthy K, Nagasree PS (2021) Microwave-absorption characteristics of polyaniline-coated multi-walled carbon nanotube composites. Plast Rubber Compos 50(4):180–188

12. Thomas P, Abdulhakim LV, Pushkaran NK, Karuvandi AC (2020) Wideband radar absorbing structure using polyaniline-graphene nanocomposite. C 6(4):72
13. Yu Q, Wang Y, Chen P, Nie W, Chen H, Zhou J (2019) Reduced graphene oxide-wrapped super dense Fe_3O_4 nanoparticles with enhanced electromagnetic wave absorption properties. Nanomaterials 9(6):845
14. Jen YM, Huang YC (2021) Improvement in tensile quasi-static and fatigue properties of carbon fiber-reinforced epoxy laminates with matrices modified by carbon nanotubes and graphene nanoplatelets hybrid nanofillers. Nanomaterials 11(12):3459
15. Gharaibeh N, AlAjlouni M, Rayes A (2024) Structural and mechanical characterization of aluminum matrix nanocomposites reinforced with SiC and Al_2O_3 Nanoparticles. J Compos Adv Mater/Revue des Composites et des Matériaux Avancés 34(6)
16. Nan Z, Wei W, Lin Z, Chang J, Hao Y (2023) Flexible nanocomposite conductors for electromagnetic interference shielding. Nano-Micro Letters 15(1):172
17. Del Bosque A, Lampropoulos G, Vergara D (2024) Nanocomposites for multifunctional sensors: a comprehensive bibliometric exploration. Nanomaterials 15(1):34
18. Cho CJ, Chung PY, Tsai YW, Yang YT, Lin SY, Huang PS (2023) Stretchable sensors: novel human motion monitoring wearables. Nanomaterials 13(16):2375
19. Khan SU, Khalid W, Atif M, Ali Z (2024) Graphene oxide–cerium oxide nanocomposite modified gold electrode for ultrasensitive detection of chlorpyrifos pesticide. RSC Adv 14(38):27862–27872
20. Lyu Y, Cheng Y, Zhao G, Wang M, Chen G, Zhang X, Han W (2023) Modification of SiBCN by Zr atom and its effect on ablative resistance of Cf/SiBCN (Zr) composites. Compos B Eng 253:110511
21. Yu A, Kolotylo I, Hashim HA, Eltoukhy AE (2025) Electronic warfare cyberattacks, countermeasures and modern defensive strategies of UAV avionics: a survey. IEEE Access
22. Tu C, Chen X, Chen F, Zhuang Y, Cai W, Li Y, Li W, Zhou C, Xie R (2024) Effect of aluminum and ammonium perchlorate particle sizes on the condensed combustion products characteristics of aluminized NEPE propellants. Sci Rep 14(1):19462
23. Pang W, Li Y, DeLuca LT, Liang D, Qin Z, Liu X, Xu H, Fan X (2021) Effect of metal nanopowders on the performance of solid rocket propellants: a review. Nanomaterials 11(10):2749
24. Lopez-Campos JE, Mojica-Gómez J, Maciel-Cerda A, Castaño VM, Hernández-Padrón G (2024) Hybrid epoxy-SiO_2/GO nanosheets anti-corrosive coating for aeronautic aluminum Al6061-T5. J Coat Technol Res 21(2):559–574
25. Dong J, Wang W, Wang X, Zhou Q, Miao R, Du M, Tan B, Wang Y, Zhang T, Li Y, Cao F (2021) Effects of nano aluminum powder on the mechanical sensitivity of RDX-based explosives. Nanomaterials 11(9):2182
26. Monteiro B, Simões S (2024) Recent advances in hybrid nanocomposites for aerospace applications. Metals 14(11):1283
27. Kumar K, Das S, Garg R, Goyat MS (2024) A comprehensive review on enhancing the strength of CFRPs through nano-reinforcements: applications, characterization, and challenges. J Fail Anal Prev 24(4):1588–1627
28. Karadimas G, Ioannou A, Kolios A, Salonitis K (2024) Techno-Economic analysis of ceramic matrix composites integration in remaining useful life aircraft engine hot section components. Int J Adv Manuf Technol 135(9):4189–4203
29. Zhu M, Zhang P, Gao F, Bai Y, Zhang H, Zu M, Liu L, Zhang Z (2024) Advanced lightweight lightning strike protection composites based on super-aligned carbon nanotube films and thermal-resistant zirconia fibers. Nanoscale Adv 6(19):4858–4864
30. Hedayati DP, Termine S, Bascucci C, Al Malak P, Bondavalli P, Semitekolos D, Clemens F, Charitidis C, Böhm R (2024) Carbon-based nanocomposites for sensing applications-a review. J Phys: Mater
31. Velmurugan R, Karthikeyan K (2005) The reinforcement effect of nanoclay on the damping characteristics of hybrid laminated composites. J Aerosp Sci Technol 345–353

32. Cestino E, Catapano J, Galvano F, Felis A, Zuccalà S, Martilla V, Sapienza V, Chesta L (2024) Effectiveness of nanotechnology treatments in composite aircraft applications. Appl Sci 14(5):1721
33. Kinloch IA, Young RJ (2018) Composites with carbon nanotubes and graphene nanoplatelets. Science 362(6414):547–548
34. Liu J, Gao Y, Cao D, Zhang L, Guo Z (2011) Nanoparticle dispersion and aggregation in polymer nanocomposites: insights from molecular dynamics simulation. Langmuir 27(12):7926–7933
35. Zare Y, Rhee KY, Hui D (2017) Influences of nanoparticles aggregation/agglomeration on the interfacial/interphase and tensile properties of nanocomposites. Compos Part B: Eng 122:12–19
36. Rahman MM et al (2025) Polymer nanocomposites with optimized nanoparticle dispersion and enhanced functionalities for industrial applications. MDPI Process 13(4):994
37. Parveez B, Kittur MI, Badruddin IA, Kamangar S, Hussien M, Umarfarooq MA (2022) Scientific advancements in composite materials for aircraft applications: a review. Polymers, 14(22):5007. https://doi.org/10.3390/polym14225007
38. Li JW, Cheng CC, Chiu CW (2024) Advances in multifunctional polymer-based nanocomposites. Polymers 16(23):3440. https://doi.org/10.3390/polym16233440

Chapter 7
Self-healing Aeronautical Nanocomposites

Himadri B. Bohidar

Abstract This chapter delves into the innovative development and characterization of self-healing nanocomposites, highlighting their potential to significantly improve the safety, longevity, and performance of aerospace materials. Self-healing nanocomposite materials are engineered to provide for real-time in situ damage repair, thereby extending the lifespan of aerospace components and reduce their maintenance cost. This review explores various self-healing mechanisms, including microencapsulation, remediation of cracks and networks, and intrinsic self-healing through the engagement of reversible chemical bonds, often driven by shape memory of materials. Key focus areas include the synthesis and integration of self-healing agents within the nanocomposite matrix. Microencapsulation techniques involve embedding healing agents within microcapsules that rupture upon damage, releasing the healing agent to fill the cracks and restore its original structural integrity. Vascular networks mimic biological systems with microchannels delivering healing agents to the damaged sites. Intrinsic self-healing utilizes materials that can reversibly bond or deform under specific conditions, thus allowing for repeated healing cycles. Mechanical testing, including tensile, flexural, and fatigue tests, assess the durability and effectiveness of the self-healing mechanisms under simulated aeronautical conditions. This in turn governs the potential applications of these advanced materials in aeronautical engineering, highlighting their promise in improving safety, reducing downtime, and enhancing the performance of aerospace components. This chapter captures the salient features of self-healing mechanisms and their potential applications in aerospace materials, in particular.

Keywords Aerospace industry · Mechanisms of self-healing · Microencapsulation · Vascular networks · Reversible chemical bonds · Shape memory effect · Self-healing mechanisms · Self-healing materials · Vascular networks · Intrinsic self-healing materials

H. B. Bohidar (✉)
Applied Physics Department, Delhi Technological University, Delhi, India
e-mail: hbbohidar@jnu.ac.in

S. K. Panda (ed.), *Nanocomposites for Defense, Aeronautical and Aerospace Applications*, Advanced Structured Materials 248,
https://doi.org/10.1007/978-3-032-11568-3_7

7.1 Introduction

The aerospace industry continually faces challenges related to material durability and maintenance, which are critical for ensuring safety and performance due their exposure to extreme heat and mechanical stress conditions [1, 2]. Traditional aerospace materials, while strong and lightweight, are susceptible to damage from mechanical stress, thermal cycling, and environmental exposure [2–6]. Such damage can lead to costly repairs, increased downtime, and in extreme cases, catastrophic failures. To address these issues, there is a growing demand for the development of advanced functional materials that possess self-healing capabilities.

Self-healing nanocomposites represent a cutting-edge approach to enhancing the resilience of aerospace materials. These materials are engineered to autonomously repair the damage in situ, thereby extending the lifespan of components and reducing the need for frequent maintenance. The concept of self-healing in materials science draws inspiration from biological systems, where damage triggers a healing response that restores functionality. By incorporating similar mechanisms into synthetic materials, researchers aim to create smarter, more durable materials for aeronautical applications [7–9].

There are several mechanisms by which self-healing can be achieved in nanocomposites. Microencapsulation involves embedding healing agents within tiny capsules that break open under mechanical stress releasing the agents to fill the cracks and restore structural integrity. Vascular networks mimic biological circulatory systems, using microchannels to deliver healing agents to damaged sites. Intrinsic self-healing leverages materials that can reversibly bond or deform, enabling repeated healing cycles without the need for external intervention [10–19].

This chapter explores the various strategies for developing self-healing aeronautical nanocomposites with a focus on the synthesis and integration of self-healing agents. Additionally, the mechanical performance of these nanocomposites under simulated aeronautical conditions is assessed through tensile, flexural, and fatigue testing. By advancing the development of self-healing aeronautical nanocomposites, this research aims to pave the way for the next generation of smarter, more resilient materials that can significantly impact the aerospace industry.

7.2 Importance of Self-healing in Aerospace Industry

The aerospace industry is continuously evolving, with increasing demands for safety, reliability, and efficiency, but with minimum downtime. Self-healing materials are emerging as a revolutionary solution to meet these challenges. These materials have the remarkable ability to autonomously repair minor damage, such as cracks or delamination, before they progress into more serious structural failures. This enhances the overall reliability of aerospace components, which must endure extreme environmental conditions, mechanical stress, and the temperature during the flight.

By self-repairing itself, these materials help maintain structural integrity, thereby reducing the likelihood of catastrophic component failures and improving operational safety.

In addition to enhancing reliability, self-healing materials offer significant cost-saving potential in aerospace maintenance. Aircrafts require regular inspections and repairs to ensure they remain in optimal condition and flightworthy. Traditional maintenance procedures can be time-consuming, expensive, and often lead to extended downtime. Self-healing materials, however, can autonomously address certain types of damage, reducing the need for frequent manual interventions. Over time, this capability can lower maintenance costs, minimize operational disruptions, and extend the service life of aerospace components.

The incorporation of self-healing materials also contributes to the overall efficiency of aerospace structures. Modern aircrafts and spacecrafts are designed with strict performance criteria including weight minimization for fuel efficiency. Self-healing materials offer the advantage of added durability without significantly increasing weight, making them a favorable option for designing lightweight structures that can withstand demanding conditions. This improves both fuel efficiency and overall performance, making aerospace systems more economical and environmentally friendly.

Safety and durability are critical in the aerospace industry, and self-healing materials provide a new level of assurance in this regard. The ability of these materials to recover from minor damage ensures that even under unexpected circumstances, such as impacts or high-stress scenarios, the structural integrity of aircraft is maintained. This is essential for high-risk operations like flight missions, where immediate repair options are limited. By integrating self-healing materials, the aerospace industry can develop more resilient designs, improving both safety and long-term performance.

Lastly, self-healing materials open the door to innovative aerospace designs. Engineers can now explore more complex, lightweight, and efficient structures, knowing that the materials used will be capable of autonomously managing the damage. This allows for greater flexibility in design and enhances the overall potential of aerospace technologies, enabling the industry to push the boundaries of what is possible in terms of performance and resilience.

7.3 Mechanisms of Self-healing

The ability of materials to autonomously repair damage has gained significant attention in the field of material science, particularly for applications in the aerospace industry where the reliability and resilience of materials are critical. Self-healing materials employ various mechanisms to detect and repair damage at the molecular or microscopic level, ensuring that the material's structural integrity is maintained over time. Understanding these mechanisms is crucial for the development and optimization of self-healing aeronautical nanocomposites. This section explores the primary mechanisms of self-healing, including microencapsulation, vascular networks, reversible chemical bonds, and shape memory effects.

7.3.1 Microencapsulation

Microencapsulation is a widely studied and implemented mechanism for self-healing materials. This approach involves embedding microcapsules filled with a healing agent within the material matrix. When damage occurs, such as a crack formation, these microcapsules rupture, releasing the healing agent into the damaged area. Once released, the healing agent undergoes a chemical reaction, like polymerization or crosslinking, which bonds the crack and restores the material's mechanical properties. This mechanism offers an efficient way to repair materials autonomously, making it highly valuable in applications where structural integrity is critical.

The self-healing process is initiated through the rupture of microcapsules when mechanical stress occurs. As a crack propagates through the material, it breaks open the capsules, allowing the healing agent to flow into the crack. Typically, this healing agent is a monomer or resin that polymerizes upon exposure to air, moisture, or a catalyst embedded within the material. This polymerization creates new bonds across the crack, effectively sealing the damage and restoring the material's mechanical strength. The simplicity of this process and its ability to react to damage automatically make microencapsulation a practical solution for maintaining material durability.

Microencapsulation is particularly effective for repairing polymers-based composites, where maintaining material integrity is crucial for continued operation. Its relatively straightforward implementation allows it to be adapted to a variety of materials, making it an ideal approach for aerospace applications where lightweight, durable materials are essential. By enabling one-time damage repair, this method can significantly enhance the lifespan and reliability of aerospace components, reducing maintenance needs.

However, there are some challenges associated with microencapsulation. One of the primary limitations is that it typically allows for only a single healing event. Once the microcapsules have ruptured and the healing agent is released and used, the material cannot heal again without external intervention. Additionally, careful control is needed during the manufacturing process to ensure that the microcapsules are evenly distributed within the material matrix and do not rupture prematurely during handling or processing. Despite these challenges, microencapsulation remains a promising self-healing strategy, particularly for critical aerospace applications. A representative demonstration of in situ repair of cracks by microencapsulation method is shown in Fig. 7.1.

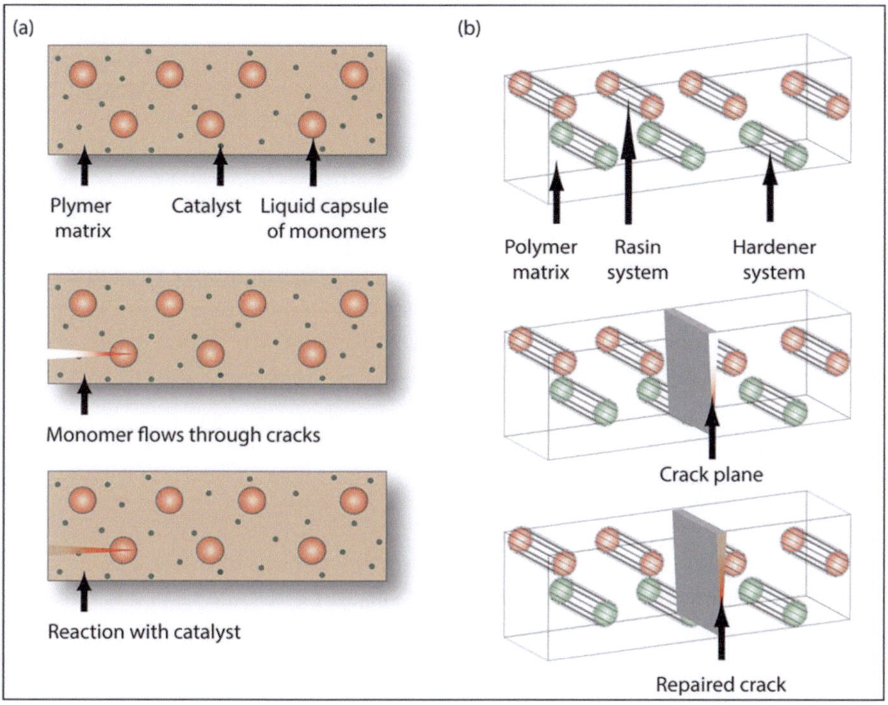

Fig. 7.1 A representative depiction of in situ crack repair by microencapsulation method. The onset of repair is trigged by the change in the mechanical material stress due to opening of the crack

7.3.2 Vascular Networks

Vascular networks for self-healing materials are inspired by biological systems, such as the circulatory system in animals, where interconnected channels distribute essential fluids throughout the body. Similarly, in self-healing materials, a network of channels supplies healing agents to damaged areas, enabling multiple cycles of repair. These channels are embedded within the material and can continuously deliver the healing agent, ensuring that the material can heal repeatedly over its lifetime.

The self-healing process in vascular networks involves microchannels or hollow fibers filled with a healing agent. These channels are often connected to an external reservoir, allowing the healing agent to be replenished as needed. When a crack forms and intersects a channel, the healing agent flows into the crack, initiating the repair process. Depending on the system design, the healing agent may cure or harden when exposed to air, heat, or a catalyst, effectively sealing the crack and restoring the material's integrity. This cycle can be repeated multiple times, as long as the reservoir remains filled and the network is functional.

One of the primary advantages of vascular networks is their ability to support continuous healing. Unlike systems that allow for only one healing event, vascular networks can repair damage multiple times, significantly extending the lifespan of the material. As long as the channels remain unobstructed and the healing agent is replenished, the material can autonomously repair itself after each damage event, making it ideal for long-term applications like those in aerospace.

However, vascular networks also present certain challenges. Creating such intricate networks within a material requires advanced fabrication techniques, such as 3D printing or microfabrication. These processes can be complex and may increase the cost of production. Additionally, over time, the channels may become blocked by cured healing agent or debris, which can reduce the effectiveness of the self-healing system. Despite these challenges, vascular networks represent a highly promising approach to developing materials capable of multiple, autonomous healing cycles, making them particularly attractive for applications requiring long-term durability.

7.3.3 Reversible Chemical Bonds

Reversible chemical bonds offer another promising approach to self-healing materials, where the material is engineered to heal itself by reforming bonds that were broken due to damage. These bonds can be either covalent or non-covalent and can break and reform under certain conditions, allowing the material to autonomously recover from damage. This method provides a dynamic and flexible means of repairing materials, offering potential for repeated healing cycles.

In materials with reversible covalent bonds, damage leads to the breaking of these bonds. However, under the right external conditions, such as the application of heat, light, or exposure to a specific chemical environment, these bonds can reform, effectively restoring the material. A notable example is the Diels–Alder reaction, in which the material heals through a reversible cycloaddition reaction (see Fig. 7.2). This process allows the material to undergo multiple damage-healing cycles, maintaining its structural integrity over time. The ability to reform covalent bonds provides a robust mechanism for healing in materials that need to maintain high strength and durability.

In contrast, non-covalent bonds, such as hydrogen bonding, ionic interactions, or van der Waals forces, facilitate self-healing at a molecular level. These weaker bonds can easily break and reform, allowing the material to recover from damage without the need for extreme external conditions. Healing through non-covalent interactions often occurs at lower temperatures, making it more efficient for certain applications. Additionally, this type of healing can be repeated multiple times without significantly degrading the material's properties, making it highly suitable for environments where the material may experience frequent damage. Figures 7.3, 7.4 and 7.5 demonstrate the self-healing of a hydrogel after it was inflicted with a cut. The mechanical strength of this gel was almost recovered during this process. Non-covalent bonds are primarily associated with such healing processes [20–24].

Despite its advantages, self-healing through reversible chemical bonds does present challenges. The efficiency of healing depends on how well the bonds can

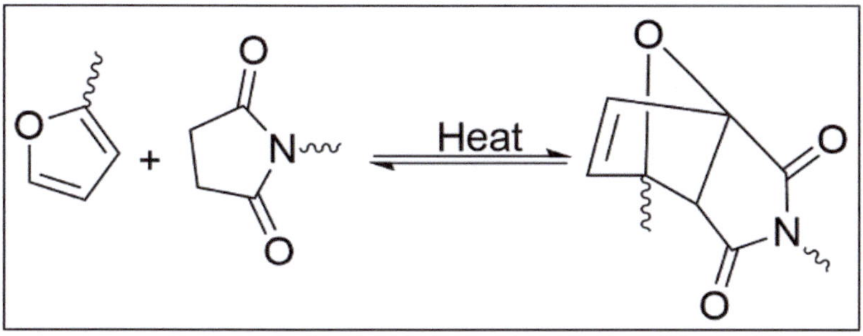

Fig. 7.2 Diels–Alder reaction. The reaction is a single step cycloaddition reaction. Here, two unsaturated molecules combine to form a cyclic adduct. There is a net reduction in bond multiplicity. All the bond formations and bond breakages happen simultaneously

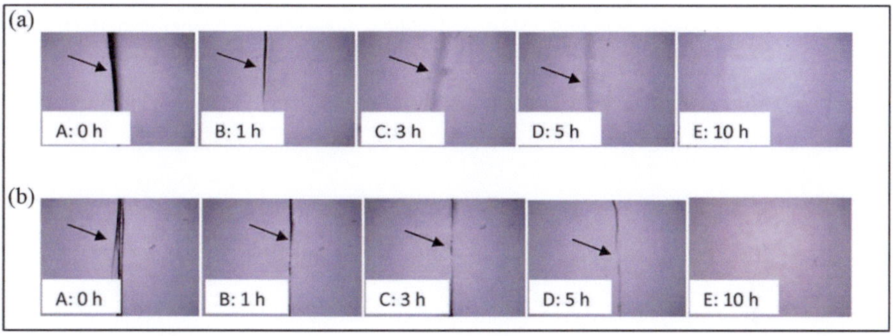

Fig. 7.3 Hourly time sequence images of ionogels, **a** top panel, IL = 1%, and **b** bottom panel, IL = 5%, indicating healing under ambient conditions. Arrows indicate the incision line which fades with time as the gel heals. Note the complete healing after 10 h. Adapted from ref. [24]

reform and the specific conditions required to trigger the process. Achieving full mechanical recovery after damage can be difficult, especially after repeated healing cycles, as some bonds may not reform completely. Furthermore, systems based on reversible bonds may be sensitive to environmental conditions, such as temperature, humidity, or UV exposure. These factors can affect the reliability and repeatability of the healing process, particularly in demanding applications like aerospace, where materials are exposed to a wide range of environmental stresses.

7.3.4 Shape Memory Effect

Shape memory materials possess the unique ability to "remember" their original shape and revert to it when exposed to a specific stimulus, such as heat. This property can be utilized for self-healing purposes, especially in materials where a shape change

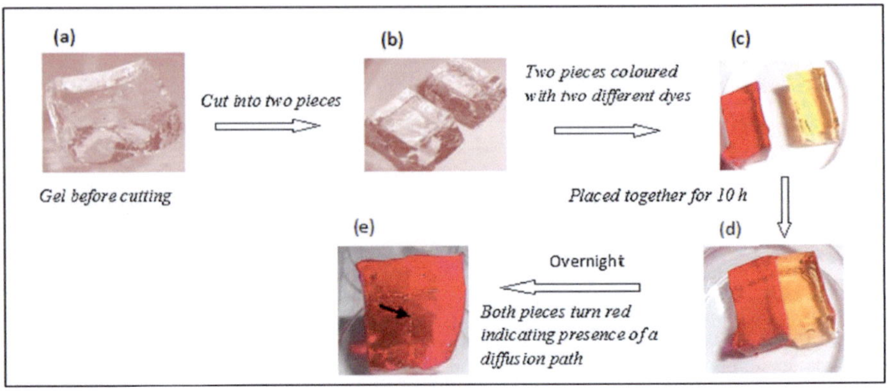

Fig. 7.4 Self-healing of a gel. **a** the original gel, **b** the gel cut into two pieces, **c** one piece was soaked in rhodamine B (0.1 mg/mL) and another in orange G (0.1 mg/mL) dye solution for 30 min each, **d** the two pieces were kept in physical contact face to face for 12 h without at 10 °C in the incubator, **e** the rhodamine G dye diffused into the gel piece containing orange G dye due to its higher diffusion coefficient. The boundary line between the two halves is clearly seen in (**e**, arrow) but the two gel pieces have fused. Adapted from ref. [24]

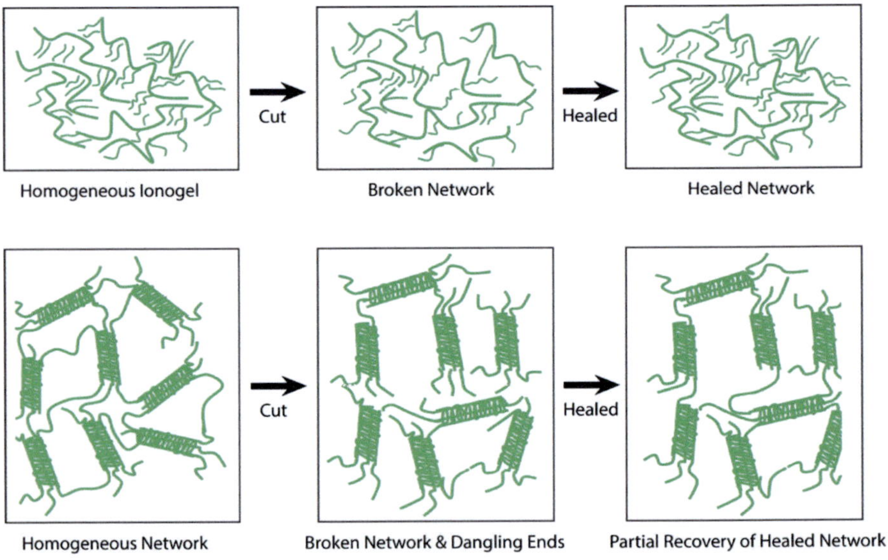

Fig. 7.5 Gel healing mechanism. Note the rearrangement of hydrophobic molecules that assist in the self-healing process. Adapted from ref. [24]

can close cracks or other forms of damage. By returning to their original shape, these materials can effectively repair damage and restore structural integrity.

Shape memory polymers (SMPs) are a prominent example of such materials. SMPs can be deformed into a temporary shape and will revert to their original configuration when heated above a certain temperature. In self-healing applications, SMPs can close cracks by returning to their original form, effectively sealing the damage. This process can be further enhanced by incorporating additional healing agents that bond the crack surfaces together as they close, resulting in a completer and more lasting repair.

Similarly, shape memory alloys (SMAs) function in a comparable way, returning to their original shape when heated. These materials are particularly useful in applications where maintaining mechanical recovery and structural integrity is critical. SMAs are often employed in structural components that must endure stress and still return to their original form, making them ideal for high-performance environments. Figure 7.6 illustrates this mechanism.

However, the use of shape memory materials in self-healing systems is not without challenges. One key limitation is the need for thermal activation. This requirement can restrict the use of these materials in environments where temperature control is difficult or where the operating temperature is close to the activation threshold. Additionally, repeated cycles of shape memory activation can lead to material fatigue, gradually reducing the effectiveness of the self-healing process over time. Despite these challenges, shape memory materials offer a promising avenue for creating smart, self-repairing systems that can enhance material durability and performance in demanding applications.

7.3.5 Comparative Analysis of Self-healing Mechanisms

Each of the self-healing mechanisms discussed offers unique advantages and is suited to different aerospace applications. Microencapsulation provides a straightforward and cost-effective approach but is limited to single-use healing. Vascular networks offer the potential for multiple healing cycles but require complex manufacturing processes. Reversible chemical bonds allow for repeated healing with minimal intervention, though they may require specific environmental conditions to activate. Shape memory effects enable mechanical recovery but are dependent on thermal triggers. Various mechanisms of shelf healing are summarized in Fig. 7.7.

When designing self-healing aeronautical nanocomposites, engineers must consider factors such as the type and frequency of damage expected, the operational environment, and the desired mechanical properties of the material. By selecting the appropriate self-healing mechanism or combining multiple approaches, it is possible to develop advanced materials that enhance the safety, reliability, and longevity of aerospace structures.

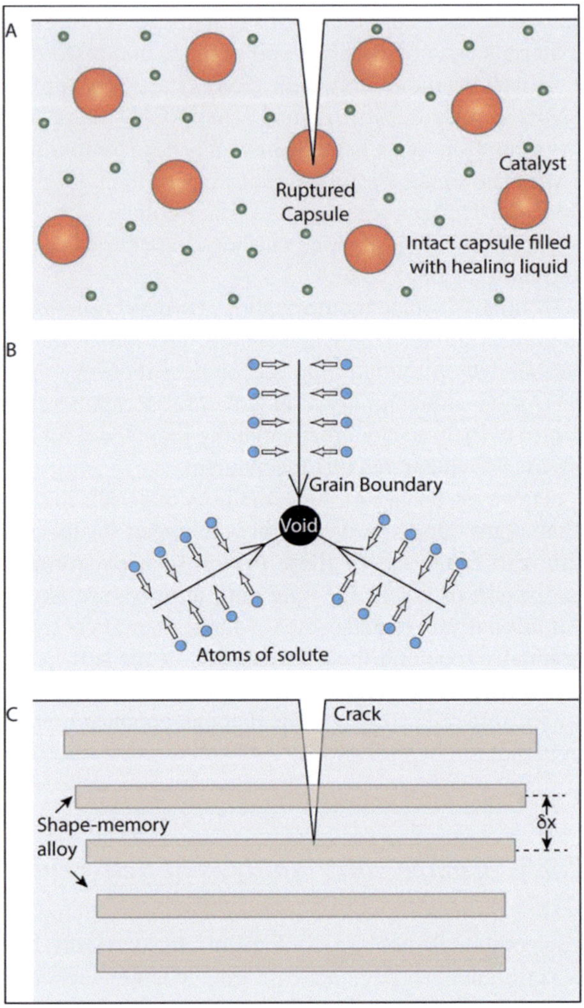

Fig. 7.6 The material integrity is stored in the structure as shape-memory with the damaged material trying to restore its original shape. **a** When a crack appears, **b** there is a considerable amount of distortion in the grain boundary that **c** promotes the rupture of capsules containing the healing agent and repair the crack. Presence of a catalyst activates this process

7.4 Self-healing Materials

The development of self-healing materials represents a significant advancement in material science, particularly in the context of aerospace applications. These materials possess the remarkable ability to autonomously repair damage without the need for external intervention, thus enhancing the durability, safety, and longevity of aerostructures. This subtopic delves into the various types of self-healing materials that are currently being explored for use in the aerospace industry, including nanomaterials and composites, polymer materials and composites, and metals and alloys.

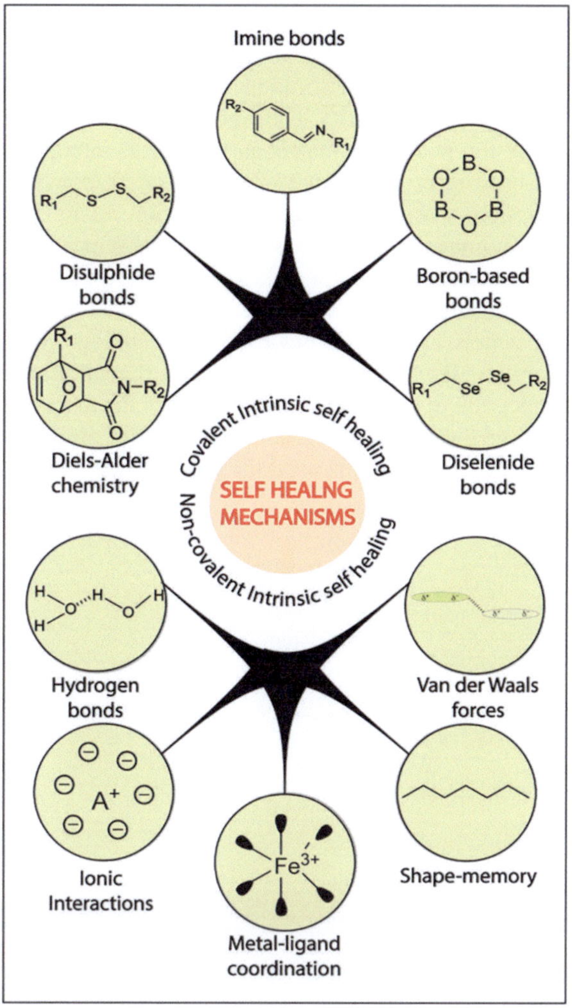

Fig. 7.7 Schematic depiction of various self-healing mechanisms

7.4.1 *Nanomaterials and Composites*

Nanomaterials, with their unique properties at the nanoscale, offer significant potential for self-healing applications in the aerospace industry. Nanocomposites combine nanoscale fillers like carbon nanotubes, graphene, or nanoclays with a polymer matrix to create materials that exhibit both enhanced mechanical properties and self-healing capabilities. These advanced materials promise to improve the durability and reliability of aerospace components, contributing to the development of more resilient and long-lasting systems.

The mechanism of self-healing in nanomaterials typically involves the incorporation of nanocapsules or nanofibers that contain healing agents. When damaged, such as cracking, occurs, these nanostructures rupture, releasing the healing agent into the affected area (Fig. 7.8 and 7.9). This agent then undergoes a chemical reaction—often polymerization or crosslinking—which fills and bonds the crack, restoring the material's structural integrity. The high surface area of nanomaterials allows for its efficient dispersion within the matrix and increases interaction with the polymer, making the self-healing process more effective and responsive to damage. A molecular level depiction of self-healing is shown in Fig. 7.10.

Nanomaterials and nanocomposites are particularly advantageous for aerospace applications due to their lightweight nature and exceptional mechanical properties. These materials are used in critical aerospace components like wing structures, fuselage panels, and turbine blades, where maintaining structural integrity is vital for safe operation. By incorporating self-healing capabilities, nanocomposites help extend the service life of these components, reduce the frequency and cost of maintenance, and improve the overall safety and reliability of aircraft and spacecraft systems. Various types of nanomaterials used for self-healing applications is listed in Table 7.1.

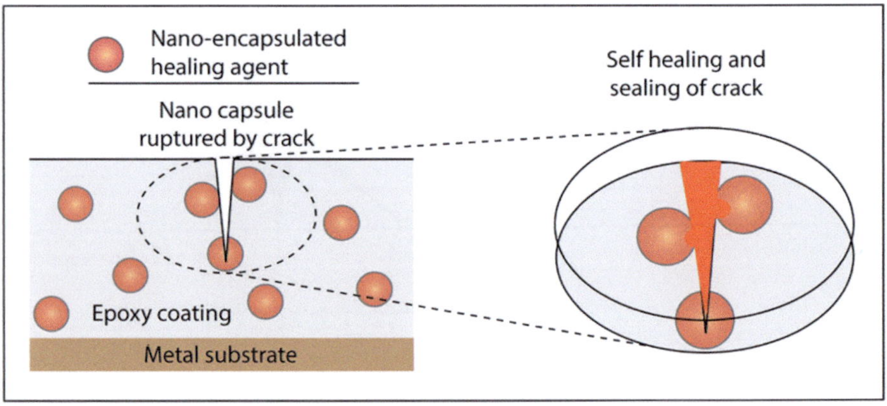

Fig. 7.8 Nanomaterial based repair process. The healing agent may be encapsulated in nanocapsules which under trigger release this cargo into the affected site for repair

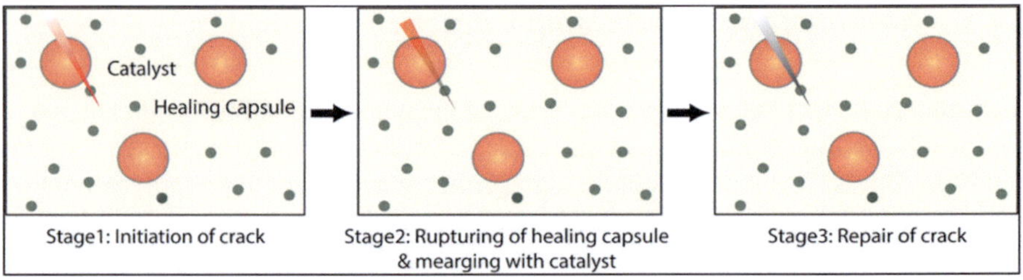

Fig. 7.9 An alternative method can be the catalyst mediated healing process where the nanocapsules releases the cargo into the affected site and the catalyst facilitates the repair process. This is a more structured process

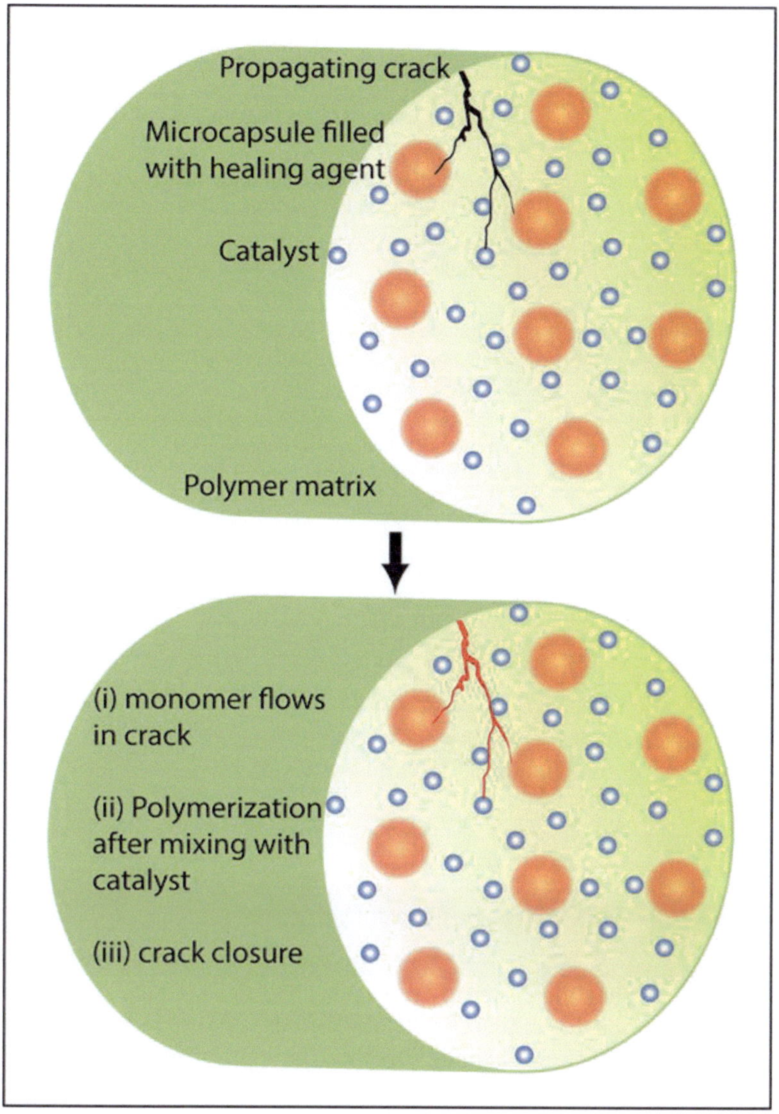

Fig. 7.10 Molecular mechanism of catalyst based self-healing

Despite the promising advantages of nanomaterials in self-healing composites, several challenges remain. One of the main difficulties is ensuring the uniform dispersion of nanomaterials within the polymer matrix to achieve consistent healing performance. Additionally, varying environmental conditions can impact the effectiveness of the self-healing process, making it essential to optimize these materials for different operational settings. Future research is focused on addressing these

Table 7.1 The various types of nanomaterials used as self-healing materials in the aerospace industry. The nanomaterials used for self-healing properties in aerospace applications, along with their mechanisms and specific uses

S. no.	Nanomaterial	Self-healing mechanism	Applications in aerospace	References
1	Nanocapsules	Encapsulated healing agents released upon damage	Composite materials, coatings	[25–30]
2	Nanofibers	Embedded fibers that reorient or reconnect	Structural components, fabrics	[31–35]
3	Nanoparticles	Triggered release of healing agents or polymers	Coatings, adhesives, matrix materials	[36–41]
4	Carbon nanotubes	Electrical conductivity aids in self-diagnosis	Structural materials, composites	[42–47]
5	Graphene	Forms conductive networks aiding in damage detection	Sensors, structural materials	[48–54]
6	Nanoclays	Clay platelets intercalate healing agents	Polymer matrix composites, coatings	[55–60]

challenges, as well as scaling up the production of nanocomposites for widespread use in aerospace, ultimately aiming to harness the full potential of nanotechnology in the industry.

7.4.2 Polymer Materials and Composites

Polymer-based self-healing materials are among the most extensively studied for aerospace applications due to their versatility and ease of processing. Polymers can be engineered to exhibit self-healing properties through various mechanisms, such as thermally activated healing, photoresponsive healing, and chemical-based healing systems. These mechanisms enable polymers to autonomously repair damage, making them highly desirable in industries where material integrity is crucial.

The self-healing process in polymer materials can be achieved through several mechanisms, including reversible covalent bonding, hydrogen bonding, or supramolecular interactions. A common approach involves embedding microcapsules filled with a healing agent within the polymer matrix. When damage occurs, these capsules rupture, releasing the healing agent into the crack. External stimuli, such as heat or light or/ and a catalyst, are often required to activate the healing agent, which then forms new bonds to repair the material. This versatile approach allows polymers to heal in response to different types of damage and environmental conditions.

Polymer composites are widely used in aerospace due to their lightweight nature and high strength-to-weight ratio. The integration of self-healing capabilities into these materials makes them even more appealing, particularly for components that are frequently subjected to stress and potential damage, such as fuselage skins, interior panels, and fuel tanks. The ability of these materials to autonomously heal reduces the need for frequent inspections and repairs, improving operational efficiency and enhancing safety in aerospace applications. A summary of polymer-based composites used for self-healing applications is provided in Table 7.2.

However, there are challenges in developing self-healing polymer composites. Ensuring the longevity and reliability of the healing process is a key concern, as is maintaining the mechanical properties of the material after repeated healing cycles. Advances in polymer chemistry and material engineering are expected to address these challenges, paving the way for the next generation of high-performance, self-healing polymer composites in aerospace, with enhanced durability and operational lifespan.

7.4.3 Metals and Alloys

While self-healing capabilities are commonly associated with polymer and composite materials, recent advancements have extended these properties to metals and alloys, which play a crucial role in aerospace engineering. This development represents a significant leap forward in material science, opening new possibilities for enhancing the durability and safety of aerospace structures.

Self-healing in metals and alloys is generally achieved through the incorporation of healing agents or by utilizing the intrinsic properties of the metal, such as diffusion or precipitation processes. One approach involves embedding microcapsules filled with liquid metal or other reactive agents within the metal matrix. When a crack forms, these capsules rupture, allowing the liquid metal to flow into the crack, where it solidifies and restores the material's integrity. Alternatively, some alloys can be engineered to undergo phase transformations or precipitation of secondary phases that fill cracks and voids, further reinforcing the structure.

In aerospace, metals and alloys with self-healing capabilities are particularly valuable in high-stress areas, such as landing gear, engine components, and load-bearing frames. These materials help prevent the propagation of cracks and other forms of damage that could lead to catastrophic failure. By enhancing the safety and reliability of critical components, self-healing metals and alloys offer the potential to extend the lifespan of aircraft. Additionally, these advancements may allow for the use of thinner, lighter materials, contributing to overall weight reduction without compromising safety. Metals and alloys used in the self-healing applications are listed in Table 7.3.

Table 7.2 The various types of polymeric materials and their composites used as self-healing materials in the aerospace industry, along with their self-healing mechanisms and applications. This table covers a range of polymeric materials and composites that exhibit self-healing properties, their mechanisms, and how they are used in the aerospace industry

S. no.	Polymer/composite	Self-healing mechanism	Applications in aerospace	References
1	Epoxy resins	Microcapsules or vascular networks with healing agents	Structural components, adhesives, coatings	[11, 61–64]
2	Polyurethane (PU)	Thermally reversible Diels–Alder reactions	Coatings, sealants, flexible components	[65–70]
3	Polydimethylsiloxane (PDMS)	Supramolecular chemistry, hydrogen bonding	Flexible electronics, sealants	[71–74]
4	Diels–Alder polymers	Reversible covalent bonding reactions	High-performance composites, coatings	[75–78]
5	Polyethylene (PE)	Thermoplastic healing via melting	Lightweight structures, impact-resistant components	[79–83]
6	Polymethyl methacrylate (PMMA)	Thermally induced phase separation	Transparent components, impact-resistant structures	[84–87]
7	Polyimides	High-temperature healing processes	High-temperature components, electronic applications	[88–91]
8	Carbon fiber reinforced polymers (CFRP)	Embedded microcapsules, fiber bridging	Structural components, fuselage, wings	[88–91]
9	Glass fiber reinforced polymers (GFRP)	Matrix cracking and fiber pull-out mechanisms	Radomes, fairings, structural components	[92–95]
10	Polymer nanocomposites	Nanoclay, carbon nanotubes, or graphene enhancement	Lightweight structural materials, coatings	[96–99]
11	Thermoplastic polymers	Heat-activated self-healing	Repairable composite structures, interior panels	[100–103]
12	Shape memory polymers (SMPs)	Shape recovery upon heating	Morphing structures, adaptive components	[104–106]

Despite their promise, the development of self-healing metals and alloys is still in its early stages, with several challenges to overcome. Ensuring consistent healing across different types of metals, understanding the potential impact on the metal's mechanical properties, and maintaining precise control over the healing process are key areas of concern. Ongoing research is focused on optimizing self-healing

Table 7.3 Various types of metals and alloys used as self-healing materials in the aerospace industry, along with their self-healing mechanisms and applications. This table outlines different metals and alloys with self-healing capabilities, their mechanisms, and their applications in the aerospace industry

S. no.	Metal/alloy	Self-healing mechanism	Applications in aerospace	References
1	Aluminum alloys	Microcapsules with liquid metal or healing agents	Structural components, fuselage, wing structures	[107–111]
2	Nickel-based superalloys	Oxidation-induced healing at high temperatures	Turbine blades, high-temperature components	[112–115]
3	Titanium alloys	Thermally activated shape memory effect	Structural components, high-stress areas	[116–120]
4	Magnesium alloys	Galvanic corrosion protection	Lightweight structural components	[121–123]
5	Zinc-based alloys	Sacrificial anode for corrosion protection	Coatings, sacrificial layers	[124, 125]
6	Copper alloys	Stress corrosion cracking resistance	Electrical components, thermal management systems	[126–128]
7	Shape memory alloys (SMA)	Phase transformation-induced healing	Actuators, sensors, adaptive structures	[129–133]
8	Steel alloys	Embedding of microcapsules with corrosion inhibitors	Structural components, landing gear	[134, 135]
9	Chromium-based alloys	Self-passivating oxide layers	High-temperature components, exhaust systems	[136–138]
10	Cobalt-based alloys	High-temperature oxidation resistance	Turbine components, high-stress areas	[139–142]
11	Gold alloys	Self-passivating surface layers	Electronics, connectors, sensors	[143–145]
12	Silver alloys	Anti-microbial and self-healing surface coatings	Electronics, connectors	[146, 147]

mechanisms in metals and alloys, with the goal of developing materials that can autonomously repair damage under the extreme conditions typical of aerospace environments.

7.5 Relative Merits of Various Self-healing Approaches

The development of self-healing materials has opened new avenues in material science, particularly for applications in the aerospace industry where maintaining structural integrity is crucial. Various self-healing approaches have been explored, each offering distinct advantages and limitations. Understanding the relative merits of these approaches is essential for selecting the most appropriate self-healing system for specific aerospace applications. This section provides an overview of the key self-healing mechanisms, including microcapsule-based systems, vascular networks, intrinsic self-healing materials, and external stimulus-based healing, and evaluates their relative merits in the context of aerospace engineering.

7.5.1 Microcapsule-Based Self-healing Systems

Microcapsule-based self-healing systems are one of the most extensively studied methods for self-healing materials. This approach involves embedding microcapsules filled with a healing agent into the material. When a crack forms, the microcapsules rupture, releasing the agent, which then reacts, typically through polymerization or crosslinking, to repair the damage and restore the material's integrity.

The key advantages of this system include its ability to autonomously heal damage without external intervention, a crucial feature in aerospace applications where manual inspection and repair can be difficult. Additionally, microcapsule-based systems are highly versatile, suitable for a range of materials including polymers, composites, and metals, which makes them adaptable for various aerospace components. Moreover, they are relatively easy to manufacture and integrate, providing a cost-effective solution for self-healing materials.

However, the main limitation of microcapsule-based systems is their ability to support only a single healing event. Once the capsules are depleted, the material cannot heal itself again. Additionally, there is a risk of the microcapsules rupturing prematurely during material processing, leading to the unintended release of the healing agent.

7.5.2 Vascular Networks

Vascular self-healing systems replicate the circulatory systems of living organisms, embedding hollow channels within materials to supply healing agents. When a crack intersects the vascular network, the healing agent flows into the damaged area and repairs it. The key advantage of vascular systems is their ability to support multiple healing cycles, as the healing agent can be replenished from an external reservoir or continuously circulated. This makes them effective for repairing larger-scale damage, as the healing agent can flow into and fill extensive cracks, providing a robust solution for maintaining material integrity.

However, vascular systems come with challenges. Manufacturing these networks is complex and costly compared to microcapsule-based systems, which can limit their scalability in aerospace applications. Additionally, the channels are prone to blockages from debris or cured healing agents, potentially reducing the system's effectiveness over time.

7.5.3 Intrinsic Self-healing Materials

Intrinsic self-healing materials rely on their inherent properties to repair damage, using reversible chemical bonds like hydrogen bonds or dynamic covalent bonds that can break and reform. This enables the material to heal itself without the need for external healing agents.

One of the key merits of intrinsic self-healing materials is their ability to undergo repeated healing cycles, making them durable and long-lasting. Additionally, since the healing mechanism is built into the material, there is no need for external agents, such as microcapsules or vascular networks, which simplifies the design and processing of these materials.

However, these materials often depend on external stimuli, such as heat, light, or pressure, to activate the healing process. In aerospace applications, where environmental conditions can be highly variable, this dependency may pose a challenge. Furthermore, some intrinsic self-healing materials may exhibit reduced mechanical strength, especially after multiple healing cycles, which can limit their use in high-stress aerospace applications.

7.5.4 External Stimulus-Based Healing

External stimulus-based healing involves materials that require a specific external trigger, such as heat, light, or an electrical field, to initiate the healing process. These materials are engineered to respond to stimuli by either activating a healing agent or inducing a physical or chemical change that repairs the damage. This method offers a controlled and reliable approach to repairing materials in critical aerospace applications.

One of the key merits of external stimulus-based healing is the precise control it offers over the healing process. The activation of the healing mechanism can be triggered only when necessary, allowing for targeted repairs. This is especially beneficial in aerospace, where controlled and deliberate healing is essential to maintain the integrity of vital components. Additionally, this method allows for tailored healing responses, enabling materials to react to specific environmental conditions or operational scenarios. Such adaptability makes these materials ideal for a variety of challenging aerospace environments.

However, there are limitations to this approach. One significant drawback is the requirement for external equipment to deliver the necessary stimulus, such as heaters or UV lamps. This adds complexity to the aerospace system and may not always be practical, particularly in remote or hard-to-reach areas of the aircraft. Furthermore, the effectiveness of the healing process can be compromised if the stimulus is not applied uniformly. In complex aerospace structures, this can lead to incomplete or uneven healing, potentially reducing the reliability of the material's self-repair capabilities.

7.5.5 Comparative Analysis and Application Considerations

In comparing the various self-healing approaches, each offers unique advantages that make them suitable for different aerospace applications. Microcapsule-based systems, with their autonomous and cost-effective healing capabilities, are ideal for components where one-time damage repair is sufficient. Vascular networks, with their ability to support multiple healing cycles, are better suited for high-stress areas subject to repeated damage. Intrinsic self-healing materials, which can heal repeatedly without external agents, offer durability and simplicity, making them suitable for long-term applications where external stimuli are available. External stimulus-based healing provides precise control and adaptability, making it an attractive option for specific, critical applications where tailored healing responses are needed.

When selecting a self-healing approach for aerospace applications, engineers must consider factors such as the expected types of damage, the operational environment, the required durability and mechanical properties, and the cost and complexity of integrating the self-healing system into the material. By carefully evaluating these factors, it is possible to select the most appropriate self-healing approach that balances performance, reliability, and cost-effectiveness in aerospace applications.

7.6 Conclusion

This chapter has explored the transformative potential of self-healing aeronautical nanocomposites, highlighting their promising applications and significant benefits for the aerospace industry. The integration of self-healing technologies into aerospace materials represents a major advancement, offering solutions to critical challenges related to material durability, maintenance, and safety.

Self-healing mechanisms, including microencapsulation, vascular networks, and intrinsic self-healing properties, provide innovative approaches to repairing damage autonomously. Microencapsulation allows for the release of healing agents upon damage, while vascular networks simulate biological repair systems to deliver healing agents precisely where needed. Intrinsic self-healing materials, such as those based on reversible chemical bonds or shape memory polymers, offer continuous repair capabilities and adaptive responses to damage.

The potential benefits of self-healing nanocomposites are substantial. Enhanced safety is achieved through the reduction of catastrophic failures, as these materials can autonomously repair minor damage and maintain structural integrity. The extended lifespan and reduced need for maintenance and repair contribute to greater operational efficiency and cost savings. By decreasing the frequency of manual inspections and repairs, self-healing materials offer a more cost-effective approach to managing aerospace components.

Case studies have demonstrated the practical applications and effectiveness of self-healing technologies. Self-healing carbon fiber composites, vascular network systems in aircraft panels, and shape memory polymers have shown promising results in experimental settings. These examples underscore the potential for self-healing nanocomposites to revolutionize aerospace engineering by enhancing the reliability and performance of critical components.

Looking ahead, ongoing research and development will focus on optimizing these materials for broader implementation. Advancements in synthesis techniques, characterization methods, and integration strategies will drive further improvements in self-healing technologies. The successful incorporation of self-healing nanocomposites into aerospace structures promises to deliver smarter, more resilient materials, ultimately advancing the field and addressing key challenges in aerospace engineering. The future of aerospace materials lies in harnessing the capabilities of self-healing nanocomposites, paving the way for safer, longer lasting, and more efficient aerospace systems.

References

1. Williams JC, Starke EA Jr (2003) Progress in structural materials for aerospace systems. Acta Mater 51(19):5775–5799
2. Zhang X, Chen Y, Hu J (2018) Recent advances in the development of aerospace materials. Prog Aerosp Sci 97:22–34
3. Boyer RR, Cotton JD, Mohaghegh M, Schafrik RE (2015) Materials considerations for aerospace applications. MRS Bull 40(12):1055–1066
4. Singh J, Srivastawa K, Jana S, Dixit C, Ravichandran S (2024) Advancements in lightweight materials for aerospace structures: a comprehensive review. Acceleron Aerosp J 2(3):173–183
5. Nagaraju SB, Priya HC, Girijappa YGT, Puttegowda M (2023) Lightweight and sustainable materials for aerospace applications. In: Lightweight and sustainable composite materials. Woodhead Publishing, pp 157–178
6. Blanco D, Rubio EM, Lorente-Pedreille RM, Sáenz-Nuño MA (2021) Lightweight structural materials in open access: latest trends. Materials 14(21):6577
7. Basheer AA (2020) Advances in the smart materials applications in the aerospace industries. Aircr Eng Aerosp Technol 92(7):1027–1035
8. Loewy RG (1997) Recent developments in smart structures with aeronautical applications. Smart Mater Struct 6(5):R11

9. Mangalgiri PD (1999) Composite materials for aerospace applications. Bull Mater Sci 22:657–664
10. Venkateswaran MR, Khosravi A, Zarepour A, Iravani S, Zarrabi A (2024) Self-healing materials in biomedicine and the circular economy. Environ Sci Nano 11(7):2771–2802
11. Shields Y, De Belie N, Jefferson A, Van Tittelboom K (2021) A review of vascular networks for self-healing applications. Smart Mater Struct 30(6):063001
12. Liu J, Urban MW (2024) Dynamic interfaces in self-healable polymers. Langmuir 40(14):7268–7285
13. Yao H, Liu T, Jia Y, Du Y, Yao B, Xu J, Fu J (2023) Water-insensitive self-healing materials: from network structure design to advanced soft electronics. Adv Func Mater 33(48):2307455
14. Obayi CS, Nnamchi PS (2020) Exploits, advances and challenges in characterizing self-healing materials. Adv Funct Mater
15. Uman S, Dhand A, Burdick JA (2020) Recent advances in shear-thinning and self-healing hydrogels for biomedical applications. J Appl Polym Sci 137(25):48668
16. Sethi SK, Verma A, Akubueze E, Maurya AK (2024) Self-healing of polymer composites: process and developments. In: Hybrid composite materials: experimental and theoretical analysis. Springer Nature Singapore, Singapore, pp 253–271
17. Zafeer M, Bhat KS (2024) Chemical approaches for fabrication of self-healing polymers. Discov. Appl. Sci. 6(7):1–42
18. Kontiza A, Kartsonakis IA (2024) Smart composite materials with self-healing properties: a review on design and applications. Polymers 16(15):2115
19. Cai P, Wang C, Gao H, Chen X (2021) Mechanomaterials: a rational deployment of forces and geometries in programming functional materials. Adv Mater 33(46):2007977
20. Taylor DL, In Het Panhuis M (2016) Self-healing hydrogels. Adv Mater 28(41):9060–9093
21. Devi VKA, Shyam R, Palaniappan A, Jaiswal AK, Oh TH, Nathanael AJ (2021) Self-healing hydrogels: preparation, mechanism and advancement in biomedical applications. Polymers 13(21):3782
22. Zhang A, Liu Y, Qin D, Sun M, Wang T, Chen X (2020) Research status of self-healing hydrogel for wound management: a review. Int J Biol Macromol 164:2108–2123
23. Wang W, Zhang Y, Liu W (2017) Bioinspired fabrication of high strength hydrogels from non-covalent interactions. Prog Polym Sci 71:1–25
24. Sharma A, Rawat K, Solanki PR, Bohidar HB (2017) Self-healing gelatin ionogels. Int J Biol Macromol 95:603–607
25. Samadzadeh M, Boura SH, Peikari M, Kasiriha SM, Ashrafi A (2010) A review on self-healing coatings based on micro/nanocapsules. Prog Org Coat 68(3):159–164
26. Sadabadi H, Allahkaram SR, Kordijazi A, Rohatgi PK (2022) Self-healing coatings loaded by nano/microcapsules: a review. Prot Met Phys Chem Surf 58(2):287–307
27. Silva ACM, Moghadam AD, Singh P, Rohatgi PK (2017) Self-healing composite coatings based on in situ micro–nanoencapsulation process for corrosion protection. J Coat Technol Res 1–29
28. Pulikkalparambil H, Siengchin S, Parameswaranpillai J (2018) Corrosion protective self-healing epoxy resin coatings based on inhibitor and polymeric healing agents encapsulated in organic and inorganic micro and nanocontainers. Nano-Struct. Nano-Objects 16:381–395
29. Malekkhouyan R, Neisiany RE, Khorasani SN, Das O, Berto F, Ramakrishna S (2021) The influence of size and healing content on the performance of extrinsic self-healing coatings. J Appl Polym Sci 138(10):49964
30. Wei H, Wang Y, Guo J, Shen NZ, Jiang D, Zhang X, Yan X, Zhu J, Wang Q, Shao L, Lin H (2015) Advanced micro/nanocapsules for self-healing smart anticorrosion coatings. J Mater Chem A 3(2):469–480
31. Gao Z, Xiao X, Carlo AD, Yin J, Wang Y, Huang L, Tang J, Chen J (2023) Advances in wearable strain sensors based on electrospun fibers. Adv Func Mater 33(18):2214265
32. Ghosh S, Nitin B, Remanan S, Bhattacharjee Y, Ghorai A, Dey T, Das TK, Das NC (2020) A multifunctional smart textile derived from merino wool/nylon polymer nanocomposites as next generation microwave absorber and soft touch sensor. ACS Appl Mater Interfaces 12(15):17988–18001

33. Tan GZ, Zhou Y (2020) Electrospinning of biomimetic fibrous scaffolds for tissue engineering: a review. Int J Polym Mater Polym Biomater 69(15):947–960
34. Dicker KT, Ratchford D, Casalini R, Thum MD, Wynne JH, Lundin JG (2020) Surfactant modulated phase transitions of liquid crystals confined in electrospun coaxial fibers. Langmuir 36(27):7916–7924
35. Moradi A, Szewczyk PK, Roszko A, Fornalik-Wajs E, Stachewicz U (2024) Unraveling the impact of boron nitride and silicon nitride nanoparticles on thermoplastic polyurethane fibers and mats for advanced heat management. ACS Appl Mater Interfaces
36. Noordin NM, Kahar NNF, Osman AF, Alosime E, Arsat N, Mohammad Azman NA, Syamsir A, Itam Z, Abdul Hamid ZA (2021) The versatility of polymeric materials as self-healing agents for various types of applications: a review. Polymers 13(8):1194
37. Zhu DY, Rong MZ, Zhang MQ (2015) Self-healing polymeric materials based on microencapsulated healing agents: from design to preparation. Prog Polym Sci 49:175–220
38. Fu X, Tian L, Fan Y, Ye W, Qiao ZA, Zhao J, Ren L, Ming W (2021) Stimuli-responsive self-healing anticorrosion coatings: from single triggering behavior to synergetic multiple protections. Mater Today Chem 22:100575
39. Kamaly N, Yameen B, Wu J, Farokhzad OC (2016) Degradable controlled-release polymers and polymeric nanoparticles: mechanisms of controlling drug release. Chem Rev 116(4):2602–2663
40. Pathan N, Shende P (2021) Strategic conceptualization and potential of self-healing polymers in biomedical field. Mater Sci Eng C 125:112099
41. Imazato S, Ma S, Chen JH, Xu HH (2014) Therapeutic polymers for dental adhesives: loading resins with bio-active components. Dent Mater 30(1):97–104
42. Wentzel DJ (2018) Development of self-diagnostic carbon fiber epoxy composite materials using Grahene flakes and carbon nanofibers. New Mexico State University
43. Vavouliotis AI, Kostopoulos V (2013) On the use of electrical conductivity for the assessment of damage in carbon nanotubes enhanced aerospace composites. In: Carbon nanotube enhanced aerospace composite materials: a new generation of multifunctional hybrid structural composites, pp 21–55
44. Han B, Yu X, Ou J (2011) Multifunctional and smart carbon nanotube reinforced cement-based materials. In: Nanotechnology in civil infrastructure: a paradigm shift. Springer Berlin Heidelberg, Berlin, Heidelberg
45. Ke K, Pötschke P, Gao S, Voit B (2017) An ionic liquid as interface linker for tuning piezoresistive sensitivity and toughness in poly (vinylidene fluoride)/carbon nanotube composites. ACS Appl Mater Interfaces 9(6):5437–5446
46. Das CM, Kang L, Yang G, Tian D, Yong KT (2020) Multifaceted hybrid carbon fibers: applications in renewables, sensing and tissue engineering. J Compos Sci 4(3):117
47. Kang Z, Aslani F, Han B (2024) Prediction of mechanical and electrical properties of carbon fibre-reinforced self-sensing cementitious composites. Case Stud Constr Mater 20:e02716
48. Tung TT, Nine MJ, Krebsz M, Pasinszki T, Coghlan CJ, Tran DN, Losic D (2017) Recent advances in sensing applications of graphene assemblies and their composites. Adv Func Mater 27(46):1702891
49. Ahmed O, Wang X, Tran MV, Ismadi MZ (2021) Advancements in fiber-reinforced polymer composite materials damage detection methods: Towards achieving energy-efficient SHM systems. Compos B Eng 223:109136
50. Carvalho AF, Kulyk B, Fernandes AJ, Fortunato E, Costa FM (2022) A review on the applications of graphene in mechanical transduction. Adv Mater 34(8):2101326
51. Dai Pang S, Gao HJ, Xu C, Quek ST, Du H (2014) Strain and damage self-sensing cement composites with conductive graphene nanoplatelet. In: Sensors and smart structures technologies for civil, mechanical, and aerospace systems, vol 9061. SPIE, pp 546–556
52. Miao P, Wang J, Zhang C, Sun M, Cheng S, Liu H (2019) Graphene nanostructure-based tactile sensors for electronic skin applications. Nano-Micro Lett 11:1–37
53. Mirabedini A, Ang A, Nikzad M, Fox B, Lau KT, Hameed N (2020) Evolving strategies for producing multiscale graphene-enhanced fiber-reinforced polymer composites for smart structural applications. Adv Sci 7(11):1903501

54. Nag A, Mitra A, Mukhopadhyay SC (2018) Graphene and its sensor-based applications: a review. Sens Actuators A: Phys 270:177–194
55. Pandey S, Zaidib MGH, Gururani SK (2013) Recent developments in clay-polymer nano composites. Sci J Rev 2(11):296–328
56. Guo F, Aryana S, Han Y, Jiao Y (2018) A review of the synthesis and applications of polymer–nanoclay composites. Appl Sci 8(9):1696
57. Peña-Parás L, Sánchez-Fernández JA, Vidaltamayo R (2018) Nanoclays for biomedical applications. Handb Ecomater 5:3453–3471
58. Kausar A, Ahmad I, Maaza M, Eisa MH (2022) State-of-the-art nanoclay reinforcement in green polymeric nanocomposite: from design to new opportunities. Minerals 12(12):1495
59. Das P, Manna S, Behera AK, Shee M, Basak P, Sharma AK (2022) Current synthesis and characterization techniques for clay-based polymer nano-composites and its biomedical applications: a review. Environ Res 212:113534
60. Shunmugasamy VC, Xiang C, Gupta N (2015) Clay/polymer nanocomposites: processing, properties, and applications. In: Hybrid and hierarchical composite materials, pp 161–200
61. Lee MW, An S, Yoon SS, Yarin AL (2018) Advances in self-healing materials based on vascular networks with mechanical self-repair characteristics. Adv Coll Interface Sci 252:21–37
62. An S, Lee MW, Yarin AL, Yoon SS (2018) A review on corrosion-protective extrinsic self-healing: comparison of microcapsule-based systems and those based on core-shell vascular networks. Chem Eng J 344:206–220
63. Zanjani JSM, Okan BS, Menceloglu Y, Yildiz M (2018) Self-healing thermosetting composites: concepts, chemistry, and future advances. In: Hybrid organic-inorganic interfaces: towards advanced functional materials, pp 121–150
64. Toohey KS, Sottos NR, Lewis JA, Moore JS, White SR (2007) Self-healing materials with microvascular networks. Nat Mater 6(8):581–585
65. Sridhar LM, Oster MO, Herr DE, Gregg JB, Wilson JA, Slark AT (2020) Re-usable thermally reversible crosslinked adhesives from robust polyester and poly (ester urethane) Diels-Alder networks. Green Chem 22(24):8669–8679
66. Samui AB (2022) Self-healing polymers. In: Smart polymers. CRC Press, pp 115–142
67. de Souza FM, Sulaiman MR, Gupta RK (2021) Materials and chemistry of polyurethanes. In: Materials and chemistry of flame-retardant polyurethanes volume 1: a fundamental approach. American Chemical Society, pp 1–36
68. Fortunato G (2020) Smart materials from reversible polymer networks
69. Larsen HP (2021) Synthesis of a self-healing graphene/polyurethane nanocomposite using the Diels-Alder reaction. Master's thesis, NTNU
70. Tremblay-Parrado KK (2020) Synthesis by click chemistry of biobased polyurethane networks with advanced properties. Doctoral dissertation, Université de Strasbourg
71. Zhang T, Shi Y, Chen W, Huang J, Li C (2024) Spider silk-inspired supramolecular polydimethylsiloxane network with prominent mechanical robustness for bifunctional flexible electronics. Eur Polymer J 213:113122
72. Li J, Niu H, Yu Y, Gao Y, Wu Q, Wang F, Sun P (2021) Supramolecular polydimethylsiloxane elastomer with enhanced mechanical properties and self-healing ability engineered by synergetic dynamic bonds. ACS Appl Polym Mater 3(7):3373–3382
73. Wen N, Zhang L, Jiang D, Wu Z, Li B, Sun C, Guo Z (2020) Emerging flexible sensors based on nanomaterials: recent status and applications. J Mater Chem A 8(48):25499–25527
74. Guo R, Li D, Qi W, Chen G, Li Q, Zhou Z (2023) Fabrication of high-strength and multi-recyclable supramolecular polysiloxane by incorporating quadruple hydrogen bonds into main-chains for adhesive and oil-water separation. Polymer 277:125954
75. Vauthier M, Jierry L, Oliveira JC, Hassouna L, Roucoules V, Bally-Le Gall F (2019) Interfacial thermoreversible chemistry on functional coatings: a focus on the Diels-Alder reaction. Adv Func Mater 29(10):1806765
76. Fortunato G, van den Tempel P, Bose RK (2024) Advances in self-healing coatings based on Diels-Alder chemistry. Polymer 294:126693

77. Ratwani CR, Kamali AR, Abdelkader AM (2023) Self-healing by Diels-Alder cycloaddition in advanced functional polymers: a review. Prog Mater Sci 131:101001
78. Karami Z, Zolghadr M, Zohuriaan-Mehr MJ (2020) Self-healing Diels–Alder engineered thermosets. In: Self-Healing Polymer-Based Systems. Elsevier, pp 209–233
79. Rizzo F, Cuomo S, Pinto F, Pucillo G, Meo M (2021) Thermoplastic polyurethane composites for railway applications: experimental and numerical study of hybrid laminates with improved impact resistance. J Thermoplast Compos Mater 34(8):1009–1036
80. Panneke N, Ehrmann A (2023) Stab-resistant polymers—recent developments in materials and structures. Polymers 15(4):983
81. Biron M (2015) Material selection for thermoplastic parts: practical and advanced information. William Andrew
82. Chandran Maniyamparambil S, Vadivel GK, Krishnan P (2023) Experimental and numerical damage analysis of low-velocity drop weight impact of polyethylene self-reinforced composite laminates for transportation applications. J Thermoplast Compos Mater 36(7):2965–2997
83. Silva F (2016) Mechanical properties and impact energy absorption of hybrid thermoplastic nanocomposite structures. Doctoral dissertation
84. Bird SA, Clary D, Jajam KC, Tippur HV, Auad ML (2013) Synthesis and characterization of high performance, transparent interpenetrating polymer networks with polyurethane and poly (methyl methacrylate). Polym Eng Sci 53(4):716–723
85. Deka N, Bera A, Roy D, De P (2022) Methyl methacrylate-based copolymers: recent developments in the areas of transparent and stretchable active matrices. ACS Omega 7(42):36929–36944
86. Sansul S, Yousif E, Zainulabdeen K (2023) A review study of the structure, properties and general application of poly (methyl methacrylate). Charact Appl Nanomater 6(1):2537
87. Huang J, Zhou H, Zhang L, Zha H, Shi W, Zhao T, Liu M (2024) Bioinspired stiff–soft gradient network structure for high-performance impact-resistant elastomers. Giant, p 100320
88. Kulkarni MV, Boppana SB (2023) Composites overview. In: Structural composite materials: fabrication, properties, applications and challenges. Springer Nature Singapore, Singapore, pp 3–21
89. García VR (2021) Towards the industrialization of bio-inspired multiscale hierarchical carbon fibre-reinforced polymer composites. Doctoral dissertation, Universidad Politécnica de Madrid
90. Veeramani N, Samikannu R, Deshpande AP, Varghese S, Moses V (2023) Effects of polymeric microcapsules on self-healing composites reinforced with carbon fibers: a comparative study. Int Polym Proc 38(4):483–495
91. Tavdishvili K (2022) Through-life support, diagnosis and prognosis in the aeronautical field: a critical literature review
92. Parmar MSS (2022) Study on physical and mechanical properties of basalt fibre preforms combination in laminates. Doctoral dissertation, Gujarat Technological University Ahmedabad
93. Chen Y (2021) Glass fiber-reinforced polymer composites for power equipment. In: Polymer composites for electrical engineering, pp 377–417
94. Uzun F (2020) Multidisciplinary investigation of C-type composite sandwich radome panels within the scope of acoustic emission based damage characterization and electromagnetic transmission performance. Doctoral dissertation
95. Wong PM (2004) Thermal, mechanical and structural performance of fibre reinforced polymer (FRP) composite structures at ambient and elevated temperatures. The University of Manchester (United Kingdom)
96. Kausar A, Rafique I, Muhammad B (2017) Aerospace application of polymer nanocomposite with carbon nanotube, graphite, graphene oxide, and nanoclay. Polym-Plast Technol Eng 56(13):1438–1456
97. Silvestre J, Silvestre N, De Brito J (2016) Polymer nanocomposites for structural applications: Recent trends and new perspectives. Mech Adv Mater Struct 23(11):1263–1277

98. Sanusi OM, Benelfellah A, Hocine NA (2020) Clays and carbon nanotubes as hybrid nanofillers in thermoplastic-based nanocomposites–A review. Appl Clay Sci 185:105408
99. Singh NP, Gupta VK, Singh AP (2019) Graphene and carbon nanotube reinforced epoxy nanocomposites: a review. Polymer 180:121724
100. Zheng M (2021) Study of recyclable and repairable dynamic covalent polymers for sustainable 3D printing development. Doctoral dissertation, The University of Western Ontario (Canada)
101. Wang TX, Chen HM, Salvekar AV, Lim J, Chen Y, Xiao R, Huang WM (2020) Vitrimer-like shape memory polymers: characterization and applications in reshaping and manufacturing. Polymers 12(10):2330
102. Jony B (2020) Characterization and modeling of dual phase thermoplastic self healing system for fiber reinforced thermoset composite structures. Doctoral dissertation, The University of Alabama
103. Hajiali F, Tajbakhsh S, Marić M (2021) Thermally reprocessable bio-based polymethacrylate vitrimers and nanocomposites. Polymer 212:123126
104. Yazik MM, Sultan MTH (2019) Shape memory polymer and its composites as morphing materials. In: Failure analysis in biocomposites, fibre-reinforced composites and hybrid composites. Woodhead Publishing, pp 181–198
105. Mu T, Liu L, Lan X, Liu Y, Leng J (2018) Shape memory polymers for composites. Compos Sci Technol 160:169–198
106. Xia Y, He Y, Zhang F, Liu Y, Leng J (2021) A review of shape memory polymers and composites: mechanisms, materials, and applications. Adv Mater 33(6):2000713
107. Das R, Melchior C, Karumbaiah KM (2016) Self-healing composites for aerospace applications. In: Advanced composite materials for aerospace engineering. Woodhead Publishing, pp 333–364
108. Paladugu SRM, Sreekanth PR, Sahu SK, Naresh K, Karthick SA, Venkateshwaran N, Ramoni M, Mensah RA, Das O, Shanmugam R (2022) A comprehensive review of self-healing polymer, metal, and ceramic matrix composites and their modeling aspects for aerospace applications. Materials 15(23):8521
109. Regassa Y, Sirabizuh B, Lemu HG (2019) Review and analysis of processing principles and applications of self-healing composite materials. Adv Manuf Autom VIII 8:159–167
110. Kadam MS, Kadam MP, Chavan S, Rahate MP, Pawar MS (2024) Study of self-healing composite materials for aerial vehicles. Fuzzy Syst Soft Comput 19(01(I)):1–8
111. Chuangchote S, Nukunudompanich M (2022) Self-healing carbon fiber–reinforced polymers for aerospace applications. In: Aerospace polymeric materials, pp 85–115
112. Gudivada G, Pandey AK (2023) Recent developments in nickel-based superalloys for gas turbine applications. J Alloy Compd 171128
113. Karabela A (2011) An experimental and computational study of damage and crack growth for a nickel-based superalloy under fatigue-oxidation conditions. Doctoral dissertation, University of Portsmouth
114. Evangelou A (2017) Oxidation-fatigue mechanisms at moderate service temperatures in single crystal turbine blade materials. Doctoral dissertation, University of Southampton
115. Evangelou A, Soady KA, Lockyer S, Gao N, Reed PAS (2019) On the mechanism of oxidation-fatigue damage at intermediate temperatures in a single crystal Ni-based superalloy. Mater Sci Eng A 742:648–661
116. Ramaiah KV, Saikrishna CN, Ranganath VR, Buravalla V, Bhaumik SK (2011) Fracture of thermally activated NiTi shape memory alloy wires. Mater Sci Eng A 528(16–17):5502–5510
117. Panton B, Zhou YN, Khan MI (2016) A stabilized, high stress self-biasing shape memory alloy actuator. Smart Mater Struct 25(9):095027
118. Hosseini E, Ghafoori E, Leinenbach C, Motavalli M, Holdsworth SR (2018) Stress recovery and cyclic behaviour of an Fe–Mn–Si shape memory alloy after multiple thermal activation. Smart Mater Struct 27(2):025009
119. Šittner P, Iaparova E, Kadeřávek L, Chen Y, Tyc O (2023) Tensile deformation of NiTi shape memory alloy thermally loaded under applied stress. Mater Des 226:111638

120. Saghaian SM, Karaca HE, Souri M, Turabi AS, Noebe RD (2016) Tensile shape memory behavior of Ni50. 3Ti29. 7Hf20 high temperature shape memory alloys. Mater Des 101:340–345
121. Gupta M, Ling SNM (2011) Magnesium, magnesium alloys, and magnesium composites. Wiley
122. Loukil N (2021) Alloying elements of magnesium alloys: a literature review. Magnes Alloy Struct Prop 17:58–78
123. Buldum BB, Sık A, Ozkul İ (2013) Investigation of magnesium alloys machinability. Int J Electron Mech Mechatron Eng 2(3):261–268
124. Ganesan S, Prabhu G, Popov BN (2013) Development of Zn-Mn alloy based sacrificial coatings. ECS Trans 50(31):405
125. Edavan RP, Kopinski R (2009) Corrosion resistance of painted zinc alloy coated steels. Corros Sci 51(10):2429–2442
126. Chae H, Wang H, Hong M, Kim WC, Kim JG, Kim H, Lee SY (2020) Stress corrosion cracking of a copper pipe in a heating water supply system. Met Mater Int 26:989–997
127. Chen H, Zheng F, Cheng W, Tao P, Song C, Shang W, Fu B, Deng T (2021) Low thermal expansion metal composite-based heat spreader for high temperature thermal management. Mater Des 208:109897
128. Jamwal A, Mittal P, Agrawal R, Gupta S, Kumar D, Sadasivuni KK, Gupta P (2020) Towards sustainable copper matrix composites: manufacturing routes with structural, mechanical, electrical and corrosion behaviour. J Compos Mater 54(19):2635–2649
129. Schick JR, Hartl DJ, Lagoudas DC (2012) Incorporation of shape memory alloy actuators into morphing aerostructures. Morphing aerospace vehicles and structures, pp 231–260
130. Mukhopadhyay R (2020) Development of autonomous truss modules utilizing shape memory alloys for actuation/sensing functionalities. North Carolina State University
131. Gotthardt R, Scherrer P, Stalmans R (2000) Smart materials based on shape memory alloys: examples from Europe. In: Materials science forum, vol 327. Trans Tech Publications Ltd., pp 83–90
132. Jani JM, Leary M, Subic A, Gibson MA (2014) A review of shape memory alloy research, applications and opportunities. Mater Des 1980–2015(56):1078–1113
133. Vanzo A (2017) Shape memory alloy for adaptive optics., Ph.D. Thesis, Department of Physics and Astronomy, University of Padova, Italy
134. Zhu H, Li J (2024) Advancements in corrosion protection for aerospace aluminum alloys through surface treatment. Int J Electrochem Sci 100487
135. Zade GS, Patil KD (2024) Advances in corrosion-resistant coatings: types, formulating principles, properties, and applications. In: Functional coatings: innovations and challenges, pp 110–152
136. Ballinger RG (2011) The development and production of a functionally graded composite for Pb-Bi service (No. DOE-ID14742). Massachusetts Institute of Technology (MIT), Cambridge, MA (United States)
137. Terentyev D, Khvan T, You JH, Van Steenberge N (2020) Development of chromium and chromium-tungsten alloy for the plasma facing components: application of vacuum arc melting techniques. J Nucl Mater 536:152204
138. Welles TS (2022) Manipulated electrochemical surface reactions induced by oscillatory electric potentials on metal based electrodes. Doctoral dissertation, Syracuse University
139. Akande IG, Oluwole OO, Fayomi OSI, Odunlami OA (2021) Overview of mechanical, microstructural, oxidation properties and high-temperature applications of superalloys. Mater Today: Proc 43:2222–2231
140. Sahay SK, Goswami B, Ray AK (2007) Development of cobalt-base alloys for high temperature applications. J Metall Mater Sci 49(1):1–38
141. Çoban O, Arslan ÖC, Karahan T (2018) Failure analysis of jet turbine engine high pressure nozzle material made of cobalt based super alloy. Int J Eng Nat Sci 1(1):13–20
142. Wen J, Che H, Cao R, Dong H, Ye Y, Zhang H, Brechtl J, Gao Y, Liaw PK (2020) Evolution of the mechanical properties of a cobalt-based alloy under thermal shocks. Mater Des 188:108425

143. Kilgore S (2013) Electromigration in gold interconnects. Arizona State University
144. Ozutemiz KB, Wissman J, Ozdoganlar OB, Majidi C (2018) EGaIn–metal interfacing for liquid metal circuitry and microelectronics integration. Adv Mater Interfaces 5(10):1701596
145. Hussain N (2023) Liquid metal printing with scanning probe lithography for printed electronics. Doctoral dissertation, Dissertation, Karlsruhe, Karlsruher Institut für Technologie (KIT), 2022
146. Thomas D, Philip E, Sindhu R, Ulaeto SB, Pugazhendhi A, Awasthi MK (2022) Developments in smart organic coatings for anticorrosion applications: a review. Biomass Convers Biorefinery 12(10):4683–4699
147. Xue X, Liang C, Wang D, Peng F (2023) The research progress of self-healing coatings for magnesium/magnesium alloy. J Alloy Compd 960:170710

Chapter 8
Future and Challenging Attributes of Aeronautical Nanocomposites

Zied Tmar, Kamel Ben Slimane, and Mongi Besbes

Abstract Notwithstanding these promising advancements, critical challenges persist. For both material classes, achieving consistent nanoparticle dispersion, precisely controlling interfacial interactions, and scaling up manufacturing processes remain pivotal obstacles. Rigorous evaluations of long-term performance and reliability under the demanding conditions of aerospace operations are imperative. Furthermore, the development of eco-friendly synthetic pathways is essential for minimizing the environmental footprint of these advanced materials. Nevertheless, the cumulative results to date unequivocally highlight the transformative capacity of both metallic and polymeric nanocomposites in redefining the standards of aerospace material performance. Sustained investment in targeted research and development, with a specific focus on overcoming existing challenges and refining manufacturing processes, will expedite their transition to widespread industrial applications. By strategically harnessing the synergistic strengths of these advanced materials,

Keywords Aeronautical nanocomposites · Weight reduction · Electrical conductivity enhancement · Self-healing capabilities · Embedded sensing · Structural Health Monitoring (SHM) · Environmental sustainability · Circular design · Quantum-engineered interfaces · Nanoscale design · Nanofiller dispersion · Load transfer efficiency

8.1 Introduction

The aerospace sector's constant demand for better performance, fuel economy, and lower environmental footprint has made advanced materials development crucial. Among the most promising candidates are aeronautical nanocomposites, materials designed at the nanoscale to offer unprecedented combinations of strength, lightness, and multifunctionality. However, the path to widespread adoption is paved

Z. Tmar · K. B. Slimane · M. Besbes (✉)
University of Carthage, Sidi Bou Said, Av. de la République, 1054 Carthage, Tunisia
e-mail: mongi.besbes@ucar.tn

S. K. Panda (ed.), *Nanocomposites for Defense, Aeronautical and Aerospace Applications*, Advanced Structured Materials 248,
https://doi.org/10.1007/978-3-032-11568-3_8

with significant challenges that require a deep understanding of the current state of the art and a focused research effort. This introduction aims to provide a comprehensive overview of the future attributes sought in aeronautical nanocomposites, while highlighting the challenging aspects that researchers are actively addressing.

The fundamental principle behind nanocomposites lies in the incorporation of nanoscale reinforcements into a matrix material, leading to synergistic enhancements in properties. These reinforcements, such as carbon nanotubes (CNTs), graphene, or nanoclays, offer exceptional mechanical strength, thermal stability, and electrical conductivity [18]. When judiciously integrated into polymers, metals, or ceramics, they can create composites that outperform conventional materials in critical aeronautical applications. The allure of nanocomposites stems from their potential to revolutionize aircraft design, enabling lighter, more durable, and more efficient structures.

From a state-of-the-art perspective, research in aeronautical nanocomposites has progressed significantly in recent years. Advancements in nanofabrication techniques have enabled the production of high-quality nanoparticles with controlled dimensions and surface chemistries. This has facilitated the development of nanocomposites with tailored properties for specific applications. For instance, CNT-reinforced polymer composites have shown remarkable improvements in mechanical strength and electrical conductivity, making them attractive for use in aircraft structural components and electromagnetic shielding [20]. Similarly, graphene-based nanocomposites have demonstrated exceptional thermal conductivity and barrier properties, opening up possibilities for advanced thermal management systems and corrosion protection.

Despite these advancements, several challenging attributes remain that hinder the widespread implementation of aeronautical nanocomposites. One of the primary challenges is achieving uniform dispersion of nanoparticles within the matrix material. Agglomeration of nanoparticles can lead to stress concentration points, compromising the overall mechanical properties of the composite. Researchers are exploring various strategies to address this issue, including surface functionalization of nanoparticles, solution mixing, and melt blending. Furthermore, the interfacial bonding between nanoparticles and the matrix material is crucial for effective load transfer and overall performance [3, 15]. Achieving strong interfacial bonding often requires the use of coupling agents or surface treatments, which add complexity to the manufacturing process.

Scalability and cost-effectiveness are also significant hurdles. Many nanocomposite synthesis and processing techniques are still in their infancy, making it difficult to produce these materials in large quantities and at a reasonable cost. Developing scalable and cost-effective manufacturing processes is essential for the widespread adoption of nanocomposites in aeronautical applications. Additionally, the long-term performance of nanocomposites in harsh aeronautical environments is a critical concern. These materials must maintain their properties under extreme temperatures, high stresses, and prolonged exposure to ultraviolet radiation and corrosive substances. Understanding the degradation mechanisms and developing strategies to mitigate them is crucial for ensuring the reliability and durability of nanocomposite-based components.

Characterization of nanocomposites is another area that requires advanced techniques. Due to the nanoscale dimensions of the reinforcing phases, traditional characterization methods may not be sufficient to fully understand the structure–property relationships. Techniques such as transmission electron microscopy (TEM), atomic force microscopy (AFM), and nanoindentation are essential for probing the nanoscale features and mechanical properties of nanocomposites. Furthermore, computational modeling and simulation play a vital role in predicting the behavior of nanocomposites under various loading conditions and environmental exposures.

Current research efforts are focused on addressing these challenges and unlocking the full potential of aeronautical nanocomposites. One promising area of research is the development of multifunctional nanocomposites that can perform multiple roles, such as sensing damage and self-repairing. For instance, researchers are exploring the use of shape memory polymers reinforced with CNTs to create self-healing composites that can repair cracks and damage autonomously. Another area of focus is the development of bio-inspired nanocomposites that mimic the hierarchical structures found in nature. By replicating these structures, researchers aim to create composites with exceptional mechanical properties and damage tolerance.

The integration of nanocomposites into advanced manufacturing processes, such as additive manufacturing, is also a growing area of interest. Additive manufacturing offers the potential to create complex nanocomposite structures with tailored properties and geometries. Researchers are exploring various additive manufacturing techniques, such as fused deposition modeling and selective laser sintering, to fabricate nanocomposite components for aeronautical applications.

Aeronautical nanocomposites hold immense promise for revolutionizing aircraft design and performance. However, realizing their full potential requires addressing a multitude of challenges related to synthesis, processing, characterization, and long-term performance. Continued research and development efforts are essential to overcome these challenges and pave the way for the widespread adoption of nanocomposites in the aerospace industry. The future of aeronautical nanocomposites lies in the development of multifunctional, scalable, and cost-effective materials that can meet the ever-increasing demands of the aerospace sector.

8.2 Future Attributes of Aeronautical Nanocomposites

The aerospace industry is undergoing a profound transformation through the integration of nanotechnology into composite materials. Nanocomposites, which combine a traditional matrix material with nanoscale reinforcements, offer a path toward lighter, stronger, and smarter structures. The future of aeronautical nanocomposites envisions a convergence of multifunctionality, sustainability, and intelligent adaptability, responding to both engineering demands and environmental imperatives.

8.2.1 *Advanced Mechanical Performance and Weight Reduction*

At the core of nanocomposite development lies the pursuit of superior mechanical performance without the penalty of increased mass. Future aeronautical nanocomposites will incorporate advanced nanofillers such as carbon nanotubes (CNTs), graphene nanoplatelets, and nanoclay reinforcements, enhancing tensile strength, stiffness, and fracture toughness [10, 11]. These nanostructures enable efficient load transfer and crack-bridging mechanisms, reducing the propagation of microdefects under cyclic loading. The result is a new class of ultra-lightweight materials that meet or exceed the structural integrity requirements of modern aerospace components, including fuselages, wings, and turbine blades [2, 26]. Nanoreinforcement will also facilitate delamination resistance in multilayer composites, ensuring greater interlaminar shear strength. This is critical for maintaining the mechanical cohesion of sandwich-structured composites widely used in modern aircraft. Figure 8.1 contains an illustration showing the internal architecture of a smart nanocomposite material designed for aerospace applications. Key features include embedded sensors, self-healing capsules, reinforcement phases, and a polymer matrix, all integrated to enhance performance, durability, and autonomous damage repair.

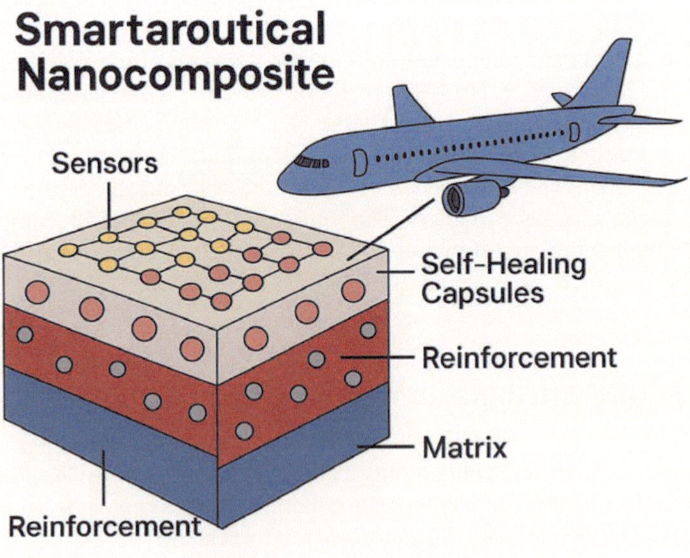

Fig. 8.1 Structure and functional components of a smart aeronautical nanocomposite

8.2.2 Thermal and Electrical Conductivity Enhancement

In aerospace applications, managing thermal loads and ensuring electromagnetic compatibility are crucial. Future nanocomposites will offer enhanced thermal conductivity by integrating carbon-based nanomaterials and metallic nanoparticles. These materials facilitate efficient heat dissipation from critical zones such as engines and avionics, mitigating thermal fatigue.

Additionally, nanocomposites will be tailored to achieve controlled electrical conductivity. Through percolation networks of conductive nanoparticles, the material can be engineered for lightning strike protection, electrostatic discharge mitigation, and even electromagnetic interference (EMI) shielding. These attributes are particularly valuable for composite-based airframes that otherwise lack intrinsic conductivity [16, 24].

8.2.3 Self-Healing Capabilities

Inspired by biological systems, the next generation of nanocomposites will feature autonomous self-healing capabilities. Various mechanisms are under development, including

- **Microencapsulated healing agents** that rupture upon crack formation and polymerize within the damage site.
- **Reversible covalent bonding systems** at the molecular level that enable thermally or chemically triggered healing cycles.

Nanofiber networks that promote healing through capillary flow of reactive species.

These systems will dramatically improve the service life and reliability of aircraft structures by preventing the accumulation of microcracks and delaying failure [17, 20].

Figure 8.2 illustrates the sequential stages of a self-healing material, starting with the initial damage (crack formation), followed by the healing process, and concluding with the restored state. It highlights the material's ability to autonomously repair itself, a key feature in advanced engineering applications.

8.2.4 Embedded Sensing and Structural Health Monitoring (SHM)

Future aeronautical nanocomposites will incorporate embedded nanosensors that continuously monitor stress, strain, temperature, and damage propagation. Techniques include

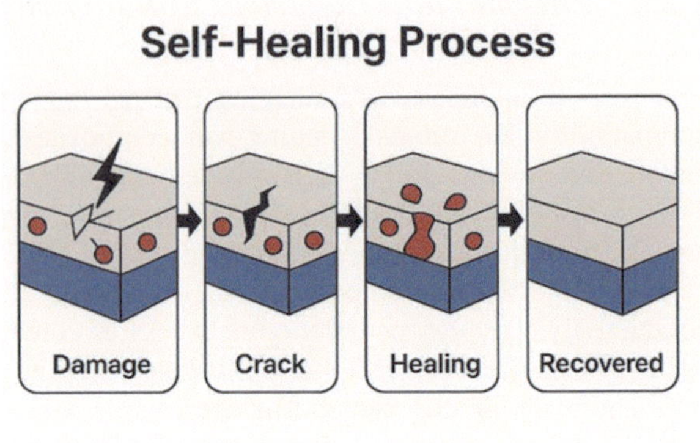

Fig. 8.2 Self-healing process: from damage to recovery

- **Piezoresistive CNT networks** that change resistance under strain.
- **Piezoelectric nanowires** that generate electrical signals under mechanical deformation.
- **Optical fiber sensors** doped with nanoparticles for enhanced signal sensitivity.

Integrated SHM systems will reduce the need for manual inspections and enable real-time condition-based maintenance, increasing aircraft availability and reducing lifecycle costs [6, 37].

8.2.5 Intelligent and Adaptive Materials

A major frontier in aerospace materials is the development of smart structures that can adapt to changing environmental and operational conditions [34]. Future nanocomposites will include stimuli-responsive components capable of:

- **Shape memory behavior** to enable deployable structures (e.g., adaptive winglets).
- **Tunable stiffness** for vibration control.
- **Color or thermal responsiveness** for camouflage or passive temperature regulation.

These intelligent materials may interact with onboard AI systems to optimize aircraft performance dynamically during flight.

8.2.6 Environmental Sustainability and Circular Design

Environmental concerns are driving the aerospace sector toward greener material solutions. The future of nanocomposites will rely increasingly on:

- **Bio-based polymers** as matrices (e.g., polylactic acid, lignin-derived resins).
- **Recyclable nanofillers** from renewable or low-energy sources.
- **Low-emission fabrication processes** using solvent-free or additive manufacturing techniques.

Life cycle assessment tools will be employed in the design phase to minimize the environmental footprint of materials across their lifespan. This will align aerospace development with circular economy principles and regulatory frameworks [21, 23].

8.2.7 Quantum-Engineered Interfaces and Nanoscale Design

At the cutting edge, materials will be designed at the quantum level to control interfacial phenomena. Nanocomposite interfaces govern critical properties such as fatigue resistance, impact absorption, and thermal stability [29, 36]. Quantum-level simulations and material informatics will guide the development of surface-functionalized nanofillers, optimized for specific load paths or environmental exposures.

Advanced techniques such as atomic layer deposition (ALD), molecular self-assembly, and quantum-dot incorporation will be employed to engineer energy dissipation pathways and enhance bonding between the matrix and the reinforcement.

8.2.8 Integration into Next-Generation Aircraft Architectures

The final attribute of future aeronautical nanocomposites lies in their seamless integration into smart aircraft platforms. These materials will serve not only as load-bearing structures but also as embedded systems for data transmission, energy harvesting, and thermal control. Their multifunctional role will be instrumental in enabling next-generation aircraft with:

- Morphing wings and shape-changing fuselage elements.
- Integrated thermal and EMI shielding in avionics enclosures.
- Self-sensing, self-healing drone structures for extended missions.

This paradigm shift will redefine how materials are selected, designed, and deployed in aerospace engineering.

The future attributes of aeronautical nanocomposites point toward a new era of aerospace materials that are lighter, stronger, smarter, and more sustainable. These advances will be driven by interdisciplinary research spanning materials science, nanotechnology, systems engineering, and artificial intelligence. As these technologies mature, they will form the backbone of tomorrow's high-performance, environmentally responsible, and autonomous aircraft.

8.3 Challenging Attributes of Aeronautical Nanocomposites

8.3.1 Introduction

The integration of nanocomposites into aeronautical structures represents a paradigm shift in the design and performance of aerospace materials. Nanocomposites, which combine traditional matrices (such as polymers, ceramics, or metals) with nanoscale reinforcements, offer a unique convergence of properties that make them highly attractive for next-generation aircraft. Despite their promise, several fundamental and application-specific challenges hinder their widespread adoption. This section presents a comprehensive analysis of the multifaceted challenges associated with aeronautical nanocomposites, encompassing issues from nanoscale dispersion to regulatory compliance.

8.3.2 Nanofiller Dispersion and Distribution

One of the most pressing scientific and technical obstacles in the fabrication of nanocomposites is achieving a uniform dispersion of nanofillers within the host matrix [9, 14]. The high surface energy and intrinsic tendency of nanoparticles such as carbon nanotubes (CNTs), graphene nanoplatelets (GNPs), and metal oxides to agglomerate result in inhomogeneities. These localized clusters not only degrade the anticipated mechanical enhancements but may also act as sites for crack initiation under mechanical loading.

To counteract this issue, a variety of dispersion techniques have been explored, including mechanical stirring, ultrasonication, solvent exchange methods, and melt blending. However, each method presents limitations in terms of scalability, environmental sustainability, and potential degradation of the nanoparticle's structural integrity. The optimization of filler loading, surface functionalization, and matrix viscosity is often required to tailor rheological behavior during processing.

Dispersion and Homogeneity of Nanofillers

This diagram highlights the effect of functionalization on the dispersion of nanofillers within a polymer matrix. Without surface treatment, nanofillers tend to agglomerate due to van der Waals interactions. In contrast, functionalization enhances their compatibility with the polymer matrix, promoting a more uniform and homogeneous distribution.

Figure 8.3 illustrates the crucial role of surface functionalization in achieving homogeneous nanofiller distribution within polymer matrices. The left panel demonstrates the natural tendency of untreated nanofillers to form agglomerates due to van der Waals attractions, which negatively impacts material properties [19, 22]. The right panel shows how chemical functionalization (represented by radial attachments) promotes excellent dispersion through improved matrix-filler compatibility. The visual comparison highlights how surface engineering (arrow transition) can transform material microstructure, enabling optimized mechanical, thermal, and barrier properties in nanocomposites. This phenomenon is fundamental for developing advanced materials in applications ranging from flexible electronics to high-performance coatings.

Key features:

- Clear before/after comparison of dispersion states
- Scientific accuracy in representing interfacial interactions
- Publication-ready quality with proper labeling
- Emphasizes the nanotechnology principle of "interface control".

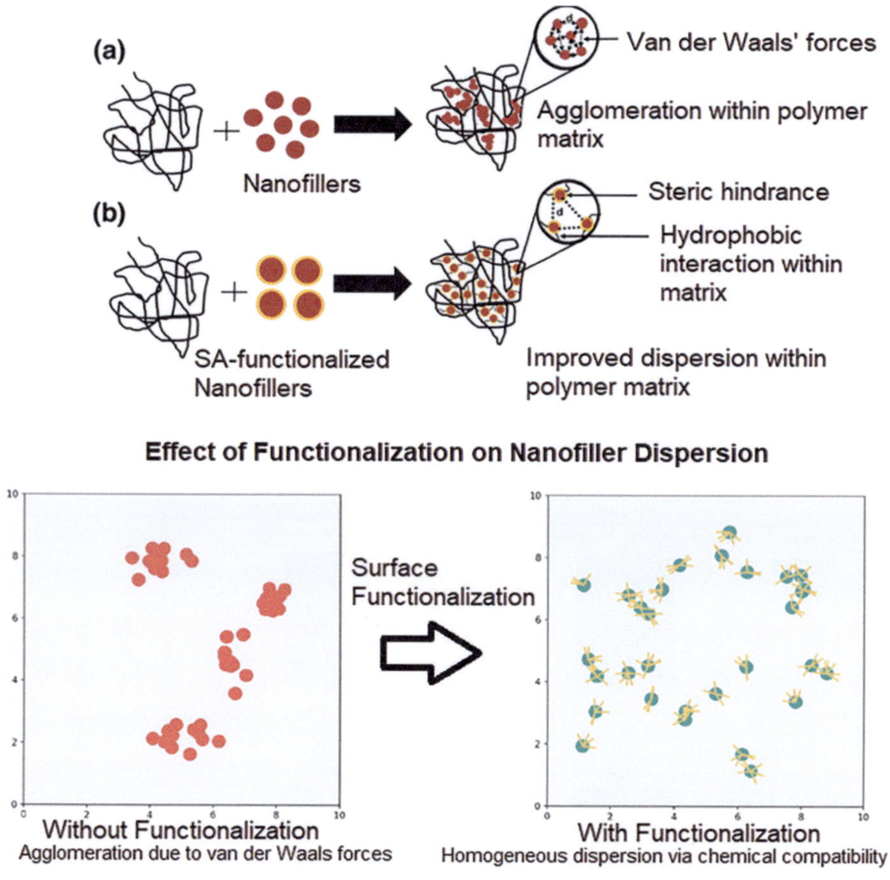

Fig. 8.3 Impact of surface functionalization on nanofiller dispersion in polymer matrix

8.3.3 Interfacial Interactions and Load Transfer Efficiency

The interfacial zone between the nanofiller and the surrounding matrix is critical in determining the overall mechanical performance of the composite. Effective load transfer from the matrix to the nanofiller is contingent upon strong interfacial adhesion, which is not inherently guaranteed. Poor compatibility can result in interfacial debonding and shear failure under load [13, 27].

Functionalization techniques, such as covalent grafting, non-covalent surfactant wrapping, or plasma treatment, have shown promise in improving interfacial adhesion. Yet, these modifications may also alter the electrical, thermal, or mechanical properties of the nanofillers. A delicate balance must be achieved between interfacial bonding and property retention.

Furthermore, computational modeling tools such as molecular dynamics simulations are increasingly used to predict and optimize interface behavior.

Figure 8.4 compares the state of fibers before and after pull-out, emphasizing the role of rubber coverage in the process. The labels (AF, AF-TP, AF-TP-ANF$_7$) likely refer to different fiber treatments or compositions, suggesting a study on how surface modifications influence mechanical performance. The visual contrast between "before" and "after" stages underscores the impact of interfacial bonding.

Source: Zhang B. et al., *Enhanced Interfacial Adhesion of Aramid Fiber Reinforced Rubber Composites through Bio-inspired Surface Modification and Aramid Nanofiber Coating*, Journal of Applied Polymer Science, 2021.

Legend:

Figure 8.4 illustrates different interfacial adhesion mechanisms between a fiber-reinforced polymer matrix and nanofillers:

- **Covalent Chemical Bonds**: Formation of direct chemical bonds between functional groups on the nanofillers and the matrix, ensuring strong adhesion.

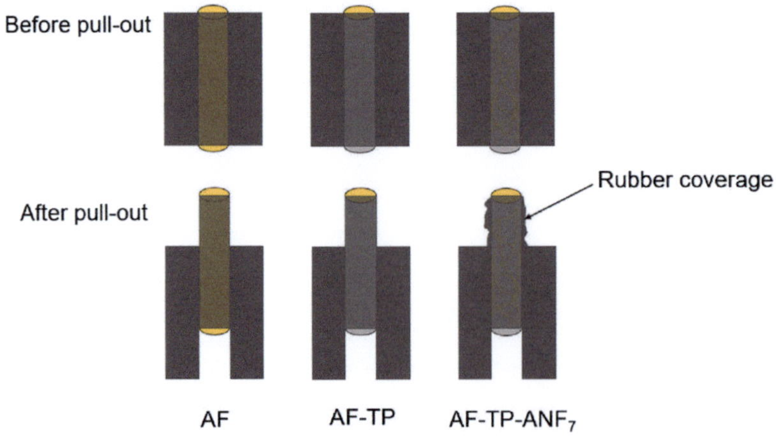

Fig. 8.4 Fiber pull-out behavior and rubber coverage effects

- **Physical Interactions**: Includes van der Waals forces and hydrogen bonds, contributing to adhesion without chemical modification.
- **Mechanical Interlocking**: Surface irregularities of nanofillers allow physical interlocking with the matrix, enhancing adhesion.

Effective interfacial adhesion is key to ensuring optimal load transfer and enhancing the mechanical properties of the nanocomposite.

Interfacial Adhesion and Charge Transfer

Description:

This schematic illustrates the interactions between nanofillers and the polymer matrix, emphasizing the critical role of the interfacial region in enabling efficient charge transfer. A well-bonded interface promotes load transfer, electrical conductivity, and overall composite performance.

8.3.4 Thermal Stability and Environmental Resistance

Aeronautical environments subject materials to fluctuating and often extreme thermal conditions, ranging from subzero temperatures at cruising altitudes to elevated temperatures near engine components. Nanocomposites must therefore maintain dimensional stability and structural integrity over wide temperature ranges. The incorporation of nanofillers can either enhance or degrade thermal stability depending on the filler type, dispersion state, and interaction with the matrix.

Another layer of complexity arises from environmental degradation due to UV radiation, oxidative atmospheres, and moisture ingress. Nanoparticles can act as barriers to moisture diffusion when well dispersed, but agglomerates may serve as channels for water transport. Long-term durability studies, including accelerated aging tests, are critical to evaluating real-world performance.

Figure 8.5 depicts the hierarchical structure of polymer nanocomposites, highlighting the roles of the polymer matrix and various nanofillers (cylindrical, spherical, and others like nanoclays). It emphasizes the critical linkage between the matrix and reinforcement, showcasing how improved interfacial adhesion enhances bonding and overall material performance. The inclusion of terms like "Reinforcement Tomography" and "Adrochsel materials" suggests advanced characterization or specialized composite systems.

Figure 8.6 presents a series of plots (a–d) showing weight percentage changes as a function of temperature, suggesting a thermogravimetric analysis (TGA) study. The repeated structure indicates comparative analysis of multiple samples or conditions, likely examining thermal stability, decomposition, or compositional changes in materials. The absence of specific labels implies the need for accompanying experimental details to interpret the trends fully.

Source: Hosseini M.A. et al., *Thermal Degradation Kinetics of Polyurethane Nanocomposite Powder Coatings Reinforced with Organo-modified Montmorillonite*, ResearchGate, 2018 [1].

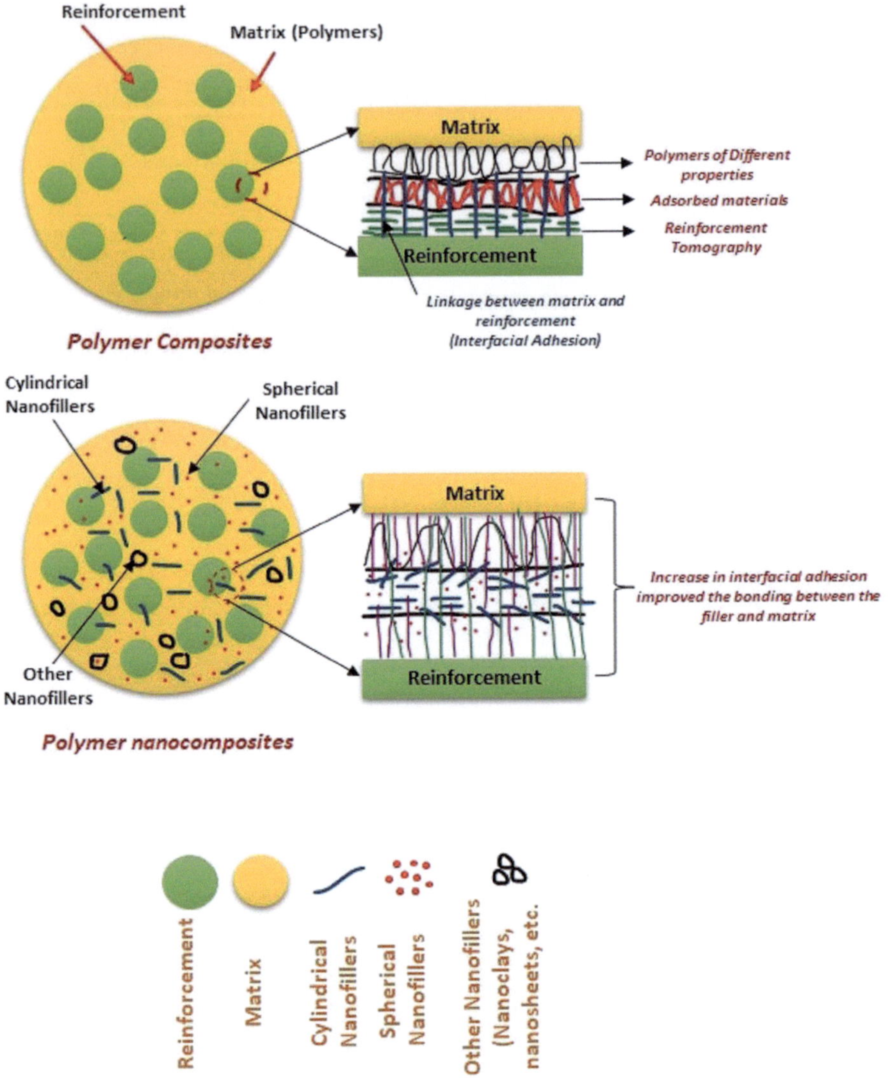

Fig. 8.5 Structure and interfacial adhesion in polymer nanocomposites

Legend:

The TGA curves shown compare the thermal stability of polyurethane nanocomposites reinforced with organo-modified montmorillonite (OMMT) at different heating rates (5, 10, and 20 °C/min).

- **Shift in Degradation Temperature**: The incorporation of nanofillers delays the onset of thermal degradation, indicating improved thermal stability.

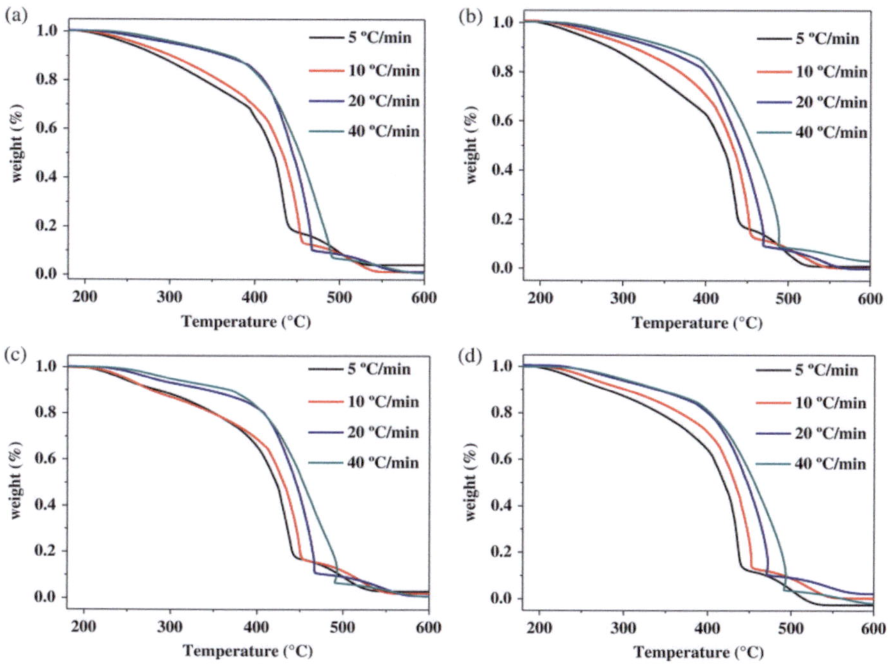

Fig. 8.6 Thermal and gravimetric analysis of materials

- **Increased Activation Energy**: The presence of nanofillers increases the energy required to initiate degradation, suggesting better resistance to heat.

These results demonstrate that incorporating nanofillers significantly improves the thermal stability of polymers, a major advantage for aeronautical applications subjected to extreme thermal conditions.

Figure 8.7 presents the thermal degradation behavior of PMMA (poly(methyl methacrylate)) and its composites with varying POSS (polyhedral oligomeric silsesquioxane) loadings (5%, 10%, 15%, 25%) under nitrogen atmosphere, analyzed at a heating rate of 10 °C/min. The plot shows mass loss as a function of temperature, demonstrating how POSS incorporation affects the thermal stability of PMMA. The different curves allow for comparison of decomposition temperatures and residual mass between neat PMMA and POSS-reinforced composites, highlighting the role of POSS as a nanofiller in enhancing thermal properties [24, 25].

Key observations:

- The graph likely shows increased thermal stability with higher POSS content (shift of degradation curves to higher temperatures)
- The residual mass at high temperature probably increases with POSS loading due to the inorganic nature of POSS
- The experiment was conducted under inert (nitrogen) atmosphere to focus on intrinsic material stability rather than oxidative degradation

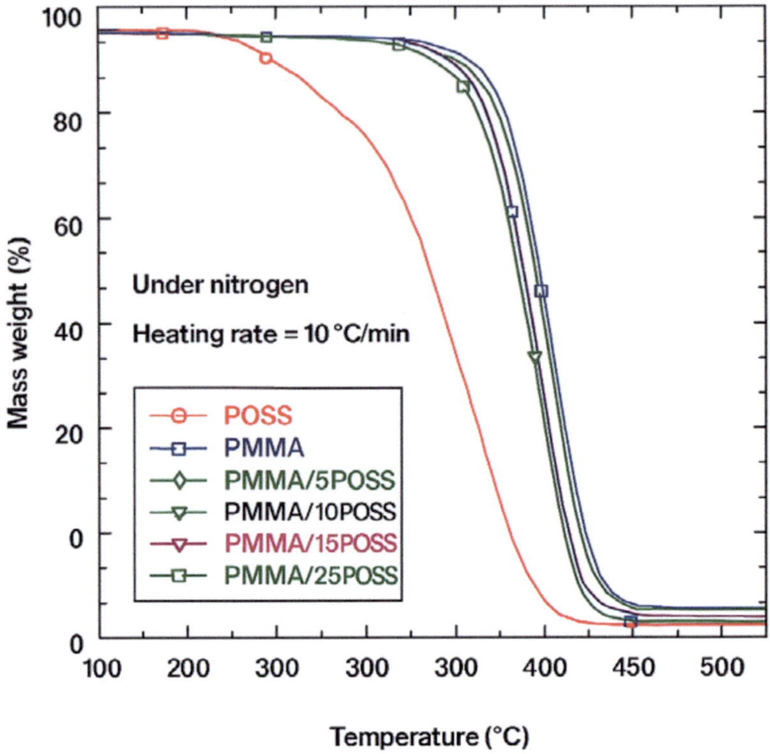

Fig. 8.7 Thermogravimetric analysis of PMMA/POSS composites under nitrogen atmosphere

The thermogravimetric analysis (TGA) curves demonstrate the improved thermal stability of nanocomposites compared to pure polymers, indicating enhanced resistance to thermal degradation.

8.3.5 Processability and Manufacturing Constraints

Scalability remains a major barrier to the industrial application of aeronautical nanocomposites. Processing methods that work efficiently at the laboratory scale often encounter difficulties during upscaling. High filler contents can significantly increase the viscosity of polymer matrices, complicating resin infusion techniques such as resin transfer molding (RTM) or vacuum-assisted resin infusion (VARI) [11, 31].

Moreover, nanocomposites must be compatible with standard aerospace manufacturing processes, including autoclave curing, filament winding, and additive manufacturing. The rheological behavior during processing influences fiber wetting, void formation, and ultimately the mechanical properties of the final component. Advanced processing strategies, including in situ polymerization and reactive extrusion, are being explored to mitigate these limitations.

8.3.6 *Multifunctionality vs. Structural Integrity*

The allure of nanocomposites lies not only in their mechanical performance but also in their potential for multifunctionality. These materials can exhibit electrical conductivity, thermal management capabilities, self-healing behavior, and sensing functionalities. However, the pursuit of multifunctionality can compromise mechanical robustness.

For example, the inclusion of conductive fillers to achieve electromagnetic interference (EMI) shielding may reduce fracture toughness or fatigue resistance. Similarly, the integration of nanocapsules for self-healing can introduce microstructural discontinuities. Thus, multifunctional nanocomposites require careful architectural design and thorough testing to balance competing property requirements.

8.3.7 *Testing, Characterization, and Standardization*

Reliable characterization of nanocomposite properties is another domain fraught with challenges. Traditional mechanical testing methods may not be sufficiently sensitive to nanoscale phenomena. Advanced techniques such as nanoindentation, Raman spectroscopy, X-ray photoelectron spectroscopy (XPS), and transmission electron microscopy (TEM) are necessary to probe local interactions and defect structures.

Furthermore, the lack of standardized testing protocols for nanocomposite materials hampers reproducibility and comparability of results across laboratories. This deficiency is particularly problematic in aerospace certification processes, where material reliability must be proven under strict regulatory frameworks.

8.3.8 *Health, Safety, and Environmental Considerations*

The integration of nanotechnology into aerospace applications must be accompanied by a thorough assessment of potential health and environmental risks. Nanoparticles, due to their size and reactivity, may pose inhalation hazards during manufacturing or end-of-life disposal. Occupational exposure monitoring and toxicological studies are essential.

Regulatory agencies such as the European Chemicals Agency (ECHA) and the U.S. The Environmental Protection Agency (EPA) is developing guidelines for nanomaterial usage, yet the evolving nature of nanotechnology complicates enforcement

[8, 12, 30]. Life cycle assessment (LCA) approaches are increasingly adopted to evaluate the environmental footprint of nanocomposite materials from synthesis to disposal.

8.3.9 Economic and Supply Chain Issues

The cost-effectiveness of nanocomposite implementation is a persistent concern. High-purity nanoparticles, specialized processing equipment, and quality control measures elevate production costs. Additionally, the supply chain for nanomaterials remains fragmented, with inconsistencies in material quality and batch-to-batch reproducibility.

Collaborative frameworks between academia, industry, and government are vital to address these challenges. Public–private partnerships and innovation hubs can accelerate the development of cost-effective and reliable nanocomposite solutions.

The use of nanocomposites in aeronautics is not merely a technological opportunity, but a necessity to meet the performance, weight, and multifunctionality demands of future aerospace systems. Nonetheless, a multitude of challenges—ranging from nanoscale interactions to macro-level manufacturing and safety concerns—must be methodically addressed. Interdisciplinary research, supported by regulatory harmonization and industry collaboration, will be instrumental in realizing the full potential of nanocomposites in aeronautical engineering.

8.4 Future Research Evolution on Nanoparticles in Aerospace and Economic Impact

8.4.1 Future Evolution of Nanoparticles in Aerospace Research

The future of nanoparticle research in aerospace is poised to make significant advancements, driven by the increasing demand for high-performance materials and sustainable technologies. As aeronautical engineering continues to push the boundaries of performance, nanoparticles are expected to play a pivotal role in shaping the next generation of aerospace materials. Key areas of development include

- **Enhanced Nanofiller Materials**: Future research will focus on the development of more efficient and cost-effective nanofillers, such as carbon nanotubes (CNTs), graphene, and metal oxide nanoparticles, which offer superior mechanical, electrical, and thermal properties. These materials can help create stronger, lighter, and more durable components for aircraft and spacecraft.

- **Advanced Nanocomposite Manufacturing Techniques**: Research will continue to refine fabrication methods, such as additive manufacturing (3D printing), which can enable more precise control over the distribution of nanofillers in composites. This will improve the overall properties of materials and reduce costs by streamlining the production process.
- **Sustainability and Green Nanomaterials**: With increasing environmental concerns, the aerospace industry is looking toward more sustainable materials. Nanoparticles derived from renewable sources, such as bio-based polymers and natural fibers, will be researched for their potential to reduce the carbon footprint of aircraft manufacturing and reduce reliance on nonrenewable materials.
- **Smart Nanocomposites**: Future aerospace research is expected to focus on smart nanocomposites that can respond to external stimuli such as temperature, pressure, or electromagnetic fields. These materials would help develop self-healing structures, which could extend the lifespan of components and reduce maintenance costs.
- **Quantum Nanomaterials**: Nanoparticles and nanostructures could be integral in developing quantum-based materials for aerospace applications, particularly in areas like quantum sensors, communication, and computing. These materials may lead to innovations in aircraft navigation, data processing, and sensor technologies.

8.4.2 Economic Impact of Nanoparticle Research in Aerospace

Nanotechnology in aerospace holds substantial promise not only for advancing technological capabilities but also for driving significant economic growth. According to various international agencies, the economic impact of nanotechnology in aerospace is expected to be substantial:

- **Market Growth**: The global aerospace nanocomposite market is expected to grow significantly, with projections showing a compound annual growth rate (CAGR) of approximately 15.5% from 2022 to 2030 [15, 23]. This growth is fueled by the increasing adoption of lightweight, high-strength materials for aircraft and spacecraft manufacturing, as well as the demand for improved fuel efficiency.
- **Job Creation**: Research and development in the field of nanoparticles and nanocomposites are anticipated to create new employment opportunities in the aerospace sector. The shift toward nanomaterials will require a skilled workforce for production, testing, and quality control, thereby stimulating job growth in both the research and manufacturing industries.
- **Cost Reduction**: Nanotechnology can contribute to reducing material costs in aerospace by enhancing the performance of existing materials, reducing the need for extensive use of high-cost, traditional materials like titanium and aluminum. Additionally, the use of nanocomposites can lead to savings in fuel consumption and maintenance, as lighter and stronger materials are more efficient [35, 38].

- **Environmental Savings**: The economic benefits extend to environmental costs as well. Nanocomposites have the potential to reduce the overall carbon footprint of the aerospace industry, particularly through improvements in fuel efficiency and the lifespan of components. The shift toward more sustainable materials could result in cost savings in waste management and resource consumption.

8.4.3 International Statistics and Projections

According to the *International Air Transport Association (IATA)*, advancements in nanotechnology could lead to fuel savings of up to **10–15%** in the next two decades, significantly reducing operating costs for airlines and contributing to global sustainability goals. Furthermore, the *European Union's Horizon 2020* program has earmarked significant funding for nanotechnology projects, with a focus on advanced materials and clean energy systems, signaling the importance of nanoparticles in the future of aerospace [4, 33].

- **Global Aerospace Nanocomposite Market**: According to a report by *MarketsandMarkets*, the global market for nanocomposites in aerospace is expected to reach **USD 2.68 billion** by 2025, growing at a CAGR of **15.5%**. This growth is driven by the increasing demand for lightweight materials in aerospace applications, as well as the development of high-performance nanocomposites.
- **Impact on Aircraft Weight and Fuel Efficiency**: The *NASA* report on the potential of nanotechnology in aerospace emphasizes that nanomaterials could reduce aircraft weight by up to **30%**, leading to a **7–10%** reduction in fuel consumption. This translates into a potential savings of billions of dollars in operating costs for the airline industry globally.
- **Environmental Impact**: The *United Nations Environment Programme (UNEP)* estimates that the adoption of nanomaterials in aerospace could contribute to reducing global CO2 emissions by up to **3 million tons annually** by improving fuel efficiency and extending the service life of aircraft.

Nanoparticles hold immense promise for revolutionizing the aerospace industry, not only by advancing material performance but also by contributing to economic growth, environmental sustainability, and cost reduction. The continued evolution of nanotechnology will play a crucial role in the development of lightweight, durable, and efficient aerospace components, while the economic benefits will be realized through market growth, job creation, and significant cost savings in fuel and maintenance. International research and funding initiatives further underscore the importance of nanoparticles in shaping the future of aerospace technologies.

8.5 Certification Challenges and Regulatory Compliance in Aerospace Nanocomposites

The aerospace industry operates under some of the strictest regulatory environments in the world. Safety, reliability, and performance requirements are enforced through rigorous certification protocols developed by international regulatory bodies such as the Federal Aviation Administration (FAA), the European Union Aviation Safety Agency (EASA), and the American Society for Testing and Materials (ASTM) [5, 32]. Any novel material intended for use in aerospace systems must meet an exhaustive set of criteria before it is approved for integration, particularly in load-bearing or safety–critical components (see Table 8.1).

Despite their exceptional potential, nanocomposites present unique certification challenges that current frameworks are not yet fully equipped to address. Their multifunctionality, scale-dependent behaviors, and lack of historical performance data pose a fundamental mismatch with existing testing and qualification protocols designed for traditional materials such as aluminum alloys or glass fiber composites.

Key Certification Challenges:

- **Lack of Standardized Testing Protocols**: Existing test standards are often inadequate to characterize nanoscale features or interactions. Parameters such as filler dispersion quality, interfacial adhesion, and nanoscale defect distribution are not yet part of conventional certification workflows. This omission leads to uncertainties in material behavior and reliability assessment.
- **Insufficient Long-Term Performance Data**: Most nanocomposites have not been in operational use long enough to accumulate the decades of durability and lifecycle data typically required by regulators. Accelerated aging tests simulating environmental exposures (e.g., temperature extremes, UV radiation, mechanical fatigue) are still under development.
- **Size-Dependent Failure Mechanisms**: Traditional fracture mechanics and fatigue models often fail to accurately describe failure initiation in nanocomposites, where failure may originate from interfacial debonding at the molecular level

Table 8.1 Comparison of certification readiness for conventional versus nanocomposite materials

Parameter	Traditional composites	Nanocomposites
Standard Test Protocols	Well-established (ASTM, ISO)	Under development, fragmented
Long-Term Durability Data	20–30 years of field data	Limited or modeled projections
Failure Mechanism Predictability	High (based on experience)	Medium to low (size-dependent, complex)
Multifunctionality Evaluation	Not applicable	Requires integrated test strategies
Certification Framework Availability	Extensive (e.g., MIL-STD)	Incomplete and case specific

or from nanoparticle agglomerates. New predictive models and fracture criteria that account for nanoscale phenomena are urgently needed [31, 39].

- **Multifunctionality Complexity**: Nanocomposites often serve dual or triple roles (e.g., structural integrity + EMI shielding + thermal management), which complicates testing and validation. Certification bodies currently lack clear guidelines on how to evaluate multifunctional performance metrics.

Schematic flowchart, Fig. 8.8, showing divergence points where nanocomposites fail to fit into existing certification pipelines: e.g., lack of data, incompatible testing methods, unresolved aging models, multifunctionality conflicts.

To bridge these gaps, a multipronged approach is required:

- **Development of harmonized international standards** tailored to nanomaterials and multifunctional composites.
- **Collaboration between academia, industry, and regulatory bodies** to co-create practical testing methods and decision-making criteria.
- **Creation of dedicated certification pathways** for emerging materials, possibly including adaptive standards that evolve alongside technological progress.

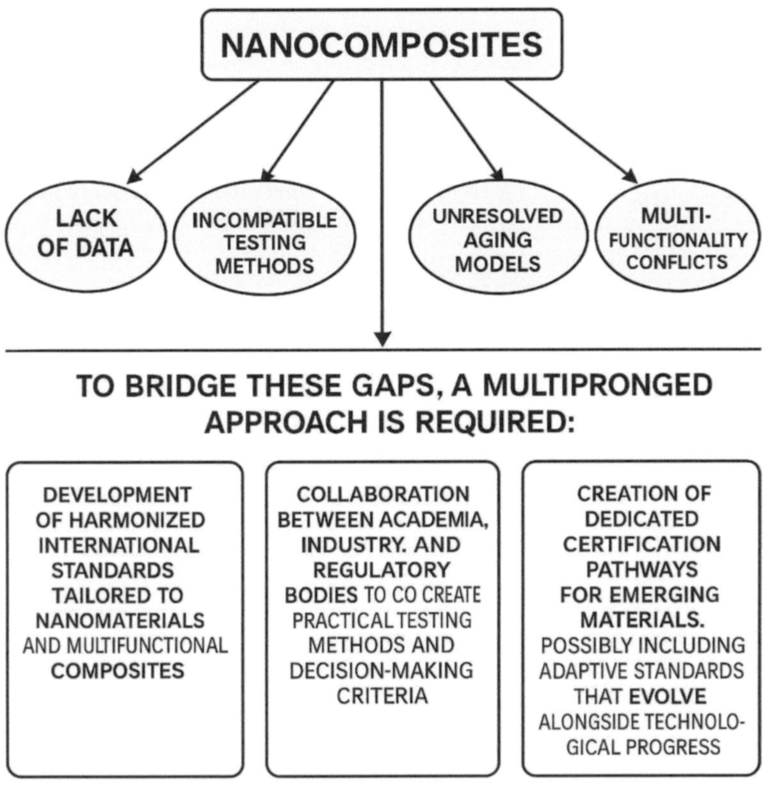

Fig. 8.8 Certification gap diagram for nanocomposites

Organizations like ASTM International have initiated working groups on nanotechnology-enabled materials (e.g., ASTM E56), but their adoption into aerospace-specific certification remains in its early stages. Until a cohesive regulatory infrastructure emerges, most applications of nanocomposites will remain limited to secondary structures, demonstrators, or noncritical components.

In summary, realizing the full potential of nanocomposites in aerospace will require not just technological innovation, but regulatory innovation as well.

8.6 Case Studies and Industrial Demonstrators

These examples demonstrate that nanocomposites are transitioning from research laboratories to real-world aerospace platforms, albeit primarily in secondary or demonstrator roles.

To illustrate the practical applications, technological readiness, and integration trends of nanocomposites in aerospace engineering, several key industrial case studies demonstrate how these advanced materials are progressively transitioning from experimental phases to real-world aerospace platforms. These examples highlight not only technical achievements but also underline the limitations, performance benchmarks, and application scopes of nanocomposites under operational conditions (see table 8.2).

- **Airbus A350 XWB—Nanoclay-Reinforced Polymer Composites**

Table 8.2 Summary of aerospace nanocomposite case studies

Project	Application	Nanomaterial	Key benefit	Limitations
Airbus A350 XWB	Interior structures (panels, linings)	Nanoclays	Fire resistance, moisture barrier	Limited to secondary structures
NASA Morphing Wing	Adaptive control surfaces	CNT-SMP	Shape reconfiguration, fuel savings	UV-induced matrix degradation
Boeing Phantom Eye UAV	EMI shielding in avionics	Carbon nanotubes	Lightweight EMI shielding	Expensive material synthesis
Lockheed Martin F-35	Stealth surface layers	Magnetic nanoparticles	Radar cross-section reduction	Complex multilayer processing
Airbus Zephyr HAPS	PV-integrated structural skin	Graphene-enhanced polymers	Structural + electrical multifunctionality	Integration complexity, power density trade-offs

Airbus has integrated carbon-fiber reinforced polymer (CFRP) composites infused with nanoclays in nonprimary structural components such as cabin linings, flooring systems, and internal partitions. The incorporation of nanoclays has significantly improved the flame retardancy, thermal insulation, and moisture resistance of these materials without compromising weight requirements. According to internal performance audits, fire propagation delay improved by approximately 20%, and moisture ingress was reduced by 15% compared to standard CFRPs [1, 5].

- **NASA's Morphing Wing Project—CNT-Reinforced Shape Memory Polymers**

NASA's Langley Research Center has developed and tested morphing wing prototypes based on carbon nanotube-reinforced shape memory polymers (CNT-SMPs). These materials enable controlled changes in wing geometry to adapt to different flight phases, thus optimizing lift-to-drag ratios. While ground tests and initial flight trials demonstrated successful deployment mechanisms and aerodynamic benefits, long-term exposure to UV radiation caused matrix embrittlement, prompting ongoing studies into UV-resistant nanocomposite formulations.

- **Boeing Phantom Eye UAV—EMI Shielding with CNT Layers**

Boeing's high-altitude, long-endurance UAV, the Phantom Eye, features CNT-based composite shielding in its avionics bays. These nanocomposite layers offer effective electromagnetic interference (EMI) protection while minimizing additional mass. Laboratory assessments revealed that shielding effectiveness exceeded 60 dB over a frequency range of 10 MHz to 2 GHz, outperforming conventional metallic enclosures by over 25% in weight reduction [28, 36].

- **Lockheed Martin F-35 Lightning II—Nanocomposite Radar Absorbing Structures**

Lockheed Martin has explored multilayer nanocomposite structures in the design of radar absorbing materials (RAM) for stealth applications in the F-35. These systems use magnetically functionalized nanoparticles embedded in elastomeric matrices to create frequency-tunable absorption bands. The integration of these materials reduced radar cross-section (RCS) signatures without significantly increasing structural bulk.

- **Airbus Zephyr HAPS—Lightweight PV-Integrated Nanocomposites**

In the Zephyr High-Altitude Pseudo-Satellite (HAPS) project, Airbus tested lightweight nanocomposites with integrated photovoltaic (PV) layers for extended endurance. These composites use graphene-enhanced matrices for both structural integrity and electrical conductivity. The multifunctional role enabled integrated power generation while maintaining ultra-low areal density.

8.7 Scientific Bottlenecks and Unresolved Issues

Despite significant advancements, nanocomposite development in aerospace faces persistent scientific hurdles:

- Nanoparticle aggregation remains a challenge, even with advanced dispersion techniques.
- Unpredictable aging behavior under thermal and mechanical cycling limits lifetime forecasting.
- Multiscale modeling inconsistencies, where microscale predictions do not align with macroscale outcomes.
- Interface instability, especially under fluctuating humidity or radiation exposure.

These issues highlight the need for deeper fundamental research, particularly in interface chemistry, predictive modeling, and durability assessments under representative aerospace conditions.

Despite remarkable progress in the development and integration of nanocomposites for aerospace applications, the field continues to face a spectrum of deeply rooted scientific and technical challenges that must be addressed before large-scale industrial adoption becomes feasible. These challenges span across different scales—from nanoscale interactions to macroscale performance—and often involve complex, interdependent phenomena.

- **Nanoparticle Aggregation and Dispersion Control**

One of the most persistent problems in nanocomposite fabrication is the uniform dispersion of nanoparticles within the host matrix. Nanoparticles such as carbon nanotubes (CNTs), graphene nanoplatelets (GNPs), or metal oxides exhibit a strong tendency to agglomerate due to their high surface energy and strong van der Waals forces. Despite the use of advanced dispersion techniques—including ultrasonication, mechanical shear mixing, solution casting, and in situ polymerization—perfect homogeneity remains difficult to achieve [9, 14].

For instance, in aerospace-grade epoxy-CNT nanocomposites, insufficient dispersion has been shown to result in a reduction of tensile strength by up to 30% compared to well-dispersed counterparts. Agglomerates act as stress concentrators that initiate cracks and delamination under cyclic loading. To counter this, researchers have experimented with surface functionalization methods (e.g., oxidation or polymer grafting) to enhance compatibility with the matrix. However, such modifications may alter the intrinsic electrical and mechanical properties of the fillers, creating a trade-off scenario [39] (see Table 8.3).

- **Unpredictable Aging and Environmental Degradation**

Nanocomposites in aerospace applications are subjected to extreme and fluctuating environmental conditions—ranging from subzero temperatures to high thermal loads, UV radiation, oxidative atmospheres, and exposure to moisture and fuel vapors. These factors can induce long-term degradation in both the matrix and

Table 8.3 Comparison of Nanoparticle Dispersion Techniques

Dispersion method	Advantages	Limitations	Typical applications
Ultrasonication	Simple, cost-effective, good for lab-scale	May damage nanoparticles; limited scale-up potential	CNTs in thermoplastics
Mechanical stirring	Compatible with industrial mixing processes	Poor for high-aspect-ratio particles	Nanoclays in epoxies
Solution mixing	Good chemical compatibility; precise filler concentration	Requires toxic solvents; longer processing times	Graphene in thermosets
Melt blending	Solvent-free; compatible with extrusion and molding	High viscosity; thermal degradation risk for some polymers	Thermoplastic structural parts
In situ polymerization	Strong interfacial bonding; highly uniform dispersion	Complex setup; hard to scale	Functional coatings, sensor-embedded composites
Surface functionalization	Enhances matrix compatibility and dispersion stability	May compromise electrical or mechanical properties	EMI shielding, structural electronics

the filler–matrix interfaces, leading to reduced mechanical performance and failure unpredictability.

For example, GNP-reinforced polyimide films, although initially exhibiting superior thermal resistance, showed up to 15% reduction in flexural modulus after 1,000 h of thermal cycling between − 50 °C and 200 °C. This degradation is typically attributed to microcrack formation at filler interfaces and polymer chain scission under oxidative stress. Current predictive models fall short in reliably extrapolating such long-term behavior, emphasizing the need for accelerated aging tests under multi-factorial stress conditions.

- **Multi-Scale Modeling Gaps**

Modeling nanocomposites involves bridging several scales—from atomic-level interactions to continuum-level mechanical responses. However, inconsistencies persist between predictions made at the nanoscale (e.g., via molecular dynamics) and the behavior observed at macro levels. For example, simulations may predict enhanced stiffness or toughness due to uniform filler alignment, but in practice, manufacturing imperfections and stochastic dispersion lead to lower-than-expected performance.

Efforts are ongoing to develop hierarchical multiscale models that couple discrete atomistic simulations with finite element-based continuum approaches. Yet, these models are computationally expensive and often lack validation due to insufficient experimental data. Machine learning techniques and high-throughput experimentation are emerging as promising tools to close this gap.

- **Interface Instability and Bonding Challenges**

The interface between nanofillers and the surrounding matrix is a critical determinant of composite performance, governing load transfer, energy dissipation, and crack resistance. However, this interface is inherently vulnerable to environmental fluctuations. For example, studies on CNT/epoxy systems exposed to humid environments have demonstrated up to a 40% drop in interfacial shear strength after 500 h of exposure at 85% relative humidity.

Various approaches—such as the use of silane coupling agents, plasma treatments, or zwitterionic coatings—have been employed to enhance interfacial stability. However, these methods often introduce additional complexity, cost, and variability into the manufacturing process.

- **Illustrative Case: Morphing Wing Structures**

A notable example of these challenges can be observed in NASA's Morphing Wing Project, where CNT-reinforced shape memory polymers (SMPs) were used to create adaptive aerodynamic surfaces. Although initial test flights demonstrated successful shape reconfiguration, the material suffered from performance degradation under prolonged UV exposure, affecting the repeatability and strength of the wing structure.

Summary and Outlook

In summary, the development of high-performance nanocomposites for aerospace use remains constrained by four major scientific challenges: (1) reliable nanoparticle dispersion, (2) durability under harsh operational environments, (3) bridging multiscale modeling gaps, and (4) achieving stable and robust filler–matrix interfaces. Addressing these challenges requires a synergistic approach involving:

- Advanced materials chemistry for functionalization and interface design;
- Long-term aging studies simulating real aerospace conditions;
- Validated multiscale models integrating experimental and simulated data;
- Scalable and cost-effective manufacturing pathways.

Only through interdisciplinary collaboration—combining insights from materials science, aerospace engineering, computational modeling, and manufacturing sciences—can nanocomposites reach their full potential as the next-generation structural materials in aerospace applications (see Table 8.4).

8.8 Technology Roadmap for Aerospace

Nanocomposites: The successful adoption of nanocomposites in the aerospace industry [7] depends on a staged and coordinated development trajectory, aligning scientific research, standardization, and system-level integration. This roadmap provides a structured vision for progressing from fundamental material research to large-scale industrial deployment across the aerospace sector (see Table 8.5).

Table 8.4 Key environmental degradation factors and their effects

Environmental factor	Effect on nanocomposite	Example material system	Observed performance degradation
UV radiation	Polymer chain scission; surface oxidation	CNT/SMP in morphing wings (NASA)	Loss of actuation reliability after 50 flight cycles
Thermal cycling	Microcracking; interfacial debonding	GNP/polyimide nanocomposites	– 15% flexural modulus after 1,000 h
Humidity/moisture	Interface weakening; filler swelling	CNT/epoxy composites	– 40% interfacial shear strength after 500 h at 85% RH
Oxidative atmosphere	Matrix embrittlement; filler oxidation	Metal-oxide/ thermoset hybrids	Discoloration, loss of ductility
Fuel vapor/solvent Attack	Swelling; matrix softening; delamination	Carbon-black filled PEEK	Delamination after 10-day exposure to jet fuel

Table 8.5 Roadmap phases and strategic focus

Timeframe	Focus area	Expected developments
2025–2027	*Fundamental research*	Investigation of interface chemistry, nanoparticle morphology, and nano-reinforcement mechanisms. Development of scalable synthesis methods
2027–2030	*Material qualification*	Establishment of standardized test protocols, accelerated aging tests, and qualification criteria for aerospace use. Small-scale integration in UAVs and satellite subsystems
2030–2035	*System integration*	Incorporation of nanocomposites in structural subcomponents (e.g., landing gear bays, fairings, control surfaces) and sensorized panels. Qualification under real flight conditions
2035 and beyond	*Full-scale deployment*	Use in primary load-bearing aircraft structures, integrated thermal-electrical systems, and multifunctional airframe elements. Full certification by regulatory bodies

This roadmap illustrates a phased and feedback-informed approach that reflects both technological complexity and regulatory constraints. Each phase builds upon the outcomes of the previous one, incorporating validation, risk mitigation, and cost-performance optimization.

By following this systematic path, stakeholders—ranging from academic researchers to aerospace manufacturers—can align their efforts and investment strategies, ensuring that nanocomposites reach maturity in a controlled, certifiable, and commercially viable manner.

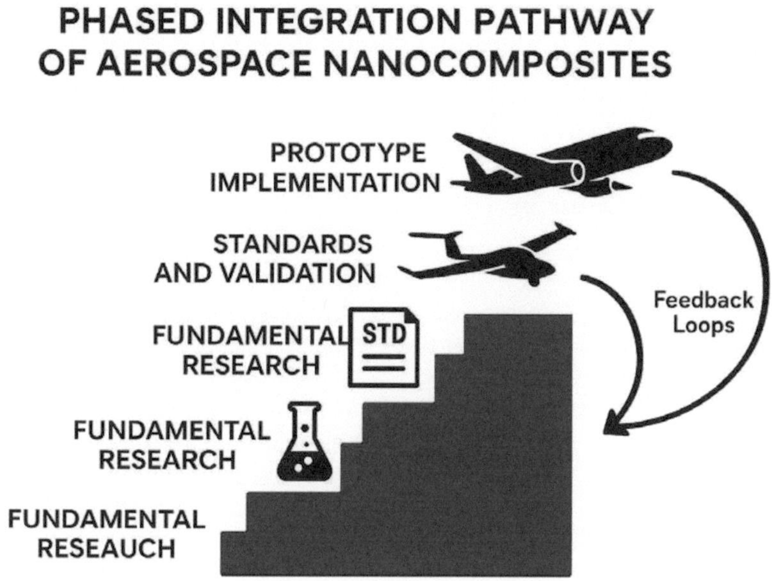

Fig. 8.9 Phased integration pathway of aerospace nanocomposites

Figure 8.9 is an illustration showing a staircase or funnel with each roadmap phase labeled, from research labs to full-scale aircraft. Icons depict test tubes, standards documents, UAVs, and airliners. Arrows represent feedback loops between stages.

8.9 Ethical and Societal Considerations

The deployment of nanocomposites in aerospace must account for ethical and societal impacts [12, 21]:

- Occupational health risks from nanoparticle inhalation during manufacturing.
- End-of-life disposal concerns, as many nanomaterials are not biodegradable.
- Environmental exposure, including nanoparticle leaching and air contamination.

To address these issues, aerospace manufacturers must implement strict workplace safety protocols, invest in green synthesis methods, and adhere to environmental regulations. Public communication and transparency will also be key in gaining societal acceptance.

The integration of nanocomposites into aerospace manufacturing raises profound ethical and societal challenges that extend far beyond technical performance. These materials—while enhancing fuel efficiency and structural resilience—introduce unique risks that demand proactive governance, transparent communication, and multidisciplinary collaboration. Below, we expand on these considerations with concrete examples, data, and mitigation strategies [13, 30].

8.9.1 Occupational Health Risks: Nanoparticle Exposure in Manufacturing

Expanded Detail:

During production processes like resin infusion, machining, or 3D printing, workers may inhale airborne nanoparticles (e.g., carbon nanotubes, graphene oxide, or nano-silica). These particles, often < 100 nm in diameter, can penetrate deep lung tissue, causing inflammation, fibrosis, or even carcinogenic effects (e.g., similar to asbestos). A 2022 *Nature Nanotechnology* study found prolonged exposure to carbon nanotubes at 0.5 mg/m^3 concentrations reduced lung function in rodents by 40% within six months.

Real-world Example:

In 2021, a Boeing composite facility reported elevated respiratory incidents among technicians sanding carbon-reinforced epoxy. Air sampling detected airborne carbon nanoparticles at 5 × the OSHA recommended limit, prompting a facility shutdown.

Mitigation Strategies:

- Engineering Controls: Closed-system manufacturing (e.g., Airbus's "NanoContain" reactors) and HEPA-filtered ventilation.
- PPE: Mandatory use of respirators with P100 filters and full-body suits.
- Monitoring: Real-time nanoparticle sensors (e.g., TSI Aerotrak) linked to automated shutdown protocols.

8.9.2 End-Of-Life Disposal: The Biodegradability Crisis

Expanded Detail:

Most aerospace nanocomposites (e.g., carbon nanotube-reinforced polymers) are non-biodegradable and may leach toxic nanoparticles when landfilled. For instance, a single commercial aircraft contains ~ 20 tons of nanocomposites, which could release heavy metals (e.g., cadmium, lead) or persistent organic pollutants over centuries. A 2023 EU study estimated that nanoparticle contamination from decommissioned aircraft could increase soil toxicity by 30% near recycling hubs by 2040.

Real-world Example:

The dismantling of the Airbus A380 prototype revealed challenges in separating nano-enhanced thermoset plastics. Traditional recycling failed, forcing ~ 60% of the material into incineration, releasing nano-copper particles into the atmosphere.

Mitigation Strategies:

- Circular Economy Models: Chemical recycling (e.g., pyrolysis at 450 °C to break down epoxy matrices while capturing nanoparticles).
- Design Innovations: BASF's development of bio-based, hydrolyzable nanocomposites for Safran's engine components.
- Regulatory Action: Adherence to the EU's SCIP database for tracking hazardous nanomaterials in products.

8.9.3 Environmental Exposure: Ecosystem Contamination

Expanded Detail:

Nanoparticles can enter ecosystems via airborne emissions (e.g., during aircraft maintenance), water runoff from manufacturing sites, or tribological wear in flight (e.g., nano-titanium debris from engine coatings). These particles bioaccumulate in flora/fauna; nano-ceramics in soil, for example, reduce microbial diversity by up to 35% (per a 2024 *Environmental Science & Technology* report).

Real-world Example:

Near a Colorado aerospace plant, groundwater testing detected nano-alumina at 50 ppb—linked to defective filtration systems—exceeding EPA limits and threatening local agriculture.

Mitigation Strategies:

- Green synthesis: Investing in water-based or solvent-free nano-production (e.g., Lockheed Martin's partnership with Green NanoTech).
- Containment systems: Double-lined wastewater ponds and electrostatic precipitators for airborne capture.
- Regulatory compliance: Enforcing ISO 14001 environmental management standards globally.

Societal Acceptance: Transparency as a Cornerstone

Expanded Detail:

Public distrust often stems from opaque risk communication. For instance, a 2023 MIT survey showed 68% of communities near aerospace sites distrusted manufacturers' safety claims about nanomaterials.

Proactive Solutions:

- Stakeholder Engagement:

  Example: Rolls-Royce's "NanoDialogue" initiative hosts open forums with NGOs and residents near its Singapore plant, publishing real-time emissions data.

- Certification & Labeling:

  Adopting ISO/TR 31321 standards for nanoparticle safety disclosure in supply chains.

- Policy Advocacy:

Collaborating with agencies like the FAA and EASA to standardize nano-lifecycle assessments.

A Framework for Responsible Innovation

Aerospace leaders must prioritize a "Safety-by-Design" ethos:

1. Preventive R&D: Funding non-toxic alternatives (e.g., cellulose nanocrystals).
2. Cross-industry Collaboration: Sharing disposal tech via consortia like the NanoSafety Cluster.
3. Global standards: Harmonizing regulations via the OECD's Working Party on Nanomaterials.

By embedding ethics into every lifecycle phase; from lab to landfill; the industry can harness nanocomposites' potential while upholding its societal license to operate.

8.10 Conclusion

In summation, the aerospace sector is poised for a profound material paradigm shift, catalyzed by the extraordinary developments in both hybrid metallic and polymeric nanocomposites. Scholarly investigations into these materials have yielded substantial breakthroughs, laying the groundwork for the creation of aerospace components that are not only more efficient and multifunctional but also inherently sustainable [26, 35].

Hybrid nanocomposites, anchored in metallic matrices such as aluminum or copper and fortified with nanostructures like carbon nanotubes, graphene, and metallic/ceramic nanoparticles, have exhibited unparalleled mechanical, thermal, and electrical attributes. Their impressive mechanical resilience, superior energy dissipation, enhanced thermal stability, and robust performance in extreme operational environments are indispensable for ensuring the safety and longevity of aerospace assets. Moreover, these materials facilitate the seamless integration of intelligent functionalities, including real-time structural health diagnostics and dynamic load adaptation, which are poised to fundamentally reshape the design and operational paradigms of aircraft and spacecraft.

Concurrently, polymeric nanocomposites are making remarkable strides, presenting viable, lightweight, corrosion-resistant, and radiation-tolerant alternatives to conventional metallic structures. Sophisticated synthesis and fabrication methodologies, coupled with meticulous characterization techniques such as AFM, TEM, and XRD, guarantee the uniform dispersion of reinforcing agents, thereby maximizing the inherent potential of these materials. Notable examples, such as Ag/Cellulose-polydopamine for electromagnetic interference shielding, SiO2-CNTs/PVDF-HFP for robust, superhydrophobic corrosion-resistant coatings, and PZT/PDMS for advanced structural health monitoring, underscore the diverse applications and significant performance enhancements achievable with polymeric nanocomposites [3, 15, 21].

Notwithstanding these promising advancements, critical challenges persist. For both material classes, achieving consistent nanoparticle dispersion, precisely controlling interfacial interactions, and scaling up manufacturing processes remain pivotal obstacles. Rigorous evaluations of long-term performance and reliability under the demanding conditions of aerospace operations are imperative. Furthermore, the development of eco-friendly synthetic pathways is essential for minimizing the environmental footprint of these advanced materials.

Nevertheless, the cumulative results to date unequivocally highlight the transformative capacity of both metallic and polymeric nanocomposites in redefining the standards of aerospace material performance. Sustained investment in targeted research and development, with a specific focus on overcoming existing challenges and refining manufacturing processes, will expedite their transition to widespread industrial applications. By strategically harnessing the synergistic strengths of these advanced materials, the aerospace industry can achieve unprecedented advancements in performance, safety, and environmental stewardship, thereby ushering in a new era of innovation and technological supremacy.

References

1. Abbasi S, Peerzada MH, Nizamuddin S, Mubarak NM (2020) Functionalized nanomaterials for the aerospace, vehicle, and sports industries. In: Handbook of functionalized nanomaterials for industrial applications. Elsevier, pp 795–825
2. Asyraf MRM, Ilyas RA, Sapuan SM, Harussani MM, Hariz HM, Aiman JM, Asrofi M et al (2022) Advanced composite in aerospace applications: opportunities, challenges, and future perspective. In: Advanced composites in aerospace engineering applications, pp 471–498
3. Bhat A, Budholiya S, Aravind Raj S, Sultan MTH, Hui D, Md Shah AU, Safri SNA (2021) Review on nanocomposites based on aerospace applications. Nanotechnology Rev 10(1):237–253. Bogue R, (2011) Nanocomposites: a review of technology and applications. Assembly Autom 31(2):106–112
4. Cabrera CR, Miranda F (eds) (2014) Advanced nanomaterials for aerospace applications. CRC Press
5. Cestino E, Catapano J, Galvano F, Felis A, Zuccalà S, Martilla V, Chesta L et al (2024) Effectiveness of nanotechnology treatments in composite aircraft applications. Appl Sci 14(5):1721
6. Del Bosque A, Vergara D, Fernández-Arias P (2025) An overview of smart composites for the aerospace sector. Appl Sci 15(6):2986
7. Dinca I, Ban C, Stefan A, Pelin G (2012) Nanocomposites as advanced materials for aerospace industry. Incas Bulletin 4(4):73
8. Edwards E, Brantley C, Ruffin PB (2017) Overview of nanotechnology in military and aerospace applications. In: Nanotechnology commercialization: manufacturing processes and products, pp 133–176
9. FENTA, Ermias Wubete et MEBRATIE (2024) Berihun Abebaw. Advancements in carbon nanotube-polymer composites: Enhancing properties and applications through advanced manufacturing techniques. Heliyon, vol 10, no 16
10. Guney Yilmaz S et al (2024) High-performance thermoplastic nanocomposites for aerospace applications. J Reinforced Plastics Compos

11. Guney Yilmaz S, Ferik E, Birak, SB et al (2024) High-performance thermoplastic nanocomposites for aerospace applications: a review of synthesis, production, and analysis. J Reinforced Plastics Compos 07316844241272035
12. Haynes H, Asmatulu R (2013) Nanotechnology safety in the aerospace industry. In: Nanotechnology safety. Elsevier, pp 85–97
13. Joshi M, Chatterjee U (2016) Polymer nanocomposite: an advanced material for aerospace applications. In: Advanced composite materials for aerospace engineering. Woodhead Publishing, pp 241–264
14. Kannan S, Siva V, Murugan A, Shameem AS, Thangamani A, Thangarasu S, Raja A (2023) Nanotechnology: nanocomposites processing. In: Innovative development in micromanufacturing processes. CRC Press, pp 372–392
15. Kausar A (2022) Future and challenging attributes of aeronautical nanocomposites. In: Polymeric nanocomposites with carbonaceous nanofillers for aerospace applications, p 317
16. Kausar A, Ahmad I, Zhao T (2023) Corrosion-resisting nanocarbon nanocomposites for aerospace application: an up-to-date account. Appl Nano 4(2):138–158
17. Kausar A, Ahmad I, Maaza M, Bocchetta P (2023) Self-healing nanocomposites—advancements and aerospace applications. J Compos Sci 7(4):148
18. Kausar A, Rafique I, Muhammad B (2017) Aerospace application of polymer nanocomposite with carbon nanotube, graphite, graphene oxide, and nanoclay. Polym-Plast Technol Eng 56(13):1438–1456
19. Kostopoulos V, Masouras A, Baltopoulos A, Vavouliotis A, Sotiriadis G, Pambaguian L (2017) A critical review of nanotechnologies for composite aerospace structures. CEAS Space J 9:35–57
20. Liu Y, Rajadas A, Chattopadhyay A et al (2013) Self-healing nanocomposite using shape memory polymer and carbon nanotubes. In: Sensors and smart structures technologies for civil, mechanical, and aerospace systems 2013. SPIE, pp 34–42
21. Malucelli G, Kausar A (2022) Green composites and nanocomposites for aerospace applications. In: Polymeric nanocomposites with carbonaceous nanofillers for aerospace applications, p 297
22. Matei A, Tucureanu V, Ţîncu BC, Mărculescu CV, Burinaru TA, Avram M (2019) Polymer nanocomposites materials for aerospace applications. In: AIP conference proceedings, vol 2071, no 1. AIP Publishing
23. Monteiro B, Simoes S (2024) Recent advances in hybrid nanocomposites for aerospace applications. Metals 14(11):1283
24. Nazarychev VM (2024) Enhanced thermal conductivity of thermoplastic polyimide nanocomposites: effect of using hexagonal nanoparticles. Polymers 16(23):3231
25. Prasad KE et al (2009) Extraordinary synergy in the mechanical properties of polymer matrix composites reinforced with 2 nanocarbons. PNAS 106(32):13186–13189
26. Ramachandran K, Boopalan V, Bear JC, Subramani R (2022) Multi-walled carbon nanotubes (MWCNTs)-reinforced ceramic nanocomposites for aerospace applications: a review. J Mater Sci 57(6):3923–3953
27. Rathod VT, Kumar JS, Jain A (2017) Polymer and ceramic nanocomposites for aerospace applications. Appl Nanosci 7:519–548
28. Romero-Fierro D et al (2022) Recent trends in magnetic polymer nanocomposites for aerospace applications. Polymers 14(19):4084
29. Romero-Fierro D, Bustamante-Torres M, Bravo-Plascencia F et al (2022) Recent trends in magnetic polymer nanocomposites for aerospace applications: a review. Polymers 14(19):4084
30. Sadiku ER, Agboola O, Mochane MJ, Fasiku VO, Owonubi SJ, Ibrahim ID, Ndamase AS et al (2019) The use of polymer nanocomposites in the aerospace and the military/defence industries. In: Polymer nanocomposites for advanced engineering and military applications. IGI Global, pp 316–349
31. She Y et al (2023) Nano-additive manufacturing and NDT of nanocomposites. Nanomaterials 13(20):2741

32. Silverman J, Astaire D, Siraeva A (2024) Applications of nanocomposites and polymeric materials in the aerospace industry. In: 2024 regional student conferences, p 85611
33. Siochi EJ, Harrison JS (2015) Structural nanocomposites for aerospace applications. MRS Bull 40(10):829–835
34. Soutter W (2012) Nanotechnology in aerospace materials. Azonano, Oct 4
35. Srinivasan V, Kunjiappan S, Palanisamy P (2021) A brief review of carbon nanotube reinforced metal matrix composites for aerospace and defense applications. Int Nano Lett 11(4):321–345
36. Trompeta AFA, Koumoulos EP, Stavropoulos SG, Velmachos TG, Psarras GC, Charitidis CA (2019) Assessing the critical multifunctionality threshold for optimal electrical, thermal, and nanomechanical properties of carbon nanotubes/epoxy nanocomposites for aerospace applications. Aerospace 6(1):7
37. Viscardi M, Arena M, Guadagno L, Vertuccio L, Barra G (2018) Multi-functional nanotechnology integration for aeronautical structures performance enhancement. Int J Structural Integrity 9(6):737–752
38. Weerasinghe J, Prasad K, Mathew J, Trifoni E, Baranov O, Levchenko I, Bazaka K (2023) Carbon nanocomposites in aerospace technology: a way to protect low-orbit satellites. Nanomaterials 13(11):1763
39. Zhao J et al (2017) Polymer nanocomposites: computer simulation understanding of properties. Polymer 133:272–287

Chapter 9
Green Composites and Nanocomposites for Aerospace Applications

Maliha Tasnim Kabir, Rudra Sarker, T. M. Nafiz Imran Dipto, and Md. Ashraful Islam Molla

Abstract The development of green and nano-enhanced composites represents a pivotal shift toward sustainable innovation in aerospace materials. Even though there are some drawbacks, there are still scopes to overcome the issues by working on them. As global priorities increasingly align with eco-conscious manufacturing, these composites are well-positioned to foster lighter, more efficient, and sustainable aircraft components. Future research should therefore focus on real-world performance validation, life cycle analyses, and scalable fabrication methods, in order to fully realize the transformative potential of these materials in aerospace engineering.

Keywords Green composites · Nanocomposite · Thermoplastic starch · Polylactic acid (PLA) cellulose · Nanofillers · Carbon nanotubes (CNTs) · Graphene · Fullerene · Inorganic clay · Metal nanoparticle · Unmanned aerial vehicles (UAVs) · Vibration damping · Noise reduction

9.1 Introduction

The aerospace industry has consistently driven innovation in advanced materials, primarily to meet demanding criteria such as high strength-to-weight ratio, thermal stability, and resistance to fatigue and corrosion. Traditionally, materials like aluminum alloys, titanium, and carbon fiber-reinforced polymers have dominated this field due to their proven performance in extreme environments [1]. In recent years, however, environmental awareness and sustainability challenges have made these materials more aware of their ecological impact, particularly their high energy consumption and limited recycling potential. Green composites have opened a new horizon for developing innovative solutions to conventional issues. These materials are composed of natural fibers and bio-based or biodegradable polymers,

M. T. Kabir · R. Sarker · T. M. Nafiz Imran Dipto · Md. Ashraful Islam Molla (✉)
Department of Applied Chemistry and Chemical Engineering, University of Dhaka, Dhaka, Bangladesh
e-mail: ashraful.acce@du.ac.bd

S. K. Panda (ed.), *Nanocomposites for Defense, Aeronautical and Aerospace Applications*, Advanced Structured Materials 248,
https://doi.org/10.1007/978-3-032-11568-3_9

as viable alternatives for certain aerospace applications. These materials also offer several advantages, including reduced environmental footprint, renewability, and lower cost, while still meeting structural and functional demands in non-primary load-bearing components [2]. Their relevance has grown particularly in the context of interior panels, seat structures, and insulation systems where sustainability can be prioritized without significant trade-offs in mechanical performance.

The integration of nanotechnology into green composites has further expanded their potential. Incorporating nanomaterials such as carbon nanotubes (CNTs), graphene, and nanoclays into natural fiber composites has been shown to significantly enhance their mechanical, thermal, and multifunctional properties. This has opened up new possibilities for hybrid green nanocomposites in aerospace applications where traditional green composites may fall short. This chapter aims to explore the evolution of sustainable material development in aerospace, highlighting the limitations of conventional materials and outlining the potential of green and nano-enhanced composites. The discussion covers processing methods, structure–property relationships, functional performance, and environmental and economic perspectives, offering a holistic view of how these innovative materials may shape the future of aerospace engineering.

9.2 Fundamentals of Green Composites and Nanocomposites

Since the earliest stages of evolution, humans have utilized available materials to improve their way of life. In ancient times, the absence of advanced technology led to a heavy reliance on natural materials, such as wood, clay, and stone. In contrast, modern society benefits from both natural and synthetic materials, which now play a pivotal role in the global techno-economic landscape. Broadly, materials are classified into three main categories: (i) metals, (ii) ceramics, and (iii) polymers [3]. Although polymers—both natural and synthetic—offer impressive properties and adaptability, they often fail to meet all the technical demands of certain specialized applications. To overcome these limitations, it became essential to integrate polymers with other materials possessing different characteristics. This approach led to the development of a new class of hybrid multi-functional materials capable of rivaling their alternative counterparts. Thus, composite materials emerged—engineered combinations that blend the strengths of their components. These composites offer the added benefit of being tailor-made, allowing for the fabrication of parts specifically optimized for unique performance requirements in various fields [4].

These polymer-based composites, including both synthetic and bio-based variants, are widely used in both industrial and domestic settings. These composites often outperform their components due to their combined or "synergistic" properties, which depend largely on the type and distribution of the reinforcing agents. In composite materials, the matrix (continuous phase) supports the reinforcement (dispersed phase), and the overall performance is influenced by the proportion of reinforcement and the strength of the fiber–matrix interface [5, 6]. Driven by their innovative nature, scientists began exploring ways to enhance polymers by combining them with ceramics and metals, not just in terms of material composition but also by modifying the shape and spatial distribution of reinforcing particles. They observed that optimizing factors such as particle concentration and geometry within the polymer matrix significantly improved composite performance. Specifically, reinforcements with high length-to-diameter (L/D) ratios were found to dramatically increase mechanical strength [7]. This led to the realization that fibrous structures, due to their geometry, offered superior stiffness and mechanical efficiency.

Despite these advancements, there was still demand for materials that were not only stronger and more efficient but also lighter and more cost-effective. Building on the benefits of high L/D ratios, researchers began experimenting with nanoscale reinforcements. This shift introduced nanocomposites—materials that could achieve enhanced strength and stiffness with minimal filler content, significantly reducing weight without compromising performance [7–9]. As a result, polymer-based materials expanded into a wide range of applications, from everyday items like paints and textiles to advanced uses in construction, electronics, transportation, aerospace, and healthcare. However, their widespread use brought new challenges regarding their environmental impact and disposal after end-of-life [5]. Natural fiber-reinforced composites are becoming increasingly attractive due to their renewable nature, ease of recycling, low cost, and lightweight properties [10]. Fibers like jute, hemp, kenaf, bamboo, banana, and sisal have historically been used as reinforcement materials. More recently, other natural fibers such as flax, wheat straw, sugarcane, coconut, silk, cotton, and oil palm have drawn interest for use in sophisticated engineering applications [11, 12]. Some of these fibers, especially when combined with biopolymers like polylactic acid (PLA), have shown promising results in thermoplastic and thermoset matrices. Their growing annual production reflects the increasing environmental consciousness driving their adoption. In addition to addressing ecological concerns, biocomposites show potential in reducing packaging waste and can be engineered for use in automotive structures, particularly in electric vehicles, where reducing overall weight is crucial for battery efficiency. These materials are also finding applications in biomedical fields, especially in creating biodegradable products.

9.3 Classification of Green Composite and Nanocomposite

Materials are considered "green" when they are both biodegradable and derived from renewable resources. The primary appeal of green composites lies in their eco-friendly nature—they are entirely degradable and support environmental sustainability throughout their life cycle. Once these materials reach the end of their functional use, they can be disposed of or composted without causing ecological harm. One of the core challenges in developing green composites is sourcing truly "green" polymers that can serve as the matrix material. For a polymer to be classified as green, it must exhibit eco-conscious characteristics, particularly the ability to degrade naturally and originate from renewable sources. Biodegradation refers to the breakdown of a polymer in a natural setting, resulting in changes to its chemical composition, loss of structural and mechanical integrity, and eventual conversion into environmentally beneficial substances [13]. Biopolymers can be derived from natural origins—such as starch, cellulose acetate, lignin, polylactic acid (PLA), polyhydroxyalkanoates (PHA), and polyhydroxybutyrate (PHB)—as well as from certain synthetic sources, including aliphatic and aromatic polyesters, polyvinyl alcohol, and modified polyolefins [14]. However, polymers obtained from synthetic processes typically lack renewability, meaning they do not fully meet the dual criteria of being renewable and biodegradable. Despite this, a wide array of natural and degradable polymers has already been integrated into green composite development, with several notable examples explored in subsequent discussions.

9.3.1 Thermoplastic Starch-Based Green Composite and Nanocomposite

Starch stands out among biopolymers as a highly promising alternative to petroleum-based plastics due to its full biodegradability and natural abundance. It is commonly derived from plant sources like corn, cassava, potatoes, and other tubers. In plants, starch exists in tightly packed granules containing both amorphous and crystalline regions. Given the growing environmental concerns, starch-based bioplastics—especially thermoplastic starch (TPS)—are increasingly used in biomedical and packaging applications. To improve their properties, bio-fillers or natural fibers are often incorporated into TPS formulations. *Da Silva* et al. report a novel Silver sulfadiazine-loaded halloysite/cassava starch (bio)nanocomposites were developed for potential use as antimicrobial wound dressings [15]. The drug was encapsulated within halloysite nanotubes and dispersed in thermoplastic starch to form the nanocomposite. Characterization confirmed strong interaction between components and uniform structure. The material showed suitable mechanical strength and moisture permeability for wound healing. Antibacterial tests revealed effective action against E. coli, P. aeruginosa, and S. aureus [15]. Bioactive PVA/starch (PSt) films were developed by *Natwat* et al. using green-synthesized silver nanoparticles (AgNPs) with oregano essential oil (OEO) as a reducing agent. Incorporating 5 wt% OEO led to effective antibacterial action against *E. coli* and *S. aureus*, showing a synergistic antimicrobial effect with AgNPs [16]. This OEO concentration was notably lower than in similar formulations. Enhanced water resistance and significant increases in tensile strength (80.9%) and modulus (88.6%) were observed.

9.3.2 Polylactic Acid (PLA)-Based Green Composites and Nanocomposites

PLA is increasingly recognized as one of the most effective eco-friendly alternatives to conventional petroleum-based polymers, thanks to its renewable origin, biodegradability, biocompatibility, and favorable thermomechanical performance. While PLA was originally used primarily in biomedical applications and short-term products like packaging due to its degradable nature, its scope has broadened. With the growing scarcity of fossil fuels, PLA is now being explored as a sustainable, bio-based material for more durable applications, including in the automotive and electronics industries. PLA was used to fabricate eco-friendly PLA/jute laminate composites through compression molding. Optimal composite quality was achieved with a layered PLA pellet–jute fabric–PLA pellet configuration, showing no voids or delamination under SEM. Adding jute layers increased stiffness by up to 145%, though ductility dropped by 60%, and strength varied with the PLA-to-jute ratio. Impact strength declined with fewer jute layers but improved slightly with higher fiber content. Enhanced storage modulus was noted even above PLA's glass transition temperature. These findings support PLA/jute systems for recyclable, sustainable composite applications [17]. On the other hand, flexible antibacterial PLA films were developed using liquidambar orientalis oil as a bio-based plasticizer and 2D Ag-decorated graphitic carbon nitride ($Ag@g\text{-}C_3N_4$) as the antimicrobial agent. The $Ag@g\text{-}C_3N_4$ nanocomposite, synthesized via thermal annealing and silver salt reduction, had a particle size of 10–30 nm and a surface area of 18 g/m^2. Incorporating 30 phr of the oil reduced PLA's glass transition to 43 °C and improved flexibility by weakening interchain interactions. The oil also enhanced $Ag@g\text{-}C_3N_4$ dispersion and boosted antibacterial effectiveness. Even at low loadings (1–4 phr), strong inhibition was observed against *S. aureus*, *E. coli*, and *A. baumannii* [18].

9.3.3 Cellulose-Based Green Composites and Nanocomposites

Cellulose, in its natural or chemically modified state, is among the most widely used polysaccharides due to its outstanding benefits, including high specific strength, biocompatibility, non-toxicity, abundant availability, and low cost. Structurally, it is a syndiotactic homopolymer composed of D-glucopyranose units connected by β-1,4-glycosidic bonds [19]. The cellulose chains feature a high density of hydroxyl groups, which engage in extensive intra- and intermolecular hydrogen bonding. This hydrogen bonding imparts key properties such as structural stability, hydrophilicity, and a high number of reactive sites for functionalization. From an application standpoint, esterification and etherification are the most important chemical modification techniques used to enhance cellulose performance [20]. Another study developed high-performance composites by incorporating nanocellulose extracted from water hyacinth via acid hydrolysis and sonication. The nanocellulose was applied to jute, cotton, and glass fabrics using dip coating. XRD showed a crystallinity index of 40.72%, while FT-IR and SEM confirmed characteristic functional groups and morphological changes. Mechanical testing revealed significant strength improvements: jute composites increased by 24.6%, cotton by 19.4%, and glass by 8.5%, with flexural strength also notably enhanced. These results surpass those of previous studies, demonstrating the effectiveness of water hyacinth nanocellulose reinforcement [21].

9.4 Nanofillers

Nanomaterials are typically described as substances with at least one dimension in the nanometer range. Their unique scale has sparked significant interest across scientific and industrial fields, leading to the rapid evolution of nanotechnology as a distinct and dynamic discipline. Much effort has been devoted to synthesizing and enhancing these materials due to their exceptional properties—such as increased surface area and distinct optical and electrical behavior—compared to their bulk counterparts. Fillers used in polymer composites are generally classified by dimensionality: 1D fibers, 2D flakes, and 3D particles. An important factor influencing composite performance is the filler's aspect ratio—the length-to-thickness ratio—which affects surface energy and therefore compatibility with polymer matrices [22, 23]. A higher aspect ratio typically enhances filler dispersion and interaction with the polymer. Reducing filler dimensions from microscale to nanoscale significantly increases the aspect ratio, leading to improved reinforcement effects within the composite material. Nanotechnology has increasingly been utilized to address the environmental challenges posed by plastic waste. The push toward sustainable materials aligned with circular economy principles and green chemistry has made this field essential. A growing focus has emerged around creating composites from biodegradable polymers reinforced with nanoscale fillers. Compared to conventional composites, nanocomposites offer enhanced performance even with minimal filler content. This is particularly important for biodegradable nanocomposites, as most nanofillers used—though not always biodegradable—are added in very small amounts (typically under 5 wt%), ensuring that the majority of the final product remains environmentally degradable. Below are some nanofillers emphasizing their unique property and applications.

9.4.1 Carbon Nanotubes (CNTs)

Carbon nanotubes (CNTs) have garnered significant attention due to their remarkable physical properties. Compared to conventional carbon fibers, CNTs exhibit much higher tensile strength (ranging from 50 to 100 GPa) and stiffness (in the terapascal range), along with superior thermal and electrical conductivity. There are two primary forms of CNTs: single-walled carbon nanotubes (SWCNTs), which consist of a single layer of graphene rolled into a thin cylindrical structure just a few nanometers in diameter but thousands of times longer; and multi-walled carbon nanotubes (MWCNTs), which are composed of several concentric graphene cylinders. While SWCNTs typically offer superior mechanical characteristics, MWCNTs are more commonly used due to their significantly lower production costs. Common synthesis techniques for CNTs include arc discharge, laser ablation, and chemical vapor deposition (CVD). CNTs suffer from interfacial adhesion and poor solubility in solvents, resulting in poor dispersion in the matrix; hence, several mechanical, physical, and chemical methods are adopted to increase the dispersity [24]. Chemical methods, the functionalization of the surface of the nanotubes involves introducing specific groups or the functionalization of the polymeric chains themselves to improve the compatibility between the load and the matrix through the interfacial interactions [25].

Biodegradable polymers like PLA have been enhanced by incorporating single-walled carbon nanotubes (SWCNTs) as reinforcing agents [26]. In a study by Wu et al., PLA/multiwalled carbon nanotube (MWCNT) composites were developed using two methods to improve polymer–filler compatibility: (1) grafting PLA with acrylic acid (PLA-g-AA), and (2) functionalizing MWCNTs with hydroxyl groups. These modifications led to significant improvements in the thermal and mechanical performance of the PLA/MWCNT composites [27]. An experimental study, explained by Poonsawat et al. 2014, shows that by increasing the concentration of the fillers, it accordingly enhances the degradation temperature, in which the highest thermal stability is attained at 5 wt% of MWCNT-COOHs (at 380 °C) (Fig. 9.1) [28]. Modifications like this enhance the quality of the composites.

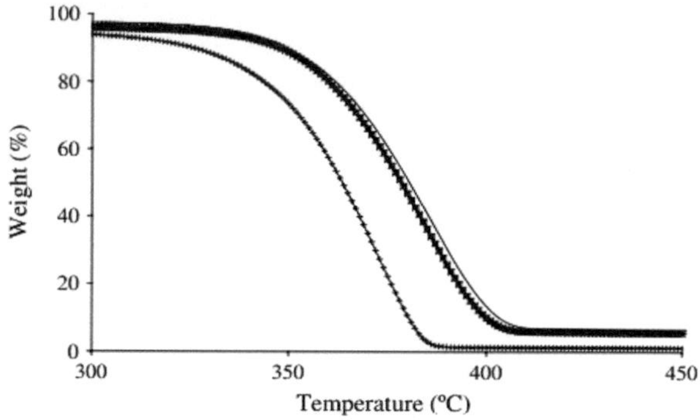

Fig. 9.1 TGA thermograms of MA-g-PLA (+) and the composites prepared with various types of MWCNTs (at 5 wt%): nonfunctionalized MWCNTs (−) [28]

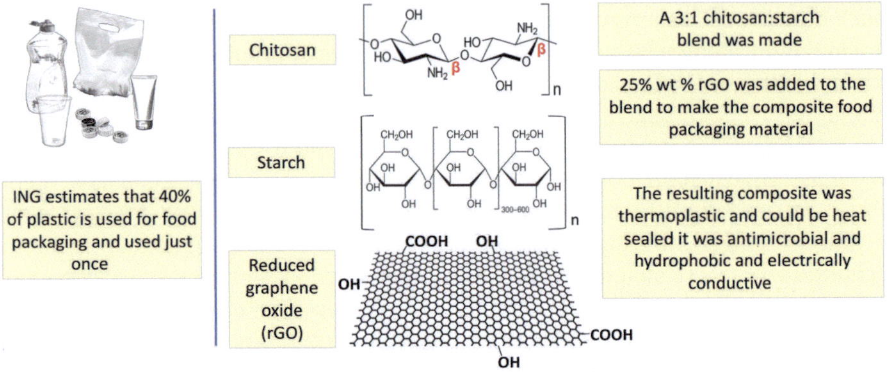

Fig. 9.2 Graphical representation of using starch-chitosan bioplastics reinforced with GO for food packaging [31]

9.4.2 Graphene

Graphene (GN) is a single atomic layer of carbon atoms arranged in a two-dimensional honeycomb lattice. Within this layer, neighboring carbon atoms form three σ bonds through sp^2 orbital overlap, while the remaining p_z orbitals contribute to the formation of valence and conduction bands via filled π and empty π* orbitals, respectively [27]. This unique sp^2-hybridized honeycomb arrangement gives graphene exceptional characteristics, including a high specific surface area, mechanical stiffness of 1 TPa, thermal conductivity up to 5000 W m^{-1} K^{-1}, optical transparency of 97.7%, and outstanding electron mobility (250,000 $cm^2 \cdot V^{-1} \cdot s^{-1}$ at room temperature) [29]. In-situ reduction of graphene oxide (GO) during polymerization or melt processing within a polymer matrix offers significant advantages for creating polymer-based nanocomposites reinforced with reduced graphene oxide (RGO). A recent study highlighted the incorporation of RGO into a biodegradable starch–chitosan matrix for use in active food packaging applications (Fig. 9.2) [30].

9.4.3 Fullerene

Fullerenes are composed of rings that form a geodesic spheroidal structure made up of sp^2-hybridized carbon atoms with significant isomeric diversity, often assembling into multi-layered formations [32]. While carbon nanotubes (CNTs) are structurally related to fullerenes and exhibit quasi-one-dimensional characteristics, fullerenes have garnered considerable interest due to their distinct geometry. Their ability to undergo endohedral, substitutional, and exohedral doping makes them valuable for applications in nanotechnology, electronics, and biotechnology, offering unique structural, magnetic, superconducting, electrochemical, and physicochemical properties [33]. Moreover, the synthesis of polymers incorporating fullerenes—including biodegradable types—has led to the development of novel materials that merge the functionalities of both components.

9.4.4 Inorganic Clay

Inorganic clays are among the most widely used fillers in polymer nanocomposites, valued for enhancing mechanical properties and reducing production costs. Common fillers include talc, silica, and various clays. A key benefit of polymer nanocomposites is that even minimal amounts of nanoparticles can significantly improve performance without negatively affecting weight or transparency [34]. Nanoclay-based composites are particularly attractive due to their affordability, availability, and ease of processing. Layered silicates, or phyllosilicates, such as montmorillonite, halloysite, hectorite, saponite, and sepiolite, are frequently used in these systems. Their structures consist of stacked tetrahedral and octahedral sheets containing cations like Al^{3+}, Mg^{2+}, or Fe^{2+}.[45] Polymer–clay interactions typically result in one of three structural arrangements: (1) agglomerated (phase-separated), (2) intercalated (polymer chains inserted between layers), or (3) exfoliated (clay layers fully dispersed within the polymer) [35–37].

9.4.5 Metal Nanoparticle

Metal and metal oxide nanoparticles have gained significant attention for their exceptional characteristics, particularly in fields like optics, catalysis, electronics, and biomedicine [38]. Typically derived from noble metals such as silver, gold, and platinum, these nanoparticles are synthesized via two main strategies: top-down and bottom-up approaches. Top-down methods reduce bulk materials into nanoscale forms using techniques like milling, laser ablation, or sputtering, whereas bottom-up approaches build nanoparticles from atoms or molecules through methods like vapor deposition, chemical reduction, or electrochemical processes [39]. In line with sustainable practices, green synthesis methods—avoiding hazardous chemicals—are increasingly favored. One such eco-friendly technique involves using plant extracts as reducing agents to convert metal ions into nanoparticles [40]. This biogenic method not only eliminates the need for toxic reagents but also proves suitable for biomedical applications. Additionally, the type of plant extract used can influence the nanoparticles' morphology, size distribution, and antibacterial effectiveness [41].

9.5 Material Processing and Fabrication Techniques

Nanocomposites can be made from polymers that contain nanoscale fillers (e.g., carbon nanotubes, graphene, and nanoclays). These materials provide enhanced mechanical, thermal, and electrical properties and can be used for versatile functions, including aerospace applications. On the other hand, green composites are materials that combine natural fibers (like flax, jute, or hemp) with biodegradable or bio-based polymer matrices. Compared to traditional composites, their lightweight nature, renewability, and reduced environmental impact make them attractive for aerospace applications.

9.5.1 Synthesis of Inorganic Nanoparticles

Different physical and chemical methods are available to synthesize nanoparticles. These particles differ in size, shape, and chemical or physical properties based on their elemental composition. Different factors like temperature, pressure, time, particle size and shape, cost of preparation, pore size etc. affect the result of the synthesis process [42]. The synthesis technique can be of three categories. Table 9.1 gives the general idea of the synthesis technique.

Mechanical method is usually done by two common ways, one is ball milling and another is melt mixing. Mechanical milling is a widely adopted top-down approach used for synthesizing nanoparticles by mechanically breaking down bulk materials into fine particles through high energy ball collisions inside rotating mills. This process is carried out using various types of mills such as planetary, vibratory, rod, and tumbler mills. This method is cost-effective, scalable, and suitable for producing nanocrystalline metals like Co, Cr, and Ag-Fe alloys [42]. On the other hand, Melt Mixing is a bottom-up method where nanoparticles are formed by rapidly mixing molten metal streams under high turbulence. The intense turbulence during mixing helps tiny particles to form, which are then quickly frozen into a disordered, glass-like structure. This glassy matrix locks the nanoparticles in place before they can grow any larger [64]. It is an ideal process for producing strong, corrosion-resistant nanocomposites. Chemical vapor deposition (CVD) is a bottom-up technique used to synthesize nanoparticles by reacting vapor-phase precursors on a heated surface. It allows precise control over particle size, purity, and composition, making it ideal for producing materials like Si, TiO_2, ZnO, and carbon-based nanostructures [65]. It is widely applied in electronics, catalysis, and aerospace due to its uniformity and scalability.

In flame spray pyrolysis (FSP), nanoparticles are formed when a precursor solution is atomized into a high-temperature flame and undergoes rapid combustion and oxidation [66]. In pulsed laser ablation (PLA) manufacturing, high-powered laser pulses are directed at solid targets, vaporizing material into plasma, which is then condensed into nanoparticles [49]. This process is ideal for synthesizing

Table 9.1 An overview of the types of methods of synthesis of nanocomposites

Method of synthesis	Type name	Typical nanoparticles	Applications	References
Physical	Mechanical method	Nanocrystalline Co, Cr, W, Ag-Fe, etc	Electro deposition, catalytic activity	[42] [43]
	Chemical Vapor Deposition	Si, TiO_2, ZnO, carbon-based NPs	Carbon-based NPs Semiconductor devices, MEMS, coatings	[44]
	Flame spray Pyrolysis	Fumed silica, Al_2O_3	Particle designing, thin film preparation, coating	[45–47]
	Pulse Laser Ablation	Au, Ag, Cu, Pt	Biomedical, electronics, nanostructured films	[48]
	Pulsed Wire Discharge method	Al, Cu, Ag, W, Ti, Fe-FeO GO	Energetic materials, powders for sintering	[49, 50]
	Ionized Cluster Beam Deposition	Ag, Au, Pt, Fe	Sensors, thin films, magnetic data storage	[51, 52]
Chemical	Sol–gel Method	SiO_2, TiO_2, ZnO, Al_2O_3	Coatings, optics, catalysts	[53]
	Hydrothermal Synthesis	TiO_2, ZnO, CeO_2, $BaTiO_3$	Batteries, piezoelectrics, photocatalysis	[54]
	Sonochemical Synthesis	Ag, Au, $NiFe_2O_4$, ZnO, TiO_2, Fe_3O_4	Catalysis, biomedical applications, sensors	[55, 56]
	Co-precipitation Method	Fe_3O_4, $ZnFe_2O_4$, MnO	Magnetic devices, pigments, biomedicine	[57]
	Inert Gas Condensation	Ag, Pt, Fe, Zn	Catalysts, magnetic materials, electronics	[58, 59]
Biological	Synthesis Using Microorganisms	Ag, Au, ZnO, TiO_2, Fe_3O_4	Antimicrobial agents, drug delivery, environmental remediation	[60, 61]
	Synthesis Using Plant Extracts	ZnO, CuO	Antimicrobial agents, environmental applications	[61]
	Synthesis Using Algae	Ag, Au, ZnO, TiO_2	Antibacterial agents, cancer therapy, environmental remediation	[62], [63]

high-purity, contamination-free nanoparticles and embedding them into polymer or ceramic matrices for optical, electronic, and biomedical nanocomposites. In sol–gel method, a metal alkoxide or salt undergoes hydrolysis and polycondensation, forms a gel that is dried and heat-treated to produce nanoparticles. This method is one of the most widely used methods for synthesizing oxide-based nanocomposites (e.g., SiO_2, TiO_2) with high homogeneity. It is applied in optical materials, sensors, membranes, and coatings [67]. The hydrothermal synthesis method involves crystallizing substances from aqueous solutions at high temperature and pressure in a sealed autoclave. It is widely used for producing metal oxides, zeolites, and layered nanomaterials. It allows for controlled size and morphology, making it suitable for electronic, catalytic, and structural nanocomposites [54].

9.6 Mechanical, Thermal, and Functional Properties

As materials science continues to evolve, green composites and nanocomposites are becoming exciting new options in aerospace design, offering not just strength and performance, but also a more sustainable approach to building the future of flight. Using natural fibers such as flax, jute, or hemp, green composites offer a favorable strength-to-weight ratio. A higher strength-to-weight ratio enables the material to withstand more stress without adding excessive mass as well as to carry more payload or passengers without compromising structural integrity.

Nanocomposites, on the other hand, include tiny particles like carbon nanotubes or graphene, and are much stronger and stiffer than regular materials (Fig. 9.3). These nanoparticles help spread out stress more evenly, making the material tougher and better at handling heavy loads. The idea of nanocomposites was first discussed in 1940, when Bower gave an idea about mixing tiny particles into metals to make them stronger. Some of the first examples were aluminum blended with materials like silicon carbide or boron nitride to form Al/SiC and Al/BN[68]. Aluminum has a very low density, which makes it perfect for corrosion resistance and can withstand moderate temperatures (~400 °C). Similarly, SiC is a low-density ceramic material that is abundant in nature and can withstand temperatures above 1500 °C. It is also extremely hard and wear-resistant, making it excellent for high-stress environments. So, the nanocomposite of Al with SiC or BN showed a good performance in aerospace applications.

Though green composites have good mechanical properties, they face challenges with their limited thermal stability, which stems from the organic nature of the materials used, particularly natural fibers and bio-based resins. These components tend to break down or lose strength when exposed to high temperatures for extended periods. Additionally, natural fibers are prone to absorbing moisture from the environment due to the presence of hydroxyl groups in their cellulose structure, which can lead to swelling, fiber degradation, and poor bonding with the matrix [70]. For instance, untreated flax fiber composites can absorb up to 13.5% moisture, leading to significant decreases in tensile strength and stiffness. This moisture uptake not only weakens the

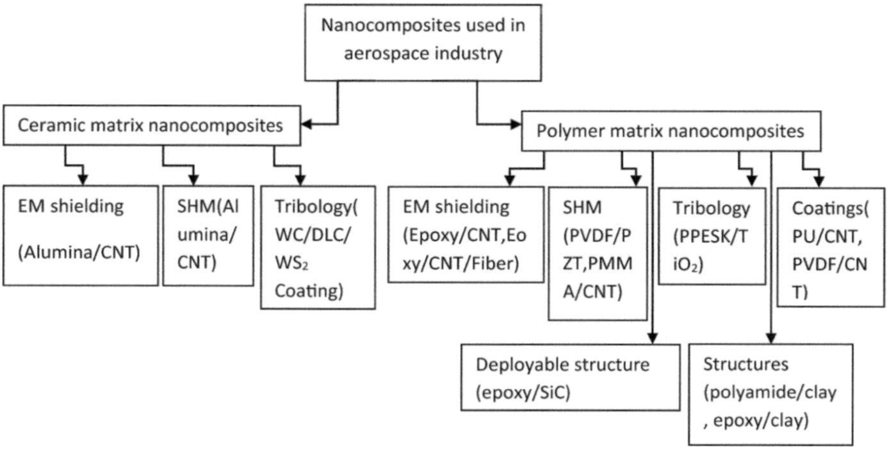

Fig. 9.3 Functional classification of nanocomposites in aerospace engineering: materials, matrices, and applications [69]

composite's mechanical properties but also further compromises its ability to perform under thermal stress by causing plasticization. As a result, green composites may not be suitable for high-temperature or humid aerospace environments without proper surface treatments or protective coatings.

On the other hand, the integration of nanoparticles into composite materials has been shown to significantly enhance their thermal conductivity and stability, making them well-suited for high-temperature aerospace applications. In high-altitude or supersonic flight, materials often face extreme temperatures, and traditional composites—especially those made with natural fibers—don't always hold up. But when nanoparticles like carbon nanotubes, graphene, or metal oxides (e.g., Al_2O_3, TiO_2) are added, they help transfer heat more efficiently and make the overall structure more thermally stable. For example, a study in *Polymer Composites* showed that adding graphene nanoplatelets to a flax fiber composite significantly improved its thermal conductivity and delayed thermal degradation, which is a key function for aerospace applications [70]. Another study found that polymer nanocomposites with fillers like carbon nanotubes exhibited enhanced thermal stability, increasing the material's glass transition temperature and resistance to heat distortion [71]. In more advanced systems, nanofluids, liquids engineered by dispersing nanoparticles into traditional coolants, have shown even greater thermal conductivity. These are being explored for use in spacecraft thermal management systems, where efficient heat dissipation is critical [72].

Green composites are biodegradable and environmentally friendly, which makes them an attractive choice for industries dedicated to reducing their ecological footprints. In aerospace, weight reduction is paramount. Green composites offer a favorable strength-to-weight ratio, making them suitable for components where high mechanical strength is not the primary requirement. For instance, they are being considered for use in interior panels, seat backs, and cabin linings, where their lightweight nature contributes to overall fuel efficiency. Moreover, the production of green composites typically involves less energy consumption compared to traditional composites, further enhancing their sustainability profile. The use of renewable resources not only reduces dependence on fossil fuels but also supports rural economies through the cultivation of natural fibers [10].

In modern aircraft, where electronics are densely packed, protecting systems from electromagnetic interference is critical. Nanocomposites infused with conductive fillers like carbon nanotubes (CNTs) or graphene create lightweight, flexible shielding layers that outperform traditional metal enclosures (Fig. 9.4). These materials help maintain signal integrity while reducing overall weight. A study by Gadtya et al. (2024) highlights the effectiveness of CNT-based polymer nanocomposites in aerospace applications, emphasizing their superior EMI shielding properties [73]. Fire safety is also an important factor in aerospace applications. Incorporating nanostructured flame retardants into polymer matrices enhances the fire resistance of composite materials. These nanomaterials improve flame retardancy through mechanisms like heat dissipation, barrier effects, and char promotion, effectively slowing down combustion and reducing smoke production. Recent advancements in nanostructured flame retardants have shown promise in enhancing the fire resistance of polymer composites [74]. In self-healing nanocomposites, healing agents are injected into microcapsules or vascular networks, and the agents are released upon damage to restore structural integrity and healing. This technology not only extends the lifespan of components but also enhances safety by preventing the propagation of micro-damage. Research has demonstrated the potential of self-healing nanocomposites in designing structural components, panels, and coatings for aerospace applications [75].

9.7 Aerospace Applications of Green Composites

Composite materials have transformed aerospace by offering exceptional strength-to-weight ratios, crucial for aircraft. Made from strong fibers in a polymer matrix, they enable lighter, more fuel-efficient, and higher-performing aircraft, becoming essential for modern wings, fuselages, and engine parts. Green composites, made from natural fibers (such as flax, hemp, and jute) and bio-based resins (like polylactic acid or epoxidized plant oils), offer a sustainable alternative to conventional aerospace materials [76–78]. Their low density, high specific strength, and inherent vibration damping make them suitable for non-critical structural applications, interior components, and UAVs. [79, 80]. Chemically, their performance depends on the composition and treatment of lignocellulosic fibers and the nature of the polymer matrix, with interface optimization playing a key role [81, 82]. While challenges such as durability and fire resistance remain, the environmental benefits and evolving material chemistry of green composites position them as promising candidates for future aerospace innovations.

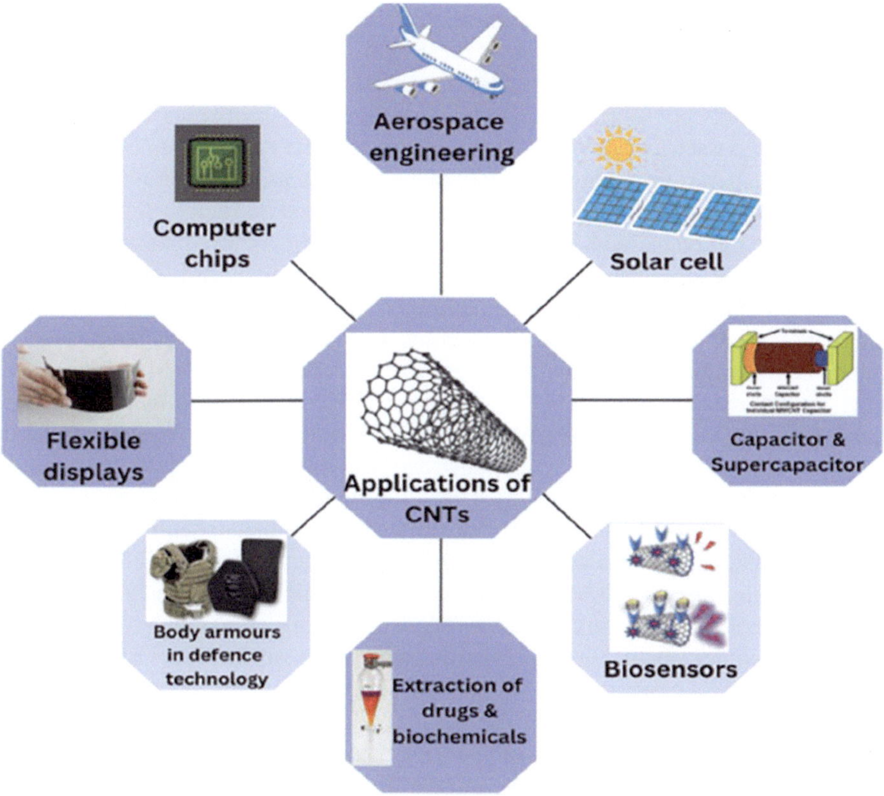

Fig. 9.4 Applications of CNT nanocomposites for aerospace [75]

9.7.1 Interior Components (Panels, Seat Frames, Insulation)

To be viable in aircraft panel, seat frames and insulation application, we need the composites to meet several performance criteria like internal bonding, thermal stability and flame retardancy. Carbon fiber-reinforced polymers (CFRP) are dominant in panels, often combined with honeycomb structures (Fig. 9.5) for lighter, stiffer fuselage and wing sections [83]. While aluminum is still used, carbon fiber composites are increasingly found in seat frames for significant weight savings [84]. For insulation, materials like melamine foams and specialized quartz fiber composites provide critical acoustic, thermal, and fire-resistant properties, ensuring passenger comfort and safety [85]. For cabin sidewall panels and ceiling panels, flax fiber/bio-epoxy composites can be used because flax have structure with high tensile strengths, stiffness and low density (~1.4 g/m^3), which reduces overall aircraft weight [86, 87]. They have a cellulose content of 60–70%, contributing to the mechanical integrity. These fibers have a porous structure due to their cellulose-based microstructure which traps air and reduces heat transfer [88]. Bio-Epoxy is derived from renewable feedstock like epoxidized soybean oil (ESO) or cardanol (from cashew nut shell liquid) which offers good adhesion, chemical resistance and cure behavior similar to petroleum-based epoxies [89]. It also has good damping properties contributing to vibration and passenger comfort which makes it suitable for seat frames.

Moreover, hemp fiber/polylactic acid (PLA) can be used in interior trim panels because Hemp contains strong bast fibers, high in cellulose and hemicellulose, readily available and cultivable with low agricultural input [90]. Hemp fibers offer ~ 350–750 MPa tensile strength, sufficient for structural support with proper design for seat frames [91]. PLA is a biodegradable thermoplastic derived from corn starch and sugarcane with tunable crystallinity and low processing temperature (~160–180 °C) and can be modified with flame retardants (e.g., ammonium phosphate) to meet aerospace safety standards [92]. PLA has a relatively low softening temperature (~60°) but this can be improved by blending with polycaprolactone (PCL) or using crosslinked PLA [93]. Fire resistance can be enhanced using phosphorus-based flame retardants chemically integrated into the matrix. Furthermore, Kenaf fiber/polyfurfuryl alcohol (PFA)/bio-PET (polyethylene terephthalate) composites can be used in insulating and acoustic panel cores because kenaf fiber is lightweight high crop with high content (~20%), which enhances thermal insulation and damping [94, 95]. It has a hollow lumen structure that reduces thermal conductivity. Adding nanoclays or aluminum trihydrates (ATH) particles into the resin improves fire resistance and decreases smoke generation [96]. Polyfurfuryl alcohol (PFA) is synthesized from furfural, a renewable chemical from agricultural waste. PFA has excellent flame resistance, high char yield and good chemical resistance [97]. Bio-PET offers durability over loading cycles which makes it suitable for seat frames. Hybrid natural fiber/

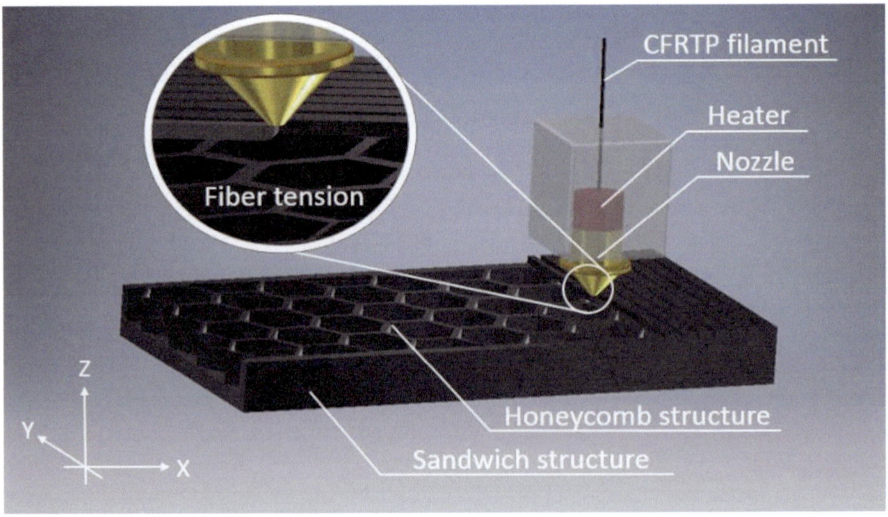

Fig. 9.5 Schematic of the 3D printing of a sandwich structure with carbon fiber composite

glass or carbon/bio-resin can be used in panels subjected to moderate mechanical loads hybridization allows combination natural fibers with small amounts of synthetic fibers that improves structural integrity while retaining sustainability [98]. Hybrid composites (e.g., flax-glass/bio-epoxy) are also useful for seat frames. They have increased strength while keeping weight and environmental impact low [99]. Also, hybrid layups can be tailored for specific load paths in seat frames jute/phenolic composites can be used in seat frames because with proper fiber alignment and layering can absorb mechanical shocks. Wool/bio-polyurethane foam composites have good fire resistance. Wool chars instead of melting, which slows flame spread. Wool's keratin structure resists ignition and limits toxic emissions, enhancing cabin safety.

9.7.2 Structural Applications (Winglets, Fairings, Secondary Structures)

Carbon fiber-reinforced polymers (CFRP) are extensively used due to their exceptional stiffness and strength, making them ideal for high-stress areas like winglets where drag reduction and fuel efficiency are critical [100]. Glass fiber-reinforced polymers (GFRP) are also employed, particularly in less critically loaded secondary structures and fairings, offering good mechanical properties at a lower cost and providing excellent electrical insulation [101]. Additionally, Aramid fiber-reinforced polymers (e.g., Kevlar-epoxy) are used where impact resistance is paramount. These composites, typically with an epoxy matrix, contribute significantly to weight reduction, corrosion resistance, and increased fatigue life, which are crucial for the overall performance, durability, and operational costs of modern aircraft [102] Flax fiber-reinforced epoxy present most promising option for winglet application. They have excellent strength-to-weight ratio, tailorable anisotropic behavior through fiber alignment and eco-friendly life cycle with potential recyclability or biodegradability [103]. Surface treatment of flax (alkaline treatment or silanization) improves fiber–matrix interfacial bonding. The crosslinking of bio-epoxy resins via an amine or anhydride curing agents ensures thermal and chemical resistance which is why flax fibers are used for fairings that are not exposed to high impact loads but need a lightweight and durable shell [104]. Flax fibers are also used in access hatches in aerospace.

Hemp fiber with bio-epoxy matrix is another choice for winglet application. It can endure cyclic loading and vibrational stress. Chemically treated hemp (e.g., with NaOH or silane coupling agents) enhances fiber–matrix compatibility, reducing interfacial voids [105]. Fiber modification (e.g., acetylation) improves compatibility with PLA and enhances composite stability, and they are feasible for UAV fairings or low-temperature environments with less mechanical stress [106]. Hybrid green composites (natural and synthetic fibers) can also be used as winglet application. Combining green fibers with minimal carbon or basalt fibers offers a synergistic effect like improved tensile and flexural strength and partial sustainability maintenance

[107]. Enhanced interfacial adhesion and hybrid toughening mechanisms increase impact resistance and fatigue life. Bio-resin matrices can be UV-stabilized and flame-retardant modified (e.g., via phosphorus-containing additives). They are ideal for outer fairings that require enhanced durability without fully sacrificing sustainability [108]. Flax/glass fiber with bio-epoxy matrix are being used in aerodynamic covers and control surfaces in UAVs. Kenaf/PLA composites are considerable option in case of winglet application. Although they have lower thermal resistance and need additives to improve ductility and temperature tolerance, kenaf has moderate tensile strength, low cost and high availability [109]. PLA is processable via compression molding or vacuum-assisted resin transfer molding (VARTM) for aerospace winglet use. Polyurethane allows tuning of hard segment-soft segment ratios, optimizing impact resistance and flexibility. Enhanced adhesion between hydroxyl-functional kenaf fibers is isocyanate-based PU chains [110]. They are used where vibration and minor impact resistance are important, such as landing gear fairings, wheel well covers, and second aerodynamic covers. Polyhydroxyalkanoate is (PHA) are more hydrophobic than PLA, offering better resistance to moisture. They are used in other secondary structures like electronic housings and cargo linings [111].

9.7.3 Use in Unmanned Aerial Vehicles (UAVs)

Composites suitable for use in unmanned aerial vehicles (UAVs) must meet critical requirements like lightweight, moderate strength, good impact resistance, thermal stability, and ideally environmental sustainability. The selection of a green composite for UAV application depends on mission profiles, flight duration and payload, environmental conditions, and manufacturing capabilities. Carbon fiber-reinforced polymers (CFRP) are considerable option in case of winglet application. Although they have lower thermal resistance and need additives to improve ductility and temperature tolerance, kenaf has moderate tensile strength, low cost and high availability [109]. PLA is processable via compression molding or vacuum-assisted resin transfer molding (VARTM) for aerospace winglet use. Polyurethane allows tuning of hard segment-soft segment ratios, optimizing impact resistance and flexibility. Enhanced adhesion between hydroxyl-functional kenaf fibers are isocyanate-based PU chains [110]. They are used where vibration and minor impact resistance are important, such as landing gear fairings, wheel well covers, and second aerodynamic covers. Polyhydroxyalkanoate is (PHA) are more hydrophobic than PLA, offering better resistance to moisture. They are used in other secondary structures like electronic housings and cargo linings [111]. Flax/bio-epoxy composites, hemp/polylactic acid composites, and kenaf/polyhydroxyalkanoate (PHA) composites are well-suited for this purpose with some modification with plasticizers or co-polymers or additives. Moreover, basalt/bio-polyester composites are suitable for load-bearing UAV components (e.g., rotor arms or landing gear). These composites can be combined with flax or kenaf for improved toughness. Natural fibers sandwich composites with high stiffness and strength with very low weight are ideal for wings, rudders, and UAV fuselages They are also useful for UAVs with sensitive payloads (e.g., sensors or cameras) [112]. By carefully tuning the fiber surface chemistry, matrix formulation, and fiber–matrix interface compatibility, these green composites can be optimized for the mechanical and environmental challenges posed by UAV deployment.

9.7.4 Vibration Damping and Noise Reduction

Composites suitable for vibration damping and noise reduction are typically made from natural fibers in bio-based polymer matrices where both constituents contribute synergistically to the material's damping behavior. The key to their effectiveness lies in their viscoelastic properties, microstructure and chemical composition, which dissipate mechanical energy as heat and attenuate sound transmission. For vibration damping, composites are engineered to absorb and dissipate energy through the inherent viscoelastic properties of their polymer matrices, frictional interfaces between layers, or through specialized designs like constrained layer damping (CLD), where viscoelastic layers are sandwiched between stiff plies. This helps reduce fatigue and improve performance [113]. For noise reduction, composites are integrated to absorb or block sound. This includes lightweight, sound-absorbing melamine foams for cabin interiors, honeycomb sandwich panels which provide both structural integrity and acoustic insulation, and specialized sound-absorbing composites designed to convert sound energy into heat [114]. Materials like quartz fiber composites also offer excellent thermal and acoustic insulation in high-temperature areas. These composite solutions significantly mitigate vibrations and noise, leading to more durable aircraft and a quieter, more comfortable experience for occupants.

Vibration damping is enhanced in flax fiber/bio-epoxy composites through microslippage at the fiber–matrix interface and through internal friction within the natural fiber structure, where viscoelastic layers are sandwiched between stiff plies. This helps reduce fatigue and improve performance [113]. Kenaf fiber/polyhydroxyalkanoate (PHA) are bacterial polyesters with semi crystalline structure which gives them balanced stiffness and damping properties [115]. Energy from vibrations is dissipated as heat during polymer chain segmental motion (in the matrix) and fiber–matrix interfacial motion. Cork-based sandwich composites have a closed cell structure, low density, and excellent damping and thermal insulation properties. Cork's regular architecture allows it to compress and rebound, absorbing and dissipating vibrational energy over a wide frequency range. They are particularly effective for low-frequency vibration damping, common in aircraft cabin noise. Coir fiber/natural rubber or bio-polyurethane composites behave like a damped spring, absorbing shock and vibrational energy in underfoot panels or vibration-isolating mounts.

9.7.5 *Performance Evaluation Under Real-World Conditions*

Evaluating the performance of green composites under real-world aerospace conditions is essential to determine their reliability and long-term viability. These materials are exposed to a range of harsh environmental and mechanical stresses, including fluctuating temperatures, high humidity, ultraviolet (UV) radiation, mechanical vibrations, and contact with aviation fluids such as hydraulic oils and fuels. Natural fibers tend to absorb moisture due to their hydrophilic nature, which can lead to swelling, loss of interfacial bonding, and degradation of mechanical properties [116]. To address this, chemical treatments such as alkali or silane modification are applied to improve fiber–matrix adhesion and moisture resistance. Thermal aging and UV exposure can cause polymer chain scission in bio-resins, reducing their strength and stability, which is commonly assessed through techniques like thermogravimetric analysis (TGA), differential scanning calorimetry (DSC), and Fourier-transform infrared spectroscopy (FTIR) [117]. Mechanical performance is tested via tensile, flexural, and impact loading under both ambient and extreme conditions. Overall, while green composites demonstrate promising behavior, particularly when protected with coatings or hybridized with synthetic components, continued evaluation and refinement are necessary to ensure consistent performance in demanding aerospace environments.

9.8 Conclusions

Green composites and nanocomposites represent a significant step toward sustainable innovation in the aerospace sector. By integrating natural fibers with bio-based or biodegradable matrices, these materials provide a viable alternative to conventional composites for noncritical and semi-structural components, offering weight reduction, lower environmental impact, and cost advantages. The strength, thermal stability, and functional capabilities of the composites improve enough to meet more demanding requirements when reinforced with nanoscale fillers such as carbon nanotubes, graphene, nanoclays, or metal nanoparticles.

Despite these advances, some challenges remain, such as moisture resistance, thermal durability, and long-term performance consistency. However, advances in fiber treatments, matrix modifications, and hybrid designs that blend small amounts of synthetic reinforcement are steadily addressing these issues. With growing pressure for greener manufacturing, these materials are well-positioned to become a regular feature in aerospace components, offering lighter structures, lower emissions, and innovative possibilities for the next generation of aircraft.

The development of green and nano-enhanced composites represents a pivotal shift toward sustainable innovation in aerospace materials. Even though there are some drawbacks, there are still scopes to overcome the issues by working on them. As global priorities increasingly align with eco-conscious manufacturing, these composites are well-positioned to foster lighter, more efficient, and sustainable aircraft components. Future research should therefore focus on real-world performance validation, life cycle analyses, and scalable fabrication methods, in order to fully realize the transformative potential of these materials in aerospace engineering.

References

1. Campbell FC (2010) Structural composite materials. ASM International
2. KG Satyanarayana GGC Arizaga F Wypych 2009 Biodegradable composites based on lignocellulosic fibers—an overview Prog Polym Sci 34 9 982 1021 https://doi.org/10.1016/J.PROGPOLYMSCI.2008.12.002
3. C Materials 2007 Science and engineering Springer New York
4. Ince JC et al (2023) Overview of emerging hybrid and composite materials for space applications. Adv Compos Hybrid Mater 6(4):1–37. https://doi.org/10.1007/S42114-023-00678-5
5. S Waghmare S Shelare K Aglawe P Khope 2022 A mini review on fibre reinforced polymer composites Mater Today Proc 54 682 689 https://doi.org/10.1016/J.MATPR.2021.10.379
6. Pokharel A, Falua KJ, Babaei-Ghazvini A, Acharya B (2022) Biobased polymer composites: a review. J Compos Sci 6(9):255. https://doi.org/10.3390/JCS6090255
7. Kontaxis LC, Kozaniti FK, Papanicolaou GC (2021) Mechanical behavior modelling and filler geometry effect of glass filler reinforced starch-epoxy hybrid matrix composites. Materials 14(21):6587. https://doi.org/10.3390/MA14216587
8. SC Tjong 2006 Structural and mechanical properties of polymer nanocomposites Mater Sci Eng R Rep 53 3–4 73 197 https://doi.org/10.1016/J.MSER.2006.06.001
9. RJ Young IA Kinloch L Gong KS Novoselov 2012 The mechanics of graphene nanocomposites: a review Compos Sci Technol 72 12 1459 1476 https://doi.org/10.1016/J.COMPSCITECH.2012.05.005
10. M Puttegowda 2025 Eco-friendly composites: exploring the potential of natural fiber reinforcement Discover Appl Sci 7 5 1 24 https://doi.org/10.1007/S42452-025-06981-8/FIGURES/10
11. Peças P, Carvalho H, Salman H, Leite M (2018) Natural fibre composites and their applications: a review. J Compos Sci 2(4). https://doi.org/10.3390/JCS2040066
12. Kurien RA, Maria Anil M, Sharan Mohan SL, Anna Thomas J (2023) Natural fiber composites as sustainable resources for emerging applications—a review. Mater Today Proc. https://doi.org/10.1016/J.MATPR.2023.04.363
13. Dintcheva NT (2024) Overview of polymers and biopolymers degradation and stabilization towards sustainability and materials circularity. Polymer (Guildf) 306:127136. https://doi.org/10.1016/J.POLYMER.2024.127136
14. Nanda S, Patra BR, Patel R, Bakos J, Dalai AK (2021) Innovations in applications and prospects of bioplastics and biopolymers: a review. Environ Chem Lett 20(1):379–395. https://doi.org/10.1007/S10311-021-01334-4
15. da Silva GLP, Morais LCA, Olivato JB, Marini J, Ferrari PC (2021) Antimicrobial dressing of silver sulfadiazine-loaded halloysite/cassava starch-based (bio)nanocomposites. J Biomater Appl 35(9):1096–1108. https://doi.org/10.1177/0885328221995920
16. N Srikhao 2021 Bioactive nanocomposite film based on cassava starch/polyvinyl alcohol containing green synthesized silver nanoparticles J Polym Environ 29 2 672 684 https://doi.org/10.1007/S10924-020-01909-2/METRICS
17. MT Zafar SN Maiti AK Ghosh 2016 Effect of surface treatments of jute fibers on the microstructural and mechanical responses of poly(lactic acid)/jute fiber biocomposites RSC Adv 6 77 73373 73382 https://doi.org/10.1039/C6RA17894D

18. Z Durmus R Köferstein A Özgen T Lindenberg AW Maijenburg A Durmus 2025 Novel and highly efficient antibacterial PLA composites prepared with liquidambar orientalis oil and Ag@g-C3N4 nanocomposite J Polym Environ 33 5 2249 2266 https://doi.org/10.1007/S10924-025-03520-9/FIGURES/11
19. Dahman Y (2017) Smart nanomaterials**by Yaser Dahman, Adil Kamil, and Daniel Baena. In: Nanotechnology and functional materials for engineers, pp 47–66. https://doi.org/10.1016/B978-0-323-51256-5.00003-4
20. Wang Y, Wang X, Xie Y, Zhang K (2018) Functional nanomaterials through esterification of cellulose: a review of chemistry and application. Cellulose (7):3703–3731. https://doi.org/10.1007/S10570-018-1830-3
21. MS Mondal SZ Hussain P Roy C Halder 2023 Development of high-performance composite via innovative route using water hyacinth extracted nanocellulose and analysis of its physical properties Heliyon 9 12 e23095https://doi.org/10.1016/J.HELIYON.2023.E23095
22. K Markandan CQ Lai 2023 Fabrication, properties and applications of polymer composites additively manufactured with filler alignment control: A review Compos B Eng 256 110661https://doi.org/10.1016/J.COMPOSITESB.2023.110661
23. Sztorch B, Brząkalski D, Pakuła D, Frydrych M, Špitalský Z, Przekop RE (2022) Natural and synthetic polymer fillers for applications in 3D printing—FDM technology area. Solids 3(3):508–548. https://doi.org/10.3390/SOLIDS3030034
24. Li H, Wang G, Wu Y, Jiang N, Niu K (2024) Functionalization of carbon nanotubes in polystyrene and properties of their composites: a review. Polymers 16(6):770. https://doi.org/10.3390/POLYM16060770
25. Nurazzi NM et al (2021) Fabrication, functionalization, and application of carbon nanotube-reinforced polymer composite: an overview. Polymers 13(7):1047. https://doi.org/10.3390/POLYM13071047
26. Gisbert Roca F, Martínez-Ramos C, Ivashchenko S, García-Bernabé A, Compañ V, Monleón Pradas M (2023) Polylactic acid nanofiber membranes grafted with carbon nanotubes with enhanced mechanical and electrical properties. ACS Appl Polym Mater 5(8):6081–6094. https://doi.org/10.1021/ACSAPM.3C00776/ASSET/IMAGES/LARGE/AP3C00776_0008.JPEG
27. CS Wu HT Liao 2007 Study on the preparation and characterization of biodegradable polylactide/multi-walled carbon nanotubes nanocomposites Polymer (Guildf) 48 15 4449 4458 https://doi.org/10.1016/J.POLYMER.2007.06.004
28. S Poonsawat S Phattanarudee 2014 Physical properties of maleated poly(lactic acid) composites containing different functionalized multiwalled carbon nanotubes J Nanosci Nanotechnol 14 4 3239 3246 https://doi.org/10.1166/JNN.2014.8614
29. Graphene—a wonder material for scientists and engineers (2022), https://doi.org/10.5772/INTECHOPEN.100684
30. Alves Z, Ferreira NM, Ferreira P, Nunes C (2022) Design of heat sealable starch-chitosan bioplastics reinforced with reduced graphene oxide for active food packaging. Carbohydr Polym 291. https://doi.org/10.1016/j.carbpol.2022.119517
31. v6 i8 Making heat sealable starch-chitosan bioplastics reinforced with reduced graphene oxide for active food packaging. Nixene Publishing. Accessed 22 Aug 2025 [Online]. Available: https://www.nixenepublishing.com/product/v6-i8-making-heat-sealable-starch-chitosan-bioplastics-reinforced-with-reduced-graphene-oxide-for-active-food-packaging/
32. M Nasrollahzadeh SM Sajadi M Sajjadi Z Issaabadi 2019 An introduction to nanotechnology Interface Sci Technol 28 1 27 https://doi.org/10.1016/B978-0-12-813586-0.00001-8
33. T Metin C Parlak Ö Alver M Tepe 2022 Hydrogen storage and electronic properties of C20, C15M5 and H2@C15M5 (M = Al, Si, Ga, Ge) nanoclusters J Mol Struct 1247 131272https://doi.org/10.1016/J.MOLSTRUC.2021.131272
34. T Druffel 2008 The role of nanoparticles in visible transparent nanocomposites Nanophotonic Mater V 7030 70300F https://doi.org/10.1117/12.795273
35. JR Beryl JR Xavier 2023 Halloysite for clay–polymer nanocomposites: effects of nanofillers on the anti-corrosion, mechanical, microstructure, and flame-retardant properties—a review J Mater Sci 58 27 10943 10974 https://doi.org/10.1007/S10853-023-08710-1

36. M Kotal AK Bhowmick 2015 Polymer nanocomposites from modified clays: recent advances and challenges Prog Polym Sci 51 127 187 https://doi.org/10.1016/J.PROGPOLYMSCI.2015.10.001
37. BKG Theng 2012 Polymer–clay nanocomposites Dev Clay Sci 4 201 241 https://doi.org/10.1016/B978-0-444-53354-8.00007-4
38. Patel N, Munjal S (2021) Metal oxide nanoparticles and their applications: a report, Dec 2021. Accessed 20 Aug 2025 [Online]. Available: https://arxiv.org/pdf/2201.04938
39. N Abid 2022 Synthesis of nanomaterials using various top-down and bottom-up approaches, influencing factors, advantages, and disadvantages: a review Adv Colloid Interface Sci 300 102597https://doi.org/10.1016/J.CIS.2021.102597
40. Vijayaram S et al (2023) Applications of green synthesized metal nanoparticles—a review. Biological Trace Element Res 202(1):360–386. https://doi.org/10.1007/S12011-023-03645-9
41. M Ghaffari-Moghaddam R Hadi-Dabanlou M Khajeh M Rakhshanipour K Shameli 2014 Green synthesis of silver nanoparticles using plant extracts Korean J Chem Eng 31 4 548 557 https://doi.org/10.1007/S11814-014-0014-6/METRICS
42. Article R et al (2021) Overview on methods of synthesis of nanoparticles. https://doi.org/10.22159/ijcpr.2021v13i2.41556
43. Y Ullal AC Hegde 2014 Electrodeposition and electro-catalytic study of nanocrystalline Ni–Fe alloy Int J Hydrogen Energy 39 20 10485 10492 https://doi.org/10.1016/J.IJHYDENE.2014.05.016
44. CN He NQ Zhao CS Shi SZ Song 2009 Fabrication of carbon nanomaterials by chemical vapor deposition J Alloys Compd 484 1–2 6 11 https://doi.org/10.1016/J.JALLCOM.2009.04.088
45. Jung DS, Bin Park S, Kang YC (2010) Design of particles by spray pyrolysis and recent progress in its application. Korean J Chem Eng 27(6):1621–1645. https://doi.org/10.1007/S11814-010-0402-5
46. LD Hafshejani 2016 Synthesis and characterization of Al_2O_3 nanoparticles by flame spray pyrolysis (FSP)—role of Fe ions in the precursor Powder Technol 298 42 49 https://doi.org/10.1016/J.POWTEC.2016.05.003
47. L Mädler HK Kammler R Mueller SE Pratsinis 2002 Controlled synthesis of nanostructured particles by flame spray pyrolysis J Aerosol Sci 33 2 369 389 https://doi.org/10.1016/S0021-8502(01)00159-8
48. Al-Kattan A, Tselikov G, Metwally K, Popov AA, Mensah S, Kabashin AV (2021) Laser ablation-assisted synthesis of plasmonic Si@Au core-satellite nanocomposites for biomedical applications. Nanomaterials 11(3):592. https://doi.org/10.3390/NANO11030592
49. W Jiang K Yatsui 1998 Pulsed wire discharge for nanosize powder synthesis IEEE Trans Plasma Sci 26 5 1498 1501 https://doi.org/10.1109/27.736045
50. Gao X, Yokota N, Oda H, Tanaka S, Hokamoto K, Chen P (2018) One step preparation of Fe–FeO–graphene nanocomposite through pulsed wire discharge. Crystals 8(2):104. https://doi.org/10.3390/CRYST8020104
51. HJ Kang MW Lee JH Kim CN Whang 1997 Molecular dynamics simulation for ionized cluster beam deposition Nucl Instrum Methods Phys Res B 121 1–4 53 57 https://doi.org/10.1016/S0168-583X(96)00559-9
52. T Takagi 1986 Ionized cluster beam technique for thin film deposition Zeitschrift für Physik D Atoms Molecules Clusters 3 2 271 278 https://doi.org/10.1007/BF01384816/METRICS
53. M Parashar VK Shukla R Singh 2020 Metal oxides nanoparticles via sol–gel method: a review on synthesis, characterization and applications J Mater Sci: Mater Electron 31 5 3729 3749 https://doi.org/10.1007/S10854-020-02994-8/FIGURES/16
54. JA Darr J Zhang NM Makwana X Weng 2017 Continuous hydrothermal synthesis of inorganic nanoparticles: applications and future directions Chem Rev 117 17 11125 11238 https://doi.org/10.1021/ACS.CHEMREV.6B00417/ASSET/IMAGES/MEDIUM/CR-2016-00417Z_0056.GIF

55. MAS Amulya HP Nagaswarupa MRA Kumar CR Ravikumar SC Prashantha KB Kusuma 2020 Sonochemical synthesis of $NiFe_2O_4$ nanoparticles: characterization and their photocatalytic and electrochemical applications Appl Surface Sci Adv 1 100023https://doi.org/10.1016/J.APSADV.2020.100023
56. K Qi C Zhuang M Zhang P Gholami A Khataee 2022 Sonochemical synthesis of photocatalysts and their applications J Mater Sci Technol 123 243 256 https://doi.org/10.1016/J.JMST.2022.02.019
57. H Al-Madhagi V Yazbik W Abdelwahed L Alchab 2023 Magnetite nanoparticle co-precipitation synthesis, characterization, and applications: mini review Bionanoscience 13 2 853 859 https://doi.org/10.1007/S12668-023-01113-1/FIGURES/4
58. LG Silva 2014 Synthesis of Fe nanoparticles functionalized with oleic acid synthesized by inert gas condensation J Nanomater 2014 1 643967https://doi.org/10.1155/2014/643967
59. FE Kruis H Fissan A Peled 1998 Synthesis of nanoparticles in the gas phase for electronic, optical and magnetic applications—a review J Aerosol Sci 29 5–6 511 535 https://doi.org/10.1016/S0021-8502(97)10032-5
60. X Li H Xu ZS Chen G Chen 2011 Biosynthesis of nanoparticles by microorganisms and their applications J Nanomater 2011 1 270974https://doi.org/10.1155/2011/270974
61. P Singh YJ Kim D Zhang DC Yang 2016 Biological synthesis of nanoparticles from plants and microorganisms Trends Biotechnol 34 7 588 599 https://doi.org/10.1016/J.TIBTECH.2016.02.006/ATTACHMENT/D7673F65-4850-4AA4-ACC4-008428450572/MMC1.PPTX
62. F Khan 2022 Prospects of algae-based green synthesis of nanoparticles for environmental applications Chemosphere 293 133571https://doi.org/10.1016/J.CHEMOSPHERE.2022.133571
63. P Khanna A Kaur D Goyal 2019 Algae-based metallic nanoparticles: synthesis, characterization and applications J Microbiol Methods 163 105656https://doi.org/10.1016/J.MIMET.2019.105656
64. Rastogi AK (2025) A mini review practice of formulations of nanoparticles. Int J Chem Synthesis Chem Reactions 3(2):1–7. Accessed 21 Aug 2025 [Online]. Available: https://chemical.journalspub.info/index.php?journal=JCSCR&page=article&op=view&path%5B%5D=432
65. MJ Hampden-Smith TT Kodas 1995 Chemical vapor deposition of metals: part 1. An overview of CVD processes Chem Vap Deposition 1 1 8 23 https://doi.org/10.1002/CVDE.19950010103;PAGEGROUP:STRING:PUBLICATION
66. R Mueller L Mädler SE Pratsinis 2003 Nanoparticle synthesis at high production rates by flame spray pyrolysis Chem Eng Sci 58 10 1969 1976 https://doi.org/10.1016/S0009-2509(03)00022-8
67. HK Schmidt E Geiter M Mennig H Krug C Becker RP Winkler 1998 The sol-gel process for nano-technologies: new nanocomposites with interesting optical and mechanical properties J Solgel Sci Technol 13 1–3 397 404 https://doi.org/10.1023/A:1008660909108/METRICS
68. A Bhat 2021 Review on nanocomposites based on aerospace applications Nanotechnol Rev 10 1 237 253 https://doi.org/10.1515/NTREV-2021-0018/ASSET/GRAPHIC/J_NTREV-2021-0018_FIG_007.JPG
69. A Iqbal A Saeed A Ul-Hamid 2021 A review featuring the fundamentals and advancements of polymer/CNT nanocomposite application in aerospace industry Polym Bull 78 1 539 557 https://doi.org/10.1007/S00289-019-03096-0/FIGURES/2
70. D Balaji 2025 A review on effect of nanoparticle addition on thermal behavior of natural fiber–reinforced composites Heliyon 11 1 e41192https://doi.org/10.1016/J.HELIYON.2024.E41192/ASSET/71AEE6B6-9897-41E2-8A17-F63781298C6F/MAIN.ASSETS/GR1.JPG
71. Barra G, Guadagno L, Raimondo M, Santonicola MG, Toto E, Vecchio Ciprioti S (2023) A comprehensive review on the thermal stability assessment of polymers and composites for aeronautics and space applications. Polymers 15(18):3786. https://doi.org/10.3390/POLYM15183786
72. Nobrega G et al (2024) A review of novel heat transfer materials and fluids for aerospace applications. Aerospace 11(4):275. https://doi.org/10.3390/AEROSPACE11040275

73. Mallik M, Saha M (2021) Carbon-based nanocomposites: processing, electronic properties and applications, pp 97–122. https://doi.org/10.1007/978-981-16-1052-3_5
74. MK Singh SM Rangappa M Misra AK Mohanty S Siengchin 2025 Recent advancements in nanostructured flame-retardants: types, mechanisms, and applications in polymer composites Nano-Structures Nano-Objects 42 101468https://doi.org/10.1016/J.NANOSO.2025.101468
75. Kausar A, Ahmad I, Maaza M, Bocchetta P (2023) Self-healing nanocomposites—advancements and aerospace applications. J Compos Sci 7(4):148. https://doi.org/10.3390/JCS7040148
76. T Cionita 2024 Overview of jute fibre as thermoplastic matrix polymer reinforcement J Renew Mater 12 3 458 https://doi.org/10.32604/JRM.2024.045814
77. Joseph G (2022) Natural fiber polymer composites. In: Sustainable plastics: environmental assessments of biobased, biodegradable, and recycled plastics, ch 13, pp 357–376
78. Subash T, Pillai SN (2015) Bast fibers reinforced green composites for aircraft indoor structures applications: review. J Chem Pharmaceutical Sci 7:305–307
79. Abdullayev S et al (2025) Improving the aerodynamic characteristics of UAVS through the use of composite materials. Int J Innov Res Sci Stud 8:721–732. https://doi.org/10.53894/ijirss.v8i5.8815
80. Chen K, Meng W, Wang J, Liu K, Lu Z (2023) An investigation on the structural vibrations of multi-rotor passenger drones. Int J Micro Air Vehicles 15. https://doi.org/10.1177/17568293231199097
81. A Oushabi 2019 The pull-out behavior of chemically treated lignocellulosic fibers/polymeric matrix interface (LF/PM): a review Compos B Eng 174 107059https://doi.org/10.1016/j.compositesb.2019.107059
82. AA Musa AP Onwualu 2024 Potential of lignocellulosic fiber reinforced polymer composites for automobile parts production: current knowledge, research needs, and future direction Heliyon 10 3 e24683https://doi.org/10.1016/j.heliyon.2024.e24683
83. Sarkar S (2008) Carbon fiber reinforced plastics in aircraft structures
84. Tang H (2025) Research progress of carbon fiber composite materials in automobile body lightweight. In: MATEC web of conferences, vol 410. https://doi.org/10.1051/matecconf/202541003011
85. D He F Jiao R Zhang 2025 Investigation on the reliability of a thermal protection system based on mullite fiber insulation tiles coated with mosi2-borosilicate glass in an aerothermal-vibration coupling environment Appl Compos Mater 32 3 955 970 https://doi.org/10.1007/s10443-025-10307-7
86. Madhu P et al (2025) Mechanical and structural optimization of flax fiber reinforced composites through controlled gamma irradiation. iScience 28(7):112531. https://doi.org/10.1016/j.isci.2025.112531
87. M Thulo W Webo MC Khoathane T Malwela 2025 Epoxy reinforced flax fibre composites for interior lining panels of an aircraft: flammability study J Eng 2025 1 e70092https://doi.org/10.1049/tje2.70092
88. Djafari Petroudy SR (2017) 3 - Physical and mechanical properties of natural fibers. In: Fan M, Fu F (eds) Advanced high strength natural fibre composites in construction. Woodhead Publishing, pp 59–83. https://doi.org/10.1016/B978-0-08-100411-1.00003-0
89. Kumar S, Samal S, Mohanty S, Nayak S (2016) Recent development of bio-based epoxy resins: a review. Polym Plast Technol Eng 57. https://doi.org/10.1080/03602559.2016.1253742
90. Padleckienė I, Stygienė L, Krauledas S, Abraitienė A, Sankauskaitė A (2025) Thermal comfort properties of biodegradable hemp and polylactide fiber knitted fabrics. Polymers (Basel) 17(7). https://doi.org/10.3390/polym17070903
91. Kovuru R, Schuster J (2025) Enhancing mechanical properties of hemp and sisal fiber-reinforced composites through alkali and fungal treatments for sustainable applications. J Manuf Mater Process 9(6). https://doi.org/10.3390/jmmp9060191
92. Henton D, Gruber P, Lunt J, Randall J (2005) Polylactic acid technology. In: Natural fibers, biopolymers, and biocomposites. https://doi.org/10.1201/9780203508206.ch16

93. Bikiaris ND et al (2023) Recent advances in the investigation of poly(lactic acid) (PLA) nanocomposites: incorporation of various nanofillers and their properties and applications. Polymers (Basel) 15(5). https://doi.org/10.3390/polym15051196
94. H Deka M Misra A Mohanty 2013 Renewable resource based 'all green composites' from kenaf biofiber and poly(furfuryl alcohol) bioresin Ind Crops Prod 41 94 101 https://doi.org/10.1016/j.indcrop.2012.03.037
95. NH Marzuki MU Wahit A Arsad N Othman NISM Yusoff 2021 The effect of kenaf loading on the mechanical properties of kenaf-reinforced recycled poly(ethylene terephthalate)/recycled poly(propylene) (rPET/rPP) composite Mater Today Proc 39 959 964 https://doi.org/10.1016/j.matpr.2020.04.333
96. Chai G, Zhu G, Gao S, Zhou J, Gao Y, Wang Y (2019) On improving flame retardant and smoke suppression efficiency of epoxy resin doped with aluminum tri-hydroxide. Adv Compos Lett 28:2633366X19894597. https://doi.org/10.1177/2633366X19894597
97. Wang H, Yao J (2006) Use of poly(furfuryl alcohol) in the fabrication of nanostructured carbons and nanocomposites. In: Industrial & Engineering Chemistry Research - IND ENG CHEM RES, vol 45. https://doi.org/10.1021/ie0602660
98. T Islam 2024 Advancements and challenges in natural fiber-reinforced hybrid composites: a comprehensive review SPE Polym 5 4 481 506 https://doi.org/10.1002/pls2.10145
99. Kumar C, Prabhakar MN, Song J (2018) Effect of interface in hybrid reinforcement of flax/glass on mechanical properties of vinyl ester composites. Polym Test 73. https://doi.org/10.1016/j.polymertesting.2018.12.005
100. Wang H et al (2025) Carbon fiber reinforced thermoplastics: from materials to manufacturing and applications. Adv Mater 37(27):2418709. https://doi.org/10.1002/adma.202418709
101. L Zheng X Yue L Yue Z Wang Y Xie Z Cui 2020 Electrical/thermal stability of GFRP under different intensities γ ray irradiation Fusion Eng Des 158 111856https://doi.org/10.1016/j.fusengdes.2020.111856
102. B Mishra S Varshney MK Gupta 2025 Effect of fillers on the performance of fibre reinforced polymer hybrid composites: a comprehensive review Polymer-Plastics Technol Mater 64 14 2179 2213 https://doi.org/10.1080/25740881.2025.2519024
103. Das S, La Rosa A, Goutianos S, Grammatikos S (2022) Flax fibers, their composites and application, pp 209–232. https://doi.org/10.1016/B978-0-12-824528-6.00017-5
104. J Zhu H Zhu K Immonen J Brighton H Abhyankar 2015 Improving mechanical properties of novel flax/tannin composites through different chemical treatments Ind Crops Prod 67 346 354 https://doi.org/10.1016/j.indcrop.2015.01.052
105. Kabir MM, Al-Haik M, Aldajah S, tak Lau K, Wang H (2020) Impact properties of the chemically treated hemp fibre reinforced polyester composites. Fibers Polym 21:2098–2110. https://doi.org/10.1007/s12221-020-9630-4
106. Moorthy SN, Chandran S (2025) A comprehensive review on the influence of surface treatment and 3D printing of natural fiber composites. Compos Interfaces 1–36. https://doi.org/10.1080/09276440.2025.2525590
107. El Bahaoui J, Hanafi I, Chairi M, Favaloro F, Borsellino C, Di Bella G (2025) Effect of manufacturing processes on basalt fiber-reinforced composites for marine applications. J Compos Sci 9(5). https://doi.org/10.3390/jcs9050233
108. A Yu J Sahouani D Webster 2018 Highly functional methacrylated bio-based resins for UV-curable coatings Prog Org Coat 122 219 228 https://doi.org/10.1016/j.porgcoat.2018.05.035
109. Abu Hassan NA, Ahmad S, Chen RS, Shahdan D, Mohamad Kassim MH (2023) Tailoring lightweight, mechanical and thermal performance of PLA/recycled HDPE biocomposite foams reinforced with kenaf fibre. Ind Crops Prod 197:116632. https://doi.org/10.1016/j.indcrop.2023.116632
110. Kojio K, Nozaki S, Takahara A, Yamasaki S (2020) Influence of chemical structure of hard segments on physical properties of polyurethane elastomers: a review. J Polym Res 27. https://doi.org/10.1007/s10965-020-02090-9
111. ZA Raza S Abid IM Banat 2018 Polyhydroxyalkanoates: characteristics, production, recent developments and applications Int Biodeterior Biodegradation 126 45 56 https://doi.org/10.1016/j.ibiod.2017.10.001

112. R Caran A Yüksel N Ercan DE Yunus A Bedeloğlu 2024 The flexural and compressive properties of sandwich composites with different 3D-printed core structures J Innov Eng Natural Sci 4 1 98 112 https://doi.org/10.61112/jiens.1355323
113. S Tian Z Xu Q Wu C Qin 2016 Dimensionless analysis of segmented constrained layer damping treatments with modal strain energy method Shock Vib 2016 1 8969062 https://doi.org/10.1155/2016/8969062
114. X-L Gai 2022 Acoustic properties of honeycomb like sandwich acoustic metamaterials Appl Acoust 199 109016https://doi.org/10.1016/j.apacoust.2022.109016
115. Thakur A, Musioł M, Duale K, Kowalczuk M (2024) Exploring the future of polyhydroxyalkanoate composites with organic fillers: a review of challenges and opportunities. Polymers (Basel) 16(13). https://doi.org/10.3390/polym16131768
116. TH Mokhothu MJ John 2015 Review on hygroscopic aging of cellulose fibres and their biocomposites Carbohydr Polym 131 337 354 https://doi.org/10.1016/j.carbpol.2015.06.027
117. Cuadri AA, Martín-Alfonso JE (2017) The effect of thermal and thermo-oxidative degradation conditions on rheological, chemical and thermal properties of HDPE. Polym Degrad Stab 141. https://doi.org/10.1016/j.polymdegradstab.2017.05.005

Chapter 10
Multifunctional Hybrid Sol–Gel Coatings in Aeronautical and Aerospace Applications

Praveen Kumar Verma and Hitesh Vasudev

Abstract The rapid evolution of aeronautical and aerospace technologies demands materials and coatings that deliver exceptional multifunctional performance under extreme environmental conditions. Among these, multifunctional hybrid sol–gel coatings have garnered significant attention owing to their superior protective properties and versatility. These coatings, derived from the sol–gel process combining organic and inorganic components at a molecular level, offer an innovative approach to enhancing corrosion resistance, wear resistance, thermal stability, and other crucial surface functionalities. Hybrid sol–gel coatings have emerged as promising candidates for addressing multiple performance requirements in aerospace applications, including corrosion resistance, thermal protection, mechanical durability, and environmental stability. This review summarizes the synthesis strategies, functional modifications, and application-specific roles of hybrid sol–gel coatings in the aeronautical and aerospace industries. Key advances, current challenges, and future outlooks are also discussed.

Keywords Hybrid sol–gel · Organic–inorganic integration · Material design · Precursors and additives · Coating techniques · Aerospace applications · Hydrophobicity · Anti-icing · Radiation shielding airframes · Exhaust systems

10.1 Introduction

The aeronautical and aerospace sectors function in among the most severe and challenging environments, where parts are constantly subjected to varying temperatures, ultraviolet (UV) radiation, corrosive environments, mechanical stress, and erosive particulate matter. For maintaining structural integrity, mission safety, and durability of these high-performance systems, the use of strong surface protection measures

P. K. Verma (✉) · H. Vasudev
School of Mechanical Engineering, Lovely Professional University, Phagwara, Punjab, India
e-mail: pk68449@gmail.com

S. K. Panda (ed.), *Nanocomposites for Defense, Aeronautical and Aerospace Applications*, Advanced Structured Materials 248,
https://doi.org/10.1007/978-3-032-11568-3_10

is essential [1, 2]. Protective coatings act as the initial line of defense against such environmental stressors, adding a service life to metallic substrates and composite materials and keeping them in their optimal operational performance. Traditional coatings such as chromate-based primers, epoxy resins, and thermal spray coatings have been used extensively for corrosion protection, thermal insulation, wear resistance, and surface functionalization on aerospace structures. Although these coatings have shown high levels of effectiveness, they are subject to many limitations that limit their versatility and longevity. Chromate-based coatings, while very effective corrosion inhibitors, are carcinogenic and ecotoxic, leading to strict regulatory control under directives like REACH and RoHS [3, 4]. Likewise, traditional polymeric coatings tend to lack satisfactory thermal stability and UV radiation and space weathering resistance, particularly in high-altitude and extraterrestrial environments. Additionally, conventional coating systems are likely to provide individual functionality, e.g., corrosion resistance or thermal protection without the ability to include more than one performance attribute in a single layer [5]. Figure 10.1 shows the benefits of advanced aerospace coatings, highlighting enhanced protection, safety, and reliability. Central circle connects to three icons representing each benefit with corresponding text labels in green, blue, and purple.

To address these issues, the recent development of multifunctional hybrid sol–gel coatings has drawn considerable interest. Hybrid sol–gel technology offers the beneficial attributes of both inorganic and organic networks, providing coatings that are light in weight, chemically resilient, heat resistant, and adjustable for varied functionalities. These coatings are prepared through low-temperature wet chemical

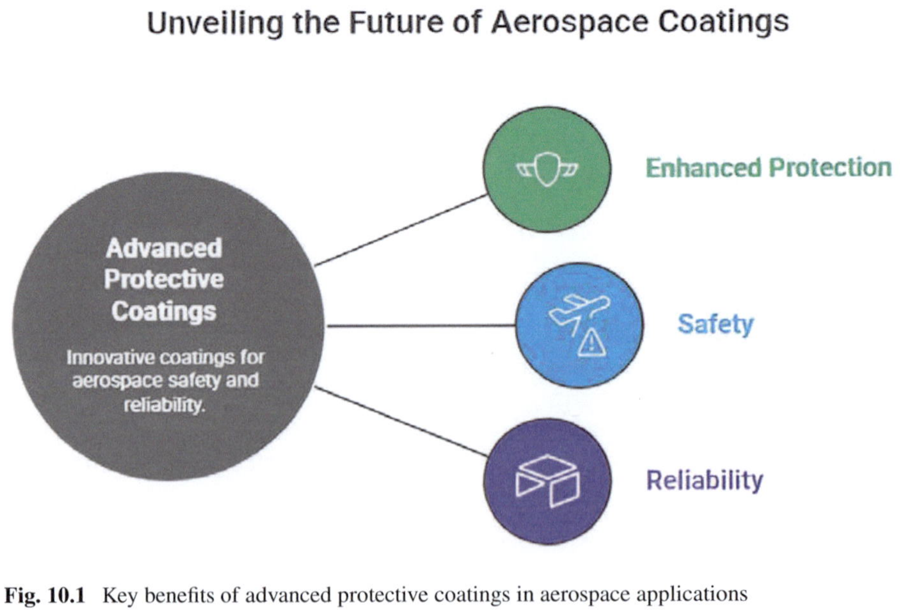

Fig. 10.1 Key benefits of advanced protective coatings in aerospace applications

routes involving metal alkoxides and organosilanes, which react through hydrolysis and condensation reactions to create a stable hybrid matrix. Sol–gel coatings can be tailored to provide a broad variety of protective properties depending on the precursors and modifiers used, such as corrosion inhibition, thermal barrier effects, hydrophobicity, UV shielding, and mechanical reinforcement [6, 7]. One of the most appealing characteristics of sol–gel coatings is that they can be designed at the molecular level, enabling nanoparticles (e.g., ceria, silica, zirconia, graphene) to be included, corrosion inhibitors, self-healing agents, and organic functionalities in a single coating system. This molecular-scale design freedom provides for the production of multifunctional surfaces that react smartly to environmental stimuli, withstand degradation due to cyclic loading, and demonstrate enhanced adhesion to aerospace-grade materials like aluminum alloys, titanium, and carbon fiber composites. Furthermore, sol–gel coatings can be deposited by scalable methods like dip-coating, spray-coating, and spin-coating, rendering them suitable for incorporation into established manufacturing and maintenance processes. With the fast-growing developments in hybrid sol–gel chemistry and increasing needs for sustainable, high-performance coatings in aerospace applications, the present review is focused on presenting an updated overview of recent trends in this area. It centers on the design, synthesis, and functional performance of hybrid sol–gel coatings for aeronautical and space use. Particular focus is given to the multifunctional properties of these coatings, such as their corrosion protection, thermal insulation, mechanical resistance, and UV and space radiation resistance [8–11]. The review also mentions the current trends, practical challenges, and future research and industrial application directions.

The aim of this chapter is to narrow the gap between material innovation and aerospace engineering demand by introducing hybrid sol–gel coatings as an exciting platform for next-generation protection systems. As researchers and engineers understand the triad of chemistry, processing, and functionality, they can collaborate to accelerate the use of sol–gel technology for aerospace performance and sustainability maximization.

10.2 Fundamentals of Hybrid Sol–Gel Coatings

10.2.1 Sol–Gel Process Overview

Sol–gel is a general and common chemical route to produce inorganic and hybrid organic–inorganic materials as powders, thin films, or monoliths. The two building block reactions are the center of this process: hydrolysis and condensation, by which molecular precursors are converted into a crosslinked network creating a gel [12]. The process starts with the choice of appropriate precursors, often alkoxysilanes such as tetraethyl orthosilicate (TEOS), tetramethyl orthosilicate (TMOS), and organically modified silanes such as 3-glycidyloxypropyltrimethoxysilane (GPTMS). For metal oxide systems, metal alkoxides such as titanium isopropoxide, zirconium n-butoxide, and aluminum sec-butoxide are the most often used [13]. These precursors contain reactive alkoxide groups (–OR) that easily undergo hydrolysis in the presence of water, often catalyzed by an acid or a base to start the sol–gel transition. Upon hydrolysis, the metal or silicon alkoxide reacts with water to produce hydroxyl groups (–OH) as the alkoxy groups are replaced. This process results in the formation of alcohol as a byproduct and leaves partially hydrolyzed species. The hydrolyzed species then proceed to undergo polycondensation reactions in which the hydroxyl groups or residual alkoxy groups condense with each other to create metal–oxygen–metal (M–O–M) or silicon–oxygen–silicon (Si–O–Si) linkages, which ultimately form an interconnected network. As the process of condensation goes on, the system evolves from being a sol (a colloidal stable suspension of polymers or particles in a liquid) to a gel (three-dimensional network that extends throughout the liquid volume), with great increase in viscosity and gelation [14–16]. This gel can be dried afterward to eliminate the solvent and aged in order to increase the network structure of the material, then cured at high temperature to promote the mechanical, thermal, and chemical properties of the coating or bulk material [17]. Figure 10.2 depicts the schematic of a multi-layered sol–gel coating system over a metallic substrate, indicating the release and migration of self-healing agents (orange circles) and corrosion inhibitors (green circles) from an OIH sol–gel layer upon weathering or mechanical damage. Arrows represent the migration of water and ions through the topcoat and the migration routes of the protective agents to the damaged zone.

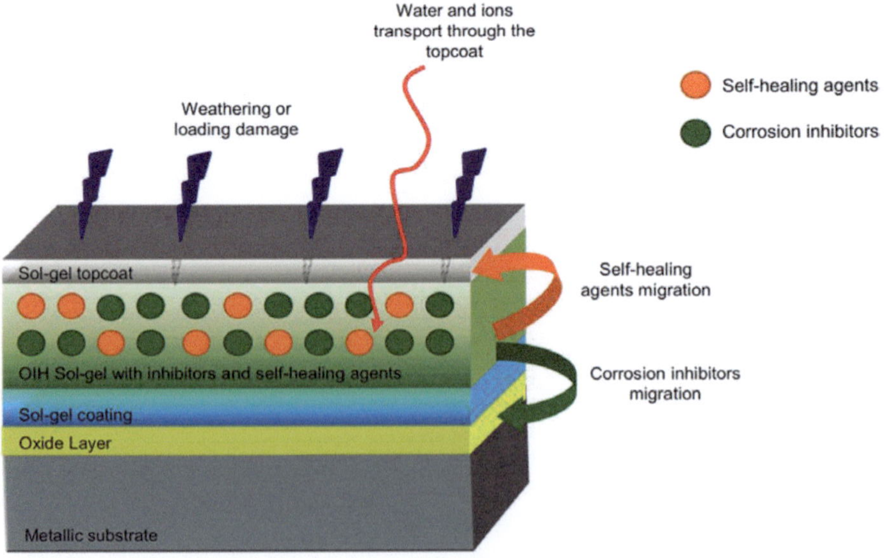

Fig. 10.2 Schematic representation of self-healing and corrosion-inhibiting mechanisms in a multi-layered sol–gel coating system [18]

The sol–gel technique offers several unique advantages over traditional high-temperature ceramic processing methods. First, it allows for molecular-level mixing of precursors, which leads to homogeneous material compositions and fine control over the final microstructure. Second, the process occurs at relatively low temperatures (often below 300 °C), making it compatible with temperature-sensitive substrates such as polymers and composites. Third, the sol–gel process facilitates the incorporation of various dopants, nanoparticles, and functional agents, enabling tailored properties such as corrosion resistance, hydrophobicity, thermal insulation, and self-healing [19, 20]. Additionally, sol–gel formulations can be easily processed using a variety of deposition techniques, including dip-coating, spin-coating, spray-coating, and inkjet printing, making them adaptable to complex aerospace component geometries. The reaction kinetics and final properties of sol–gel coatings depend on several factors, including the type and concentration of precursors, water-to-alkoxide ratio, catalyst type and concentration, solvent type, and processing parameters such as pH, temperature, and time. Acidic conditions tend to favor the formation of linear or weakly branched structures suitable for thin films, whereas basic conditions promote extensive crosslinking and rapid gelation, ideal for bulk monoliths or porous materials. The control of these parameters is essential to achieve coatings with optimal adhesion, transparency, porosity, thickness, and mechanical integrity [21–23].

In aerospace applications, sol–gel-derived coatings are valued for their ability to act as multifunctional barrier layers on light metals such as aluminum, magnesium, and titanium alloys, which are prone to corrosion and oxidation under aggressive environmental conditions. By adjusting the sol–gel chemistry, it is possible to achieve dense, crack-free, and adherent coatings with strong interfacial bonding to metallic substrates. Moreover, the low-temperature processing characteristic of sol–gel coatings makes them suitable for integration with sensitive structural components and composite materials used in modern aircraft and space vehicles. The sol–gel process represents a powerful and flexible route for developing high-performance coatings with controlled composition, structure, and functionality. Its compatibility with a wide range of materials and ease of processing make it an attractive technique for next-generation protective coatings in aeronautical and aerospace systems [24–27].

10.2.2 Hybrid Systems: Organic–Inorganic Integration

Hybrid sol–gel systems represent an advanced class of materials where organic and inorganic components are integrated at the molecular or nanoscale level to exploit the advantageous properties of both domains [28]. These systems are typically categorized into two main classes based on the nature of the interaction between the organic and inorganic phases: Class I and Class II hybrids. In Class I hybrids, the organic and inorganic components are held together through weak interactions such as van der Waals forces, hydrogen bonding, or physical entanglement. These systems are relatively easy to synthesize and often provide flexibility and processability, but they may suffer from phase separation or limited long-term stability under demanding service conditions. In contrast, Class II hybrids exhibit strong covalent or ionic-covalent bonding between the organic and inorganic phases [29]. This results in more robust networks with superior thermal stability, mechanical strength, and chemical resistance, making them particularly attractive for aerospace applications where durability is critical. The successful design of hybrid sol–gel coatings relies on the careful selection and functionalization of precursor molecules to achieve compatibility and synergistic performance. Organically modified silanes (ORMOSILs), such as GPTMS, MTES (methyltriethoxysilane), and APTES (aminopropyltriethoxysilane), are widely used to introduce organic functionalities into the siloxane network. These molecules possess both hydrolyzable alkoxy groups, which participate in the sol–gel network formation, and organic moieties, which impart specific functional properties such as flexibility, hydrophobicity, or chemical reactivity. The incorporation of polymer chains—either linear or branched—into the sol–gel matrix further enhances the toughness and elasticity of the coating, reducing the risk of cracking under mechanical stress or thermal cycling [30–32]. Figure 10.3 shows five elements of hybrid sol–gel systems: core technology, Class I and II hybrids, precursor molecules, polymer chains, and aerospace applications, each with color-coded markers and descriptions.

Nanoparticles and nanostructured additives also play a vital role in hybrid sol–gel systems by enhancing key functional attributes. For instance, metal oxide nanoparticles like ceria (CeO_2), titania (TiO_2), and silica (SiO_2) can be embedded into

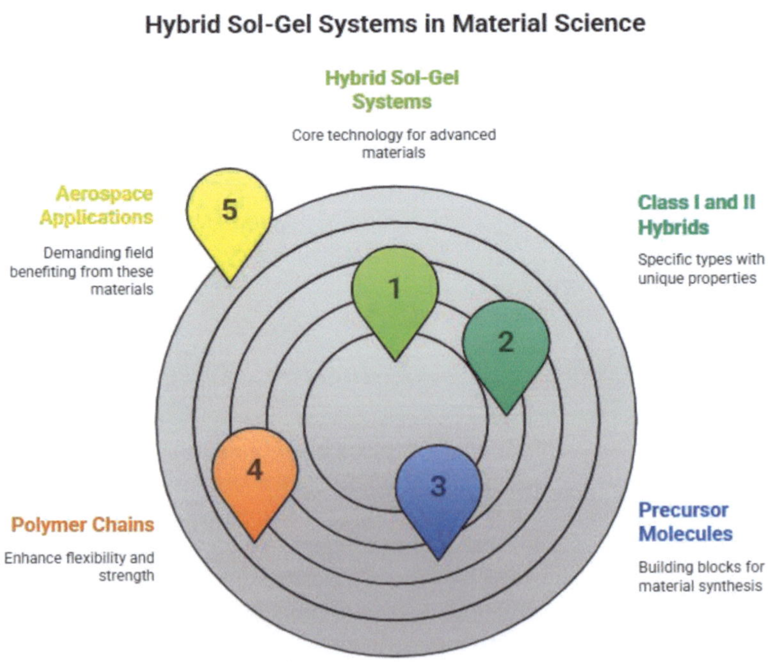

Fig. 10.3 Key components and applications of hybrid sol–gel systems in material science

the hybrid matrix to improve hardness, UV resistance, and corrosion inhibition. Carbon-based nanomaterials, such as graphene oxide and carbon nanotubes, have been explored for improving electrical conductivity, EMI shielding, and thermal dissipation [33]. In many cases, the surface of these nanomaterials is functionalized with silane coupling agents to improve dispersion and interfacial bonding within the sol–gel matrix. This ensures a uniform distribution of nanofillers and maximizes the synergistic effects between the organic–inorganic domains. Another important aspect of hybrid sol–gel design is the use of functional molecules and additives that endow the coating with stimulus-responsive or self-healing behavior [34]. Microencapsulated corrosion inhibitors, pH-sensitive dyes, or redox-active compounds can be embedded within the hybrid matrix to create coatings that respond dynamically to environmental changes. For example, upon exposure to corrosive electrolytes, embedded inhibitors may be released to protect exposed metal sites, thereby extending the service life of aerospace components. Similarly, incorporating fluorinated silanes or long-chain alkylsilanes can impart hydrophobic or superhydrophobic surface characteristics, which are beneficial for anti-icing, dewetting, and drag-reduction in aviation environments. The hybrid sol–gel approach also enables the fine-tuning of mechanical properties by controlling the crosslinking density and phase morphology. A balanced design ensures that the coating remains sufficiently elastic to accommodate substrate deformation while maintaining high hardness and abrasion resistance. This is particularly important in aerospace applications, where coatings are exposed to vibration, temperature fluctuations, and high-speed particulate erosion [35, 36].

The hybrid organic–inorganic sol–gel systems offer a platform for engineering multifunctional coatings with superior structural and functional attributes. By leveraging molecular-level integration of diverse chemistries, it is possible to create smart, durable, and lightweight coatings tailored to meet the complex performance demands of aeronautical and aerospace environments. The continued development of hybrid sol–gel systems promises to unlock new opportunities for next-generation aerospace materials and protective technologies [37].

10.3 Synthesis Strategies and Material Design

10.3.1 Precursors and Additives

The preparation of hybrid sol–gel coatings starts with the judicious choice of precursors and additives, which define the chemical structure, morphology, and functional properties of the resultant coating [38]. The most widely utilized sol–gel precursors are tetraethyl orthosilicate (TEOS), 3-glycidyloxypropyltrimethoxysilane (GPTMS), and methyltriethoxysilane (MTES). TEOS, a pure inorganic silicon alkoxide, is the backbone of silica-based networks because it can completely undergo hydrolysis and condensation to build compact Si–O–Si structures. Its strong crosslinking capability makes it a building block precursor for the formation of mechanically strong and heat-resistant coatings. Although, TEOS-based networks are of inherently brittle and inflexible nature, which limits their use in dynamic loading or thermal cycling applications prevalent in the aerospace environment. To rectify this constraint, organosilanes like GPTMS and MTES are introduced that incorporate organic functionality and enhance mechanical compliance. GPTMS incorporates an epoxide moiety that not only facilitates secondary chemical crosslinking with amine or hydroxyl-functional polymers but also imparts flexibility and adhesion-promoting characteristics. MTES, possessing a methyl group on the silicon center, decreases the hydrophilicity of the silica network, increasing water resistance and aiding in hydrophobic surface character. These organofunctional silanes play a key role in the synthesis of hybrid networks that bridge the rigidity of inorganic silica with the elasticity and tunability of organic moieties [34, 35, 38–41]. Figure 10.4 compares TEOS, GPTMS, and MTES based on chemical structure, mechanical properties, and use case in sol–gel processes for coatings, adhesion promotion, and hydrophobic surface creation.

Apart from silicon-based precursors, metal alkoxides like zirconium n-propoxide, titanium isopropoxide, and aluminum sec-butoxide are usually added to introduce specific functional properties. Zirconia precursors are prized for improving corrosion resistance and thermal stability owing to their high dielectric constant and chemical inertness. Titania precursors are extensively employed for photocatalytic activity and UV-blocking properties, which are well-suited to aerospace surfaces subjected to extreme ultraviolet exposure. Alumina precursors add hardness, wear resistance, and electrical insulation and hence are well-suited to avionic or sensor protection. To achieve even greater multifunctionality in sol–gel coatings, a vast array of nanomaterials is incorporated into the sol–gel matrix. Graphene and graphene derivatives,

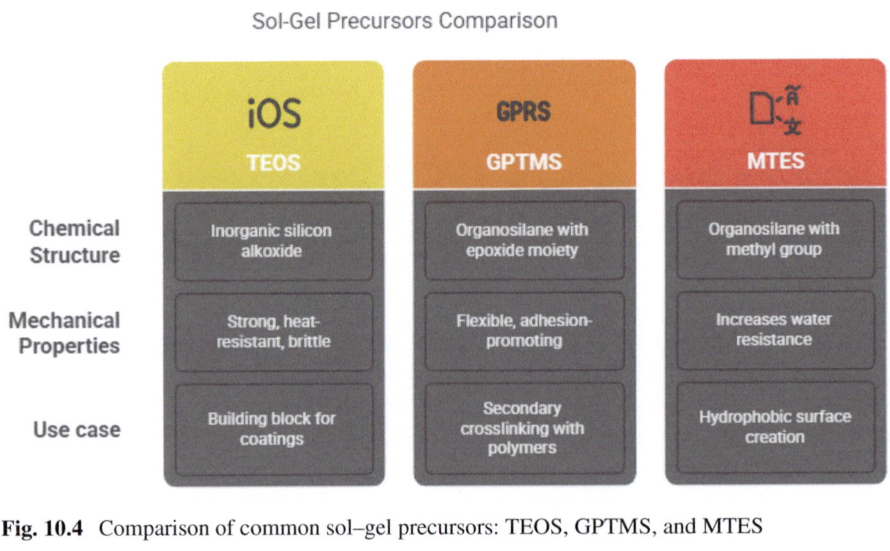

Fig. 10.4 Comparison of common sol–gel precursors: TEOS, GPTMS, and MTES

including graphene oxide (GO) and reduced graphene oxide (rGO), are incorporated to enhance electrical conductivity, mechanical properties, and barrier characteristics. The two-dimensional nature of graphene provides a tortuous path for the corrosive species, significantly improving corrosion protection. Carbon nanotubes (CNTs), single-walled, and multi-walled are utilized for their superior mechanical strength, thermal conductivity, and electrical characteristics. Their inclusion enhances the hardness and electrical conductivity of the hybrid coating and plays a role in structural strengthening [34, 35, 39–41].

Metal oxide nanoparticles like ceria (CeO_2), titania (TiO_2), and silica (SiO_2) play versatile functions varying from corrosion inhibition to UV blocking and mechanical hardening. Ceria, especially, is an oxygen buffer, which facilitates redox reactions capable of inhibiting oxidative degradation and self-healing of coatings. These nanoparticles may be functionalized using silane coupling agents to ensure better dispersion in the sol–gel matrix and increase interfacial adhesion, averting agglomeration and phase separation. In general, the judicious blending of sol–gel precursors with functional additives allows for the development of sophisticated coatings for high-performance applications in aeronautics and aerospace industries [38, 39].

10.3.2 Coating Techniques

The deposition process used is crucial in determining the uniformity, thickness, adhesion, and functional properties of films deposited through sol–gel coatings. Aerospace applications have large surface areas, complex geometries, and high-performance demands being common approaches to coating requiring versatility, scalability, and precision. Four of the most widely used deposition techniques for sol–gel coatings are dip-coating, spin-coating, spray-coating, and electrophoretic deposition, each with its relative advantages and limitations [34, 40]. Dip-coating is one of the most frequent and simplest sol–gel film deposition techniques, best suited for coating flat as well as curved substrates. In dip-coating, the substrate is initially dipped into a solution of sol and then withdrawn at a controlled rate, allowing an uniform liquid film to be formed on its surface. Thickness of the liquid film is predominantly regulated by withdrawal rate, which includes viscosity of the sol and the surface tension. After deposition, the coating is subjected to gelation, drying, and curing. Dip-coating gives good thickness control over the coating and creates very uniform films with good adhesion. It is also well-suited for coating internal and external surfaces of tubular components and is thus well-suited for aerospace fuel lines, engine components, and landing gear components [35, 41].

Spin-coating is particularly ideally suited for producing thin and uniform coatings on planar substrates and is widely employed in laboratory-scale production and sensor fabrication. In spin-coating, a small amount of sol is placed on the center of the substrate and rapidly spun. Centrifugal force drives the sol radially out onto the surface to form a thin film. The thickness of the final coating is a function of the spin rate, viscosity of the sol, and duration. While spin-coating yields coatings with excellent uniformity and reproducibility, it is less suitable for intricate geometries and big parts, limiting its use in large-scale aerospace manufacturing. Still, it remains extremely handy for functional coating testing in small device applications such as aerospace sensors, dielectric films, or EMI-shielding coatings [38, 39].

Spray-coating delivers a more flexible and scalable method that is especially useful in coating large surface areas and complex geometries that are typical in aircraft structures. Sol atomizing into fine droplets and subsequently spraying it on the substrate with air pressure (pneumatic spraying) or electrostatic forces (electrospraying) are the steps. The process parameters of spray pressure, nozzle diameter, and distance to the surface can be controlled with high accuracy to control thickness and coverage of the coating. Spray-coating has the advantage of suitability for field applications, such as maintenance, repair, and overhaul (MRO) in aerospace environments. It also allows multilayer coatings to be deposited and can handle coatings containing nanoparticles or corrosion inhibitors without becoming clogged [34, 40, 42].

Electrophoretic deposition (EPD) is a highly controlled and effective method for the coating of conductive substrates. In EPD, colloidal particles charged and suspended in a liquid medium are transported to an oppositely charged electrode (the substrate) by the force of an electric field. The process enables the formation of defect-free and uniform coats with thickness controlled and good adhesion. EPD is particularly valuable for the deposition of nanocomposites and multilayered systems and is thus of interest to aerospace coatings with layer-specific functionality, e.g., thermal barriers with embedded corrosion inhibitors or conductive layers above insulating films. However, electrically conductive substrate requirements may limit its application in certain aerospace uses [38, 41]. Each deposition method must be uniquely selected based on the intended application, substrate material, and coating properties. For instance, dip- and spray-coating are appropriate methods to deposit corrosion-resistant sol–gel coatings on large structural parts such as aluminum fuselage panels and magnesium-based fasteners, while spin-coating is better for high-precision components and sensors. Hybrid methods are being developed in advanced manufacturing through the integration of spray-coating with local laser sintering or printing technologies to produce smart coatings. Further, automation and robotics are increasingly integrated into aerospace coating processes to achieve higher consistency, reduced labor cost, and better compliance with performance and safety specifications [39].

10.3.3 Curing and Post-treatment

Following the deposition of sol–gel coatings, curing and post-treatment are necessary in order to transform the original soft gel film into a dense, adherent, and resistant coating. The post-deposition procedures eliminate unreacted species and residual solvents, promote network crosslinking, and enhance the mechanical, thermal, and chemical stability of the coating. Thermal curing and ultraviolet (UV) curing are generally the curing processes for sol–gel coatings, and the preference between them depends upon the substrate thermal susceptibility and sol formulation chemistry. Thermal curing is the most frequently applied sol–gel post-treatment process. The substrate is subjected to controlled heat, typically 100–400 °C, to volatilize volatile organic components and water and enable additional silanol (Si–OH) condensation into stable siloxane (Si–O–Si) bonds. The densification process hardens the coating matrix and strengthens adhesion onto the substrate. The curing time and temperature must be optimized based on the composition of the precursor; for example, metal alkoxide-containing coatings would be cured at reduced temperatures to avoid unwanted crystallization. For aerospace-grade aluminum or magnesium substrates, curing is normally done below 250 °C to reduce thermal distortion or microstructural damage [38, 40, 43, 44].

UV curing, in contrast, offers a low-temperature pathway, particularly useful for heat-sensitive materials such as polymers or fiber-reinforced composites. In UV curing, photoactive functionalities—such as acrylates or epoxy groups—in the sol–gel system get crosslinked upon exposure to ultraviolet radiation, typically between 200 and 400 nm. The process is conducive to rapid curing in seconds or minutes and enables high-throughput processing. UV-curable sol–gel coatings are increasingly used in aerospace electronics where temperature-sensitive components require localized curing without overall warming [41].

Surface densification techniques are also critical to promote functional performance of the coating. These include additional thermal annealing, laser sintering, and microwave-assisted curing, which reduce porosity, drive away trapped gases, and create smoother and denser surfaces. Laser sintering, for instance, is advantageous because it has high-precision localized heating, which is useful in spot treatment or patterning. Microwave cure is increasingly being recognized for its capability to achieve volumetric curing, reducing cure time, and enabling consistency of crosslinking across coating thickness [34, 45, 46]. Figure 10.5 illustrates sol–gel coating steps: deposition, thermal and UV curing, surface densification, and post-treatment, highlighting methods like laser sintering and UV exposure to enhance coating performance.

Beyond healing, post-treatment can entail plasma surface treatment, chemical functionalization, or top-layer deposition to incorporate the desired functionalities including hydrophobicity, anti-fouling behavior, or enhanced scratch resistance. Such post-treatments extend further the multifunctionality of sol–gel coatings, particularly in aerospace environments requiring high-performance surfaces able to withstand UV radiation, salt spray, humidity, and mechanical erosion. Curing and post-treatment are critical steps in sol–gel coating development, converting the deposited film into a strong, multifunctional layer. The optimal choice and optimization of these processes guarantee that the final coating possesses the mechanical strength, chemical resistance, and long-term stability needed for challenging aeronautical and aerospace applications [47, 48].

Fig. 10.5 Sequential steps in the sol–gel coating process

10.4 Multifunctional Properties in Aerospace Applications

10.4.1 Corrosion Resistance

One of the most important roles played by hybrid sol–gel coatings in aerospace technology is to defend lightweight metals like aluminum and magnesium alloys against corrosion. These alloys, though crucial for lightening aircraft and saving fuel, are extremely prone to environmental degradation, especially in humid or saline environments. Hybrid sol–gel coatings present powerful physical barriers to suppress the entry of corrosive species such as water, oxygen, and chloride ions. The impervious inorganic–organic network developed by hydrolysis and condensation reactions imparts low porosity and high substrate adhesion, which minimizes under-film corrosion. Adding silanes like TEOS and GPTMS causes strong siloxane bonds, which strongly affix the coating to the metallic surface, so enhancing barrier qualities [49].

Additionally, the addition of metal oxide nanofillers like ceria, zirconia, and titania integrates active corrosion inhibitors. Ceria nanoparticles, e.g., will capture and liberate oxygen to promote the build-up of shield oxide layers (passivation) on the support substrate. Certain state-of-the-art hybrid coatings also contain corrosion inhibitors or healing agents encapsulated, which upon damage in the coating, are released, allowing self-healing and longer service life. Such responsive action is especially beneficial in aerospace repair, where inspection and repair of inaccessible areas may be difficult. Generally, the self-healing and dual barrier mechanisms of hybrid sol–gel coatings are superior to corrosion resistance, particularly for lightweight fuselage panel alloys, structural frames, and fasteners [47].

10.4.2 Thermal Stability and Insulation

Thermal stability is a key requirement for aerospace coatings, especially those in high-temperature environments, such as engine parts, leading edges, and hypersonic flight surfaces. Hybrid sol–gel coatings are very resistant to heat because of their ceramic-like inorganic network, which is stable even under high temperatures. Addition of thermally stable oxides such as zirconia and alumina to the sol–gel matrix greatly improves heat shielding performance and thermal degradation resistance. These oxides have good melting points and low thermal conductivity, such that they are well suited to reducing heat transfer to the base metal substrate [43]. Figure 10.6 shows Venn diagram comparing basic sol–gel, traditional aerospace, hybrid sol–gel, and zirconia-based hybrid coatings in terms of oxidation resistance and thermal stability for aerospace applications.

Hybrid sol–gel coatings also retain performance under thermal cycling conditions, where repeat heating and cooling are incurred in flight operations. The hybrid organic–inorganic architecture is able to relax thermal stresses and avoid cracking or delamination, thereby enhancing the part life of coated items. In addition, the coatings may be designed to withstand high-temperature oxidation, which is a critical issue in aerospace propulsion systems. The creation of an impervious oxide layer at high temperatures avoids further oxidation of the metal and minimizes material loss with time. Such thermal protection is not merely critical for performance but also for maintaining the structural integrity and safety of aerospace components in harsh environments [41].

10.4.3 Mechanical Durability

The mechanical characteristics of hybrid sol–gel films have a direct impact on the long-term viability of such coatings in aerospace applications, where mechanical parts are constantly exposed to vibrations, impacts, abrasion, and cyclic stress. The aim of such films is to exert good wear resistance and scratch through the incorporation of hard ceramic nanoparticles of alumina and silica, resulting in surface stiffening and safeguarding against abrasion and erosive damage via particulate abrasion. The

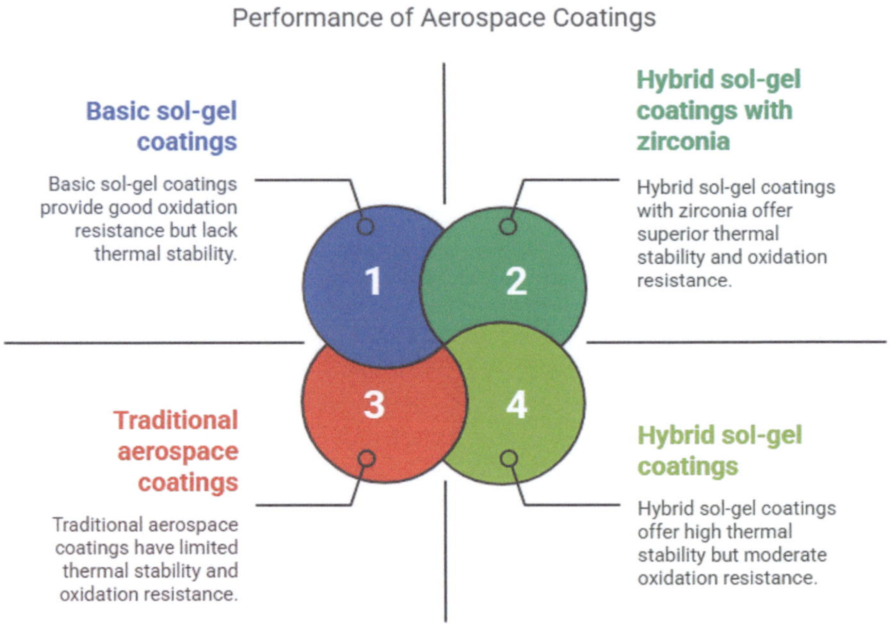

Fig. 10.6 Comparative performance of aerospace coating types

inorganic network derived by sol–gel exhibits inherent stiffness, and introduction of organic moieties contributes flexibility, enabling the coating to absorb mechanical shocks without fracture. More significantly, mechanical flexibility of hybrid coatings allows them to remain intact and adhesive under dynamic loading and vibrational fatigue, prevalent in aircraft at takeoff, flight, and landing. Utilization of flexible silanes like GPTMS and MTES enables deformed substrate to be tolerated within coatings without causing delamination. Further, enhanced mechanical performance may also be brought by using multi-layer architectures or graded coatings where surface hardness may be traded-off against inner layer flexibility. By virtue of that, hybrid sol–gel coatings remain both esthetically and structurally unblemished despite parts subject to cumulative mechanical stressing [34, 46, 50].

10.4.4 Hydrophobicity and Anti-icing

Hydrophobic and anti-icing performance is extremely valued in aerospace for drag reduction, ice accretion prevention, and sensor and window clarity preservation. Hybrid sol–gel thin films can be functionalized to become superhydrophobic through surface texturing and chemical modification with low-surface-energy molecules like fluorosilanes or long-chain alkylsilanes. The produced micro- and nanostructured surfaces lead to beading up and rolling off of water droplets, reducing surface wetting

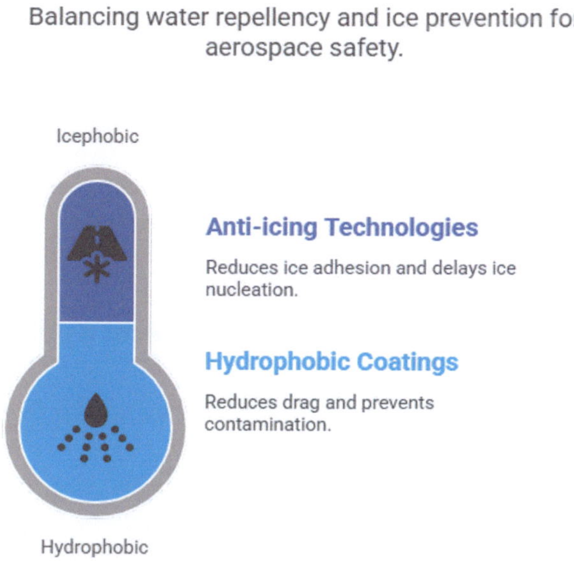

Fig. 10.7 Water-repellent and anti-icing strategies for aerospace safety

and contamination [51]. Figure 10.7 shows the graphical illustration of the balance between hydrophobic coatings, which reduce drag and contamination, and anti-icing technologies that delay ice formation, enhancing safety and performance in aerospace environments.

In cold and wet flight conditions, icephobicity becomes crucial in order to prevent ice accumulation on wings, propellers, or sensors. Superhydrophobic sol–gel coatings have been shown to reduce ice adhesion and retard ice nucleation, thus preventing the hazards of icing, including aerodynamic drag and system failure. Certain coatings include phase-change materials or thermally conductive nanomaterials to aid passive or active ice shedding. Moreover, hybrid sol–gel coatings can be hydrophobic even after exposure to UV radiation and mechanical abrasion and thus are ideal for long-term use outdoors in aviation and space vehicles [52].

10.4.5 UV and Radiation Shielding

Aerospace vehicles, especially those undertaking high-altitude or space travel, are subjected to extreme ultraviolet (UV) radiation and ionizing radiation, which cause degradation of polymers, paints, and delicate instrumentation. Hybrid sol–gel coatings can effectively shield against UV radiation by the use of UV-absorbing compounds like titania, ceria, or organic chromophores. Titania nanoparticles, in fact, are effective in blocking UVA and UVB radiation, inhibiting photodegradation of the substrate material. For radiation shielding, particularly for space use, hybrid coatings can be doped with heavy metal oxides or boron-based materials capable of absorbing gamma rays and cosmic rays. Hybrid coatings are also resistant to embrittlement and yellowing due to extended exposure to high-energy radiation. The hybrid architecture, in which flexible organic domains are embedded with radiation-tolerant inorganic phases, provides long-term stability and shielding for spacecraft surfaces, optical sensors, and solar panels. Therefore, hybrid sol–gel coatings not only prolong the service life of aerospace materials but also ensure the safety and performance of aerospace systems in radiation-rich environments [53].

10.5 Aerospace and Aeronautical Applications

10.5.1 Structural Components and Airframes

In modern aeronautical and aerospace engineering, structural members and airframes are the primary load-carrying components of an aircraft or spacecraft, and their performance is directly associated with safety, weight efficiency, and overall mission success. Such components are frequently composed of lightweight metals such as aluminum and magnesium alloys and fiber-reinforced polymer (FRP) composites due to their high strength-to-weight ratios. However, the resistivity of such materials against environmental degradation, including corrosion, damage from ultraviolet radiation, thermal stress, and abrasion wear, is not good. Hence, surface protection in the form of high-end coatings is unavoidable. Hybrid sol–gel coatings have emerged as excellent multifunctional coatings to meet the diverse demands of structural protection owing to their chemistry tunability, excellent adhesion, and capability of accommodating multiple performance-enhancing features [54]. Aluminum alloys find widespread applications in fuselage panels, wing skins, and internal frame structures due to their formability and high strength. However, their susceptibility to pitting and galvanic corrosion in chloride-containing and humid conditions requires the use of robust corrosion protection systems. Traditional coatings such as chromate-based primers have functioned satisfactorily but are susceptible to environmental and health concerns. Hybrid sol–gel coatings offer a safer alternative by virtue of superior barrier protection by way of adherent, dense siloxane networks formed from hydrolysis and condensation of precursors like TEOS, GPTMS, and MTES. These can be combined with nanoparticles like ceria or graphene oxide for active corrosion prevention and improved mechanical toughness. Magnesium alloys, although lighter than even aluminum, are more reactive and thus prone to surface degradation. They are also being applied increasingly in gearboxes, interior structural panels, and landing gear components for weight savings. Hybrid sol–gel coatings specifically developed for magnesium surfaces have been found to be highly promising in improving corrosion resistance through the formation of chemically bound, passivating oxide films. The low processing temperatures associated with sol–gel methods also make them ideally suited to coating magnesium without inducing thermal damage or distortion

[39]. At the same time, polymer fiber-reinforced composites such as carbon-fiber-reinforced epoxy (CFRP) or glass-fiber-reinforced polymer (GFRP) are commonly used in airframe structures such as vertical stabilizers, nacelles, fairings, and winglets.

Even though these composites show excellent fatigue life and low weight, their surfaces are prone to UV degradation, water absorption, and erosion due to foreign object impact. FRP hybrid sol–gel coatings can be engineered to provide UV resistance, water barriers, and abrasion protection, thereby providing long-term composite component durability. Interestingly, organic–inorganic hybrid nature of such coatings provides good affinity with polymer matrices and prevents interfacial stress or delamination during service [55]. The ability to tailor sol–gel coatings to specific substrates and application conditions makes them especially valuable in structural aerospace components. By combining corrosion protection, mechanical strength, thermal stability, and environmental protection, hybrid sol–gel coatings offer a synergistic solution to protect complex structures under harsh aerospace conditions. Their low mass, low-toxicity profile also improves sustainability goals and regulatory acceptability, making them ideal candidates for future aircraft and spacecraft applications [34].

10.5.2 Engine Components and Exhaust Systems

In advanced aerospace engine components, parts are often subjected to severe thermal, oxidative, and mechanical loading. Such unfavorable operating conditions, frequently involving temperatures higher than 1200 °C and accelerated thermal cycling, require tough surface protection schemes. Thermal barrier coatings (TBCs) made of ceramic materials have been extensively utilized in turbine blades, exhaust nozzles, and combustion chambers to thermally insulate structural substrates from harmful heat flux. Yet, traditional plasma-sprayed or electron-beam physical vapor deposition (EB-PVD) coatings, though effective, are afflicted with high processing costs, low conformity to complex geometries, and delamination susceptibility due to thermal expansion mismatches. Sol–gel derived ceramic hybrids provide an alternative solution in this situation. Sol–gel TBCs use organo-silicate or metal alkoxide precursors (e.g., tetraethyl orthosilicate or zirconium propoxide) to deposit homogeneous ceramic networks by controlled hydrolysis and condensation reactions. By doping with thermally stable ceramics like yttria-stabilized zirconia (YSZ), hafnia, or mullite, hybrid coatings from them show low thermal conductivity ($\sim 1^{-2}$ W/m K), high phase stability, and excellent adhesion to metallic substrates. Addition of polymeric or hybrid inorganic–organic binders enhances flexibility as well as thermal shock resistance, overcoming the brittle fracture modes characteristic of monolithic ceramics [56].

These coatings can be deposited by dip-coating, spin-coating, or spray-coating methods to achieve uniform deposition on intricate engine geometries without the requirement of line-of-sight processing. Thermal treatments after deposition (e.g., sintering at 500–800 °C) facilitate densification and crystallization, increasing the mechanical resistance and environmental durability of the coating. Additionally, the sol–gel technique allows for nanostructured porosity engineering, which not only minimizes effective thermal conductivity by phonon scattering but also inserts strain tolerance by allowing thermal expansion mismatches. Yet another benefit of sol–gel hybrid TBCs is the ability to be functionally graded. By designing the compositional gradient using repeated deposition cycles, e.g., from a compliant organo-silica-rich substrate to a refractory zirconia-rich topcoat interfacial stresses are reduced, thereby improving coating life. In addition, sol–gel coatings frequently exhibit self-healing tendencies through viscous flow of silicate matrices or oxidation of embedded metal particles that close microcracks upon thermal cycling [41].

10.5.3 Spacecraft and Satellite Surfaces

Low Earth orbit (LEO) and deep space's harsh environment poses severe threats to the structural integrity and longevity of spacecraft and satellite surface property. The most powerful environment-generated stressors on spacecraft surfaces are atomic oxygen (AO), ultraviolet (UV) radiation, vacuum thermal cycling, and high-energy particle fluxes. Individually and collectively, they produce surface erosion, embrittlement, outgassing, and optical degradation of spacecraft materials, particularly polymers, thermal blankets, and optical sensors. Sol–gel derived hybrid coatings offer a next-generation material solution to minimize these degradation pathways with maintaining lightweight, processable, and multifunctional protective coatings. Atomic oxygen, generated through the photodissociation of O_2 at altitudes from 200 to 700 km, is very reactive and oxidize organic surfaces at up to rates of 10^{-24} cm^3/O-atom. Traditional polymeric materials such as polyimide films (e.g., Kapton) experience severe erosion and mass loss on AO exposure. Sol–gel hybrids—most often silica or titania networks—are effective AO barriers due to their highly crosslinked, inorganic network that is not cleavable by oxidation. The incorporation of silsesquioxanes or organically modified silicates further enhances flexibility and adhesion of the coating to resist cracking and spallation during thermal cycling [43].

These coatings are usually deposited through low-temperature processes (< 200 °C), which are compatible with thermally sensitive polymeric substrates. Plasma or UV curing can then be used after deposition to enhance crosslinking and densification. By optimizing the balance of organic–inorganic components, sol–gel coatings can demonstrate rigidity as well as elasticity, accommodating the thermal expansion of underlying substrates upon day-night orbital transitions (~ ± 100 °C per cycle). Sol–gel films also demonstrate resistance against high-energy UV as well as vacuum-outgassing-induced damage. Inorganic oxides such as TiO_2 and ZnO dispersed in the sol–gel matrix have double roles: as UV absorbers to prevent photodegradation, and as photocatalytic self-cleaning agents which degrade contaminants on spacecraft surfaces. Vacuum stability of these films is critical, especially for optics and sensors, where outgassed volatile organics condense and blur optics. Well-designed sol–gel hybrids demonstrate good low total mass loss (TML) and volatile condensable material collected (CVCM) in line with NASA requirements (< 1% TML, < 0.1% CVCM) [49].

Another aspect in which sol–gel hybrids have an advantage is radiation shielding. By including high-Z fillers such as ceria, tungsten oxide, or bismuth nanoparticles, the coatings can be employed to absorb gamma rays and charged particles and protect sensitive electronics from single-event upsets and dielectric breakdown. For thermal control, the emissivity and solar absorptance of sol–gel coatings may be engineered with high accuracy through dopant chemistry and addition of nanoparticles to enable passive thermal control of satellite skins. Next-generation sol–gel technology innovations include multi-functional coatings that heal in-orbit, sense damage, and even regulate gas permeability. For example, micro-encapsulated healing agents in sol–gel matrices may be released repair chemicals on impact from AO or abrasion by micro-meteorites to restore barrier function independently. In situ re-coating methods using UV-cured sol–gel sprays are also being investigated for long-duration missions such as Lunar Gateway and Mars missions. In short, sol–gel hybrid coatings provide spacecraft and satellite surface with a lightweight, tunable, and multifunctional AO erosion, vacuum degradation, and radiation damage protection. Their processability, environmental robustness, and versatility make them one of the main components of the next-generation space-qualified materials [47].

10.5.4 Sensors and Avionics

In contemporary aerospace systems, sensor and avionics technologies are of utmost importance in providing flight control, navigation, communication, and instantaneous diagnostics. Such systems, however, work under adverse environments that are prone to high temperatures, mechanical vibrations, and high levels of electromagnetic interference (EMI). The widespread applications of wireless data transfer, high-frequency signals, and small circuit layouts also increase the vulnerability to signal crosstalk and electronic failure. Sol–gel processed dielectric and conductive hybrid coatings are a latest solution for EMI protection and sensor guarding with improved control of the dielectric properties, thermal stability, and multifunctionality [57].

Dielectric sol–gel coatings are prepared from metal alkoxides like $Si(OC_2H_5)_4$, $Al(OC_3H_7)_3$, or $Ti(OC_4H_9)_4$, which undergo hydrolysis to develop oxide networks (SiO_2, Al_2O_3, TiO_2). These networks have high dielectric strength, low permittivity (ε_r), and excellent thermal stability perfect for insulating layers in microelectronic packaging and high-voltage components in aerospace power systems. The inclusion of polymeric modifiers like polyimides, polyvinyl alcohol, or epoxy silanes in the sol–gel matrix facilitates flexible films with crack resistance and enhanced adhesion to metallic and polymeric substrates [58].

These dielectric coatings shield embedded sensors and circuits against voltage spikes and dielectric breakdown, especially at low-pressure, high-altitude environments where electrical arcing is more common. Their nanostructure morphology also can be engineered to inhibit corona discharge and reduce capacitive coupling among devices. Notably, sol–gel coatings can be deposited in ultra-thin configurations (< 10 μm), which are suitable for weight-constrained aerospace applications. To combat EMI shielding, sol–gel hybrids are reinforced with conductive nanofillers like carbon nanotubes (CNTs), graphene nanoplatelets (GNPs), silver nanowires, or metal-coated polymers. These fillers create percolative networks in the dielectric matrix, reflecting and absorbing incoming electromagnetic waves over an extensive frequency range (MHz to GHz). The challenge is to provide tunable conductivity with retained mechanical flexibility and environmental resistance. Sol–gel processing is used to precisely disperse and align conductive fillers, creating gradient structures in which the surface provides EMI shielding and the inner layer maintains insulation [59]. In Fig. 10.8, the flowchart shows sol–gel coating applications in aerospace, covering engine and spacecraft needs like thermal barriers, atomic oxygen resistance, UV protection, and radiation shielding using specific material systems.

The hybrid coatings have two functions: acting as Faraday cages for sensitive electronics and serving at the same time as thermal barriers or corrosion inhibitors. For instance, sol–gel films containing indium tin oxide (ITO) or aluminum-doped zinc oxide (AZO) nanoparticles are used as transparent EMI shielding for optical sensors and displays in avionics platforms. Current developments emphasize self-healing conductive networks through reversible bonds or encapsulated conductive

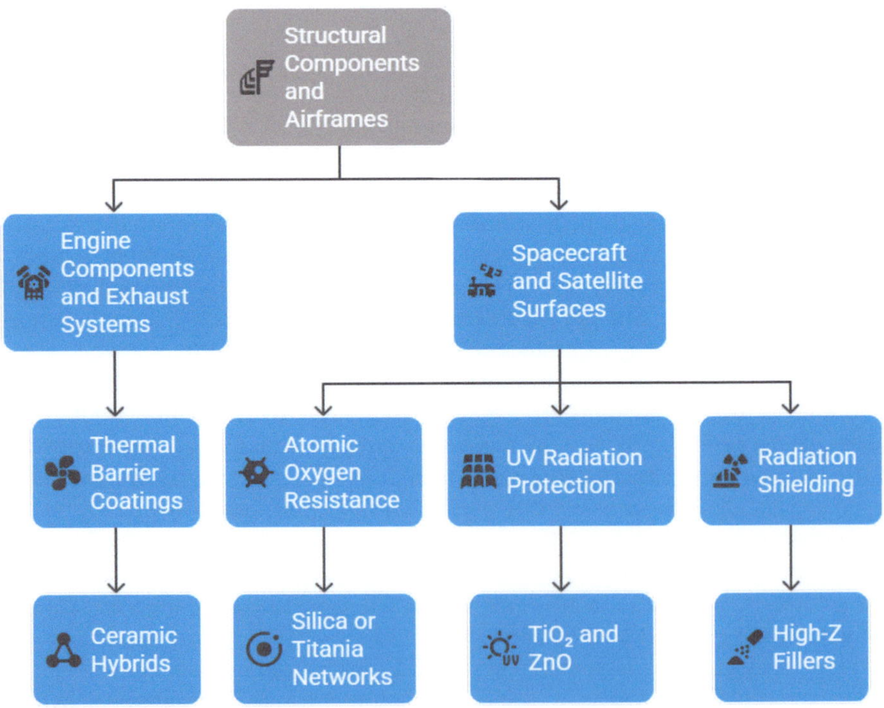

Fig. 10.8 Functional applications of coatings for aerospace structures and systems

species that regain functionality upon microcrack development. Shape-memory polymers and temperature-sensitive domains are also incorporated into sol–gel matrices to create adaptive coatings that are able to respond to thermal or vibrational stresses. In addition, sol–gel derived conformal coatings enhance MEMS sensor, fiber-optic gyroscope, and pressure transducer environmental durability by providing protection against moisture ingress, particulate contamination, and chemical attack. Hydrophobic sol–gel topcoats in avionics enclosures and printed circuit boards (PCBs) reduce icing and surface fouling and provide long-term reliability [60].

10.6 Challenges and Future Directions

Despite the promising multifunctionality and versatility of sol–gel hybrid coatings for aerospace applications, a number of crucial issues must be overcome to make them viable for widespread use in real components. Some of these key issues include scalability and reproducibility, long-term stability in repetitive aerospace conditions, embedding with smart capability, and compatibility with digital manufacturing concepts such as additive manufacturing (AM) and digital twins. Scalability and reproducibility are key obstacles to moving sol–gel coatings from laboratory-scale model to full industrial production. The sol–gel processing itself is inherently sensitive to certain processing conditions including pH, solvent composition, precursor loading, and environmental humidity. Subtle variations can introduce inhomogeneity in gelation kinetics, microstructure, and ultimate properties, thus making batch-to-batch reproducibility difficult. Furthermore, uniform deposition over extensive or complex aerospace geometries is problematic, especially for high-aspect-ratio structures or internal features. This is mitigated through greater emphasis on automated deposition methods (e.g., robotic spray, inkjet printing, or dip-coating systems) and real-time process monitoring using rheological sensors or optical coherence tomography (OCT) to offer assurance [61].

Long-term stability to aerospace-relevant cyclic stresses, i.e., thermal shock, vibrational loading, UV exposure, and oxidative environments represents another major challenge. Although sol–gel hybrids have excellent short-term performance, their organic–inorganic interpenetrating networks can be weakened upon aging through hydrolysis, UV-catalyzed chain scission, or thermal expansion coefficient mismatch between substrate and coating. Multiscale fatigue testing, accelerated aging test procedures, and in situ characterization tools need to be employed to quantify and enhance the lifecycle reliability of the coating. Graded structures and crack-arresting interfaces can potentially enhance durability even further. The emerging frontier is embedding smart functionalities into sol–gel systems. Thermally, mechanically, or chemically activated self-healing coatings are being investigated for their ability to heal autonomously microcracks or restore conductivity. Incorporation of phase-change materials, mechano-responsive polymers, or microencapsulated healing agents in the sol–gel matrix enables such action. Self-cleaning surfaces, also often relying on photocatalytic TiO_2 domains or superhydrophobic topographies, are under development to reduce maintenance cycles and maximize optical clarity of sensor windows. Responsive coatings that can alter emissivity, color, or barrier properties upon response to environmental stimuli would enable real-time adjustability to mission environments [62].

Inkjet, aerosol jet, or DIW accessible sol–gel formulations are essential to next-generation aero platforms. Spatially resolved deposition of multifunctional coatings on 3D-printed or digitally formed structures can be achieved through sol–gel formulations that are compatible with inkjet, aerosol jet, or DIW systems. In addition, the integration of sol–gel coatings with digital twins enables predictive maintenance, in which model-driven analytics and sensor feedback estimate the degradation state of the coating, allowing for condition-based reparation or reapplication strategies [63].

10.7 Conclusion

Hybrid sol–gel coatings present a powerful platform for multifunctionality in aerospace and aeronautical systems. Their tunable structure, compatibility with diverse substrates, and customizable properties make them indispensable for next-generation aircraft and space vehicles. Continued research and collaborative industry efforts are essential to address current limitations and accelerate practical implementation. Multifunctional hybrid sol–gel coatings embody a transformative innovation for aeronautical and aerospace applications, combining corrosion protection, wear resistance, thermal stability, and other surface properties into a single, versatile material platform. Their tailored chemical composition, controlled synthesis, and sophisticated structural design confer outstanding protective performance under harsh environmental and operational aerospace conditions. Recent advances in hybrid sol–gel technologies demonstrate promising pathways toward environmentally benign, high-performance coatings that address current industry challenges and anticipate future aerospace needs. Continued research and development will further unlock their full potential, enabling longer-lasting, safer, and more efficient aircraft and spacecraft components for the advancing frontiers of aerospace engineering.

References

1. Sanchez C, Julián B, Belleville P, Popall M (2005) Applications of hybrid organic–inorganic nanocomposites. J Mater Chem 15(35–36):3559–3592
2. Dey S, Mukherjee S, Saha B et al (2009) Development and implementation of sol-gel coatings for aerospace applications. SAE Technical Paper 2009-01-3208
3. Santos A, Ferreira MG, Montemor MF (2020) Design of corrosion protective and antistatic hybrid sol-gel coatings for aluminium alloys. Coatings 10(5):441
4. Guglielmi M, D'Amico M, Montesperelli G et al (2013) Potentiality of UV-cured hybrid sol–gel coatings for aeronautical applications. Surf Coat Technol 235:553–560

5. Saji VS, Cook R, Subramanian B (2021) Hybrid sol-gel coatings applied on anodized AA2024-T3 for active corrosion protection. Surf Coat Technol 421:127405
6. Saji VS, Cook R, Subramanian B (2020) Hybrid sol–gel coatings for corrosion mitigation: a critical review. Coatings 10(5):441
7. Lakshmi RV et al (2020) Sol-gel coating for corrosion protection of AA 2024. In: Proceedings of the 31st congress of the international council of the aeronautical sciences (ICAS)
8. Oubaha M, Varma PC, Duffy B et al (2014) Development of a novel hybrid aluminum-based sol-gel material: application to the protection of AA2024-T3 alloys in alkaline environment. Adv Mater Phys Chem 4(5):75–84
9. González C, Vilatela JJ, Molina-Aldareguía JM et al (2017) Structural composites for multifunctional applications: current challenges and future trends. Compos Sci Technol 153:1–14
10. Tejero-Martin D, Rezvani Rad M, McDonald A, Hussain T (2019) Beyond traditional coatings: a review on thermal sprayed functional and smart coatings. Surf Coat Technol 370:1–14
11. Theodoratou A, Bonnet L, Dieudonne-George P et al (2019) Vegetable oil hybrid films cross-linked at the air-water interface: formation kinetics and physical characterization. J Colloid Interface Sci 540:1–10
12. Wang B, Zhong S, Lee TL et al (2020) Non-destructive testing and evaluation of composite materials/structures: a state-of-the-art review. Compos Struct 236:111–117
13. Zhang L, Zhang J, Liu R (2016) Organic–inorganic hybrid sol–gel coatings for corrosion protection of aluminum alloys. Surf Innov. 4(2):81–90
14. Figueira RB (2016) Hybrid sol-gel coatings: smart and green materials for corrosion mitigation. Coatings 6(1):12
15. Fontinha RI, Salta MM, Zheludkevich ML, Ferreira MGS (2013) Effect of inorganic content on the performance of anticorrosive hybrid sol-gel coated EN AW-6063 alloy. Mater Sci Forum 730–732:745–750
16. Schottner G (2001) Hybrid sol−gel-derived polymers: applications of multifunctional materials. Chem Mater 13(10):3422–3435
17. Guin AK, Nayak S, Rout TK, Bandyopadhyay N, Sengupta DK (2011) Corrosion resistance nano-hybrid sol–gel coating on steel sheet. ISIJ Int 51(3):435–440
18. Figueira RB, Fontinha IR, Silva CJ, Pereira EV (2016) Hybrid sol-gel coatings: smart and green materials for corrosion mitigation. Coatings 6(1):12
19. Rodríguez-Alonso L, López Sánchez J, Serrano A, Rodríguez de la Fuente O, Galván JC, Carmona N (2023) Evaluation of low-toxic hybrid sol-gel coatings with organic pH-sensitive inhibitors for corrosion protection of AA2024 aluminium alloy. Gels 9(4):294
20. Figueira RB (2018) Hybrid sol-gel coatings: erosion-corrosion protection. In: Production, properties, and applications of high temperature coatings. IGI Global, p 47–73
21. Hegde M, Kavanagh Y, Duffy B, Tobin EF (2020) Multifunctional hybrid sol-gel coatings for marine renewable energy applications: synthesis, characterization and comparative analysis with organically modified silicon precursor coatings. MRS Adv 5(28):1757–1764
22. Latella BA, Barbé CJ, Cassidy DJ (2016) Investigation of the mechanical and thermal fatigue properties of hybrid sol–gel coatings applied to AA2024 substrates. J Coat Technol Res 13(6):1051–1062
23. Zheludkevich ML, Serra R, Montemor MF, Salvado IMM, Ferreira MGS (2006) Corrosion protective properties of nanostructured sol-gel hybrid coatings to AA2024-T3. Surf Coat Technol 200(9):3084–3094
24. Poznyak SK, Zheludkevich ML, Raps D, Gammel F, Yasakau KA, Ferreira MGS (2008) Preparation and corrosion protective properties of nanostructured titania-containing hybrid sol–gel coatings on AA2024. Prog Org Coat 62(3):226–235
25. Rosero-Navarro NC, Pellice SA, Castro Y, Aparicio M, Durán A (2009) Improved corrosion resistance of AA2024 alloys through hybrid organic–inorganic sol–gel coatings produced from sols with controlled polymerisation. Surf Coat Technol 203(15):1897–1903
26. Pirhady Tavandashti N, Sanjabi S, Shahrabi T (2009) Corrosion protection evaluation of silica/epoxy hybrid nanocomposite coatings to AA2024. Prog Org Coat 65(2):182–186

27. Wu KH, Li MC, Chang C, Yang CC (2006) Characterization and corrosion resistance of organically modified silicate/MO_2 (M = Zr, Ti, or Ce) hybrid coatings on a 6061–T6 aluminum alloy. J Polym Sci Part A: Polym Chem 44(2):335–342
28. Zheludkevich ML, Serra R, Montemor MF, Yasakau KA, Salvado IMM, Ferreira MGS (2005) Nanostructured sol–gel coatings doped with cerium nitrate as pre-treatments for AA2024-T3 corrosion protection performance. Electrochim Acta 51(2):208–217
29. Shchukin DG, Zheludkevich M, Yasakau K, Lamaka S, Ferreira MGS, Möhwald H (2006) Layer-by-layer assembled nanocontainers for self-healing corrosion protection. Adv Mater 18(13):1672–1678
30. Rosero-Navarro NC, Figiel P, Jedrezejewski R, Biedunkiewicz A, Castro Y, Aparicio M, Pellice SA, Durán A (2010) Influence of cerium concentration on the structure and properties of silica-methacrylate sol–gel coatings. J Sol-Gel Sci Technol 54(2):301–311
31. Feng Z, Liu Y, Thompson GE, Skeldon P (2010) Sol–gel coatings for corrosion protection of 1050 aluminium alloy. Electrochim Acta 55(11):3518–3527
32. Olivier MS, Blohowiak KY, Dauskardt RH (2010) Molecular structure and fracture properties of ZrOx/epoxysilane hybrid films. J Sol-Gel Sci Technol 55(3):360–368
33. Ivanova Y, Gerganova TS, Dimitriev Y, Salvado IMM, Fernandes MHV (2006) Nanostructured hybrid materials as precursors for synthesis of nanocomposites in Si–O–C–N–Zr system. Thin Solid Films 515(1):271–278
34. Blohowiak KY, Osborne JH, Seebergh JE (2009) Development and implementation of sol-gel coatings for aerospace applications. SAE Technical Paper 2009-01-3208
35. Moreau N, Ni L, Croutxé-Barghorn C, Chemtob A (2014) UV-cured hybrid sol-gel coatings for aeronautical and direct-to-metal applications. Surf Coat Int 97(2):116–120
36. Tejero-Martin D, Rezvani Rad M, McDonald A, Hussain T (2018) Beyond traditional coatings: a review on thermal sprayed functional and smart coatings. arXiv preprint arXiv:1811.05289
37. Verma PK, Vasudev H, Singh G, Singh S (2024) A review on the microstructural and biomedical properties of implants manufactured using additive manufacturing. J Inst Eng (India): Ser D 18 May 2024, pp 1–5
38. González C, Vilatela JJ, Molina-Aldareguía JM, Lopes CS, Llorca J (2017) Structural composites for multifunctional applications: current challenges and future trends. arXiv preprint arXiv: 1703.09917
39. Androulidakis C, Kotsidi M, Gorgolis G, Pavlou C, Sygellou L, Paterakis G et al (2021) Multi-functional 2D hybrid aerogels for gas absorption applications. arXiv preprint arXiv: 2109.12829
40. Theodoratou A, Bonnet L, Dieudonne-George P, Massiera G, Etienne P, Robin JJ et al (2018) Vegetable oil hybrid films cross-linked at the air-water interface: formation kinetics and physical characterization. arXiv preprint arXiv:1811.06903
41. Tejero-Martin D, Rezvani Rad M, McDonald A, Hussain T (2018) Beyond traditional coatings: a review on thermal sprayed functional and smart coatings. arXiv preprint arXiv:1811.05289
42. Moreau N, Ni L, Croutxé-Barghorn C, Chemtob A (2014) UV-cured hybrid sol-gel coatings for aeronautical and Direct-To-Metal (DTM) applications. Surf Coat Int 97(2):116–120
43. Gasiorek J, Gasiorek A, Babiarczuk B, Jones W, Simka W, Detyna J et al (2023) Anticorrosion properties of silica-based sol-gel coatings on steel—the influence of hydrolysis and condensation conditions. arXiv preprint arXiv:2303.14250
44. Shao G, Lu Y, Hanaor DAH, Cui S, Jiao J, Shen X (2019) Improved oxidation resistance of high emissivity coatings on fibrous ceramic for reusable space system. arXiv preprint arXiv: 1902.03943
45. Subasri R, Soma Raju KRC (2019) Multifunctional sol-gel nanocomposite coatings for aerospace, energy, and strategic applications: challenges and perspectives. In: Handbook of advanced ceramics and composites. Springer, pp 1–31
46. Moreau N, Ni L, Croutxé-Barghorn C, Chemtob A (2014) Potentiality of UV-cured hybrid sol–gel coatings for aeronautical and direct-to-metal applications. Surf Coat Technol 239:116–120
47. Zheludkevich ML, Salvado IM, Ferreira MGS (2005) Sol-gel coatings for corrosion protection of metals. J Mater Chem 15(48):5099–5111

48. Catauro M, Bollino F, Papale F, Marciano S, Pacifico S (2014) Influence of PCL on mechanical properties and bioactivity of ZrO_2-based hybrid coatings synthesized by sol–gel dip coating. Mater Sci Eng C 40:1–7
49. Subasri R, Soma Raju KRC (2020) Multifunctional sol-gel nanocomposite coatings for aerospace, energy, and strategic applications: challenges and perspectives. In: Mahajan YR, Johnson R (eds) Handbook of advanced ceramics and composites. Springer, pp 1413–1442
50. Lakshmi RV, Aruna ST, Sampath S (2021) Silica-alumina based sol-gel coating containing cerium oxide nanofibers as a potent alternative to conversion coating for AA2024 alloy. Surf Coat Technol 411:127007
51. Lakshmi RV, Aruna ST, Sampath S (2016) Ceria nanoparticles vis-à-vis cerium nitrate as corrosion inhibitors for silica-alumina hybrid sol-gel coating. Appl Surf Sci 393:397–404
52. Lakshmi RV, Aruna ST, Anandan C, Bera P, Sampath S (2017) EIS and XPS studies on the self-healing properties of Ce-modified silica-alumina hybrid coatings: evidence for Ce(III) migration. Surf Coat Technol 309:363–370
53. Shao G, Lu Y, Hanaor DAH, Cui S, Jiao J, Shen X (2019) Improved oxidation resistance of high emissivity coatings on fibrous ceramic for reusable space system. arXiv preprint arXiv: 1902.03943
54. González C, Vilatela JJ, Molina-Aldareguía JM, Lopes CS, Llorca J (2017) Structural composites for multifunctional applications: current challenges and future trends. arXiv preprint arXiv: 1703.09917. arxiv.org
55. Theodoratou A, Bonnet L, Dieudonne-George P, Massiera G, Etienne P, Robin JJ et al (2018) Vegetable oil hybrid films cross-linked at the air-water interface: formation kinetics and physical characterization. arXiv preprint arXiv:1811.06903. arxiv.org
56. Moreau N, Ni L, Croutxé-Barghorn C, Chemtob A (2014) UV-cured hybrid sol-gel coatings for aeronautical and Direct-To-Metal (DTM) applications. Surf Coat Int 97(2):116–120. researchgate.net
57. Catauro M, Bollino F, Papale F, Marciano S, Pacifico S (2014) Influence of PCL on mechanical properties and bioactivity of ZrO_2-based hybrid coatings synthesized by sol–gel dip coating. Mater Sci Eng C 40:1–7
58. Lakshmi RV, Aruna ST (2020) Recent advances in sol–gel derived coatings for corrosion protection of aerospace alloys: a critical review. Prog Org Coat 147:105858
59. Saji VS, Thomas S, Cook R (2021) Corrosion protection of aerospace magnesium alloys using hybrid sol–gel coatings. J Mater Sci Technol 79:50–65
60. Figueira RBS, Fontinha IR, Silva CJR, Pereira ECV (2016) Hybrid sol–gel coatings: smart and green materials for corrosion mitigation. Coatings 6(1):12
61. Pinna A, Montanari R, Varone A, Balbo A, Marino A, Manzoni S et al (2020) Sol–gel coating systems for aerospace applications: performance evaluation after salt spray and thermal cycling tests. Mater Today Commun 24:101172
62. Srivastava M, Khanna AS (2018) Evaluation of corrosion and mechanical properties of nanocomposite sol–gel coatings on aluminium alloy. Surf Eng 34(3):233–241
63. Montemor MF (2014) Functional and smart coatings for corrosion protection: a review of recent advances. Surf Coat Technol 258:17–37

Chapter 11
Nanocomposites for Defense, Aeronautical, and Aerospace Applications: Challenges

Amrinder Mehta and Hitesh Vasudev

Abstract Nanocomposites have brought about revolutionary changes in the defense, aeronautical, and aerospace industries owing to their lightweight, strong, and multifunctional properties. These higher technology materials are composed of a continuous matrix phase (polymer, metal, or ceramic) filled with dispersed nanoscale particles (nanoparticles, nanofibers, nanoplatelets, or nanotubes). Nanoscale materials undergo interfacial interactions and exhibit an exponentially large surface-to-volume ratio, triggering the amplification of fundamental interactions and bonding processes like conventional composites. The most interesting materials produced are carbon-based nanomaterials such as carbon nanotubes and graphene derivatives, which have unique optical, electrical, structural, and chemical properties. Another way to achieve synergies is through the design of hybrid nanocomposites, where two or more types of nanoscale fillers of different natures are used. Nanocomposites can be used for electromagnetic shielding, stealth, sensors, actuators, energy storage, and energy harvesting. Nevertheless, synthesis scalability, the ability to homogenously disperse, interfacial bonding, quality validation, duration of stability, safety, and environmental aspects are all considered challenges. Future research will focus on bio-inspired hierarchical structures, the use of machine learning to identify the mechanism of reinforcement(s), and in situ monitoring with real-time process control. Overcoming these technical, operational, and regulatory issues is important, and it would only be achieved by an integrated industrial research effort to help achieve the full potential of nanocomposites in modifying defense, aeronautical, and aerospace engineering.

Keywords Nanocomposites · Nanoscale fillers · Aerospace applications · Defense applications · Aeronautical applications · Multifunctionality · Interfacial interactions

A. Mehta (✉) · H. Vasudev (✉)
Research and Development Cell, Lovely Professional University, Phagwara, India
e-mail: amrindermehta896@gmail.com

H. Vasudev
e-mail: hiteshvasudev@yahoo.in

School of Mechanical Engineering, Lovely Professional University, Phagwara, India

S. K. Panda (ed.), *Nanocomposites for Defense, Aeronautical and Aerospace Applications*, Advanced Structured Materials 248,
https://doi.org/10.1007/978-3-032-11568-3_11

11.1 Introduction

Nanocomposites are a novel category of multiphase solids in which one of the phases has an extension of less than 100 nm in one, two, or three dimensions. These advanced materials are formed by a continuous matrix phase, which is normally a polymer, metal, or ceramic; the continuous phase; and dispersed nanoscale reinforcements that may be nanoparticles, nanofibers, nanoplatelets, or nanotubes. The basic rule of nanocomposites lies in their nanoeffect, wherein the exponentially large surface-to-volume ratio of nanoscale materials results in a heightened change in interfacial interactions and bonding processes, in contrast with that of conventional composites [1].

Space-based and aerospace systems have an especially severe environment with extreme temperature swings, vacuum ambiance, bombardment by micrometeoroids and space debris, and unsteady radiation owing to sunspot intensities. These play a critical role in material selection and analysis system design, because exposed surfaces are subjected to large, rapid temperature fluctuations, often hundreds of degrees. However, exposed surfaces must not be destroyed by atomic oxygen (AO), which readily eats many conventional materials. Advanced cladding and coatings have been developed to reduce their thermal stability, erosive scale rates, and AO resistance. This type of material plays a critical role in safeguarding sensitive parts and ensuring the structural integrity and mission duration in harsh space environments [2].

Figure 11.1 illustrates how the energy and voltage requirements for water electrolysis to produce hydrogen change with the temperature. The total demand for energy (ΔH) is depicted by the solid blue line, the electrical energy demand (ΔG) by the black line, and the heat demand (ΔQ) by the red line. The tradeoff between the demand for electrical energy and the demand for heat is indicated by a higher temperature, which leads to a lower electric energy demand and a higher heat demand. The blue shaded dashed line is the total energy required of water vapor and the vertical dashed line is where water turns to vapor. In general, the warmer it is, the more the shift in energy consumption changes to heat requirements [3]

It has been observed that nanocomposites are superior in terms of mechanical strength, lightweight, and thermal stability, and multifunctionality has revolutionized products in defense, aeronautical, and aerospace industries, with the newest materials integrating nanometer-scale structures and/or particles into a matrix and enhancing their mechanical strength, lower weight, and altered thermal stability and multifunctionality. The high strength-to-weight ratio of nanocomposites also makes them suitable for applications in aircraft and spacecraft components, where the weight-to-performance ratio plays a crucial role [4].

This also makes them more useful in extreme temperature conditions as their stability toward temperature is enhanced, making them perform better in aerospace and defense applications. Moreover, nanocomposites are multifunctional, and thus, it is possible to integrate many other features, such as electrical conductivity or the self-healing effect, within the same material. These abilities have led to more

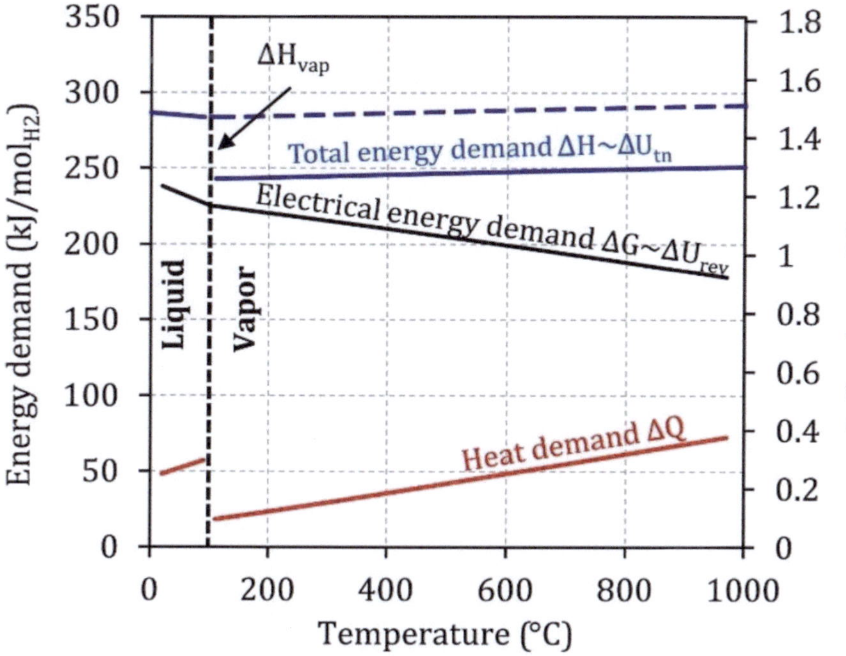

Fig. 11.1 Total enthalpy change (ΔH), thermal energy (Q), and Gibbs free energy change (ΔG) requirements of an ideal electrolysis process as a function of temperature [3]. Permission Done

efficient and competent systems within high-performance industries that have experienced increased sensation and technological development [5]. These properties included corrosion resistance, increased flexibility, increased surface-to-volume ratio, improved electrical conductivity, reduced filler concentration, and improved mechanical properties [6].

Figure 11.2 illustrates the synthesis and characterization of a nanocomposite film combining MXene and PBO nanofibers. It begins with the chemical etching and delamination of Ti_3AlC_2 to obtain $Ti_3C_2T_x$ MXene nanosheets and the transformation of PBO microfibers into nanofibers using acid treatment. These components are mixed to form a composite sol, which undergoes gelation, solvent exchange, and drying to yield a flexible, layered PBO/MXene film. Electron microscopy images reveal the film's porous, interconnected nanostructure and cross-sectional lamellar arrangement, while various spectroscopic and diffraction analyses confirm the successful integration and molecular interactions within the composite [7].

It can be applied to structural parts, as well as coating, thermal protection, and smart materials, and can be used in applications based on car nanomaterials such as carbon nanotubes (CNT), graphene, nanoclays, and advanced ceramics, which has transformed almost all industries, providing unseen opportunities in the field of innovation as well as revolutionizing performance. The most advanced parts of this technology are carbon nanotubes (CNTs), graphene, nanoclays, and advanced ceramics.

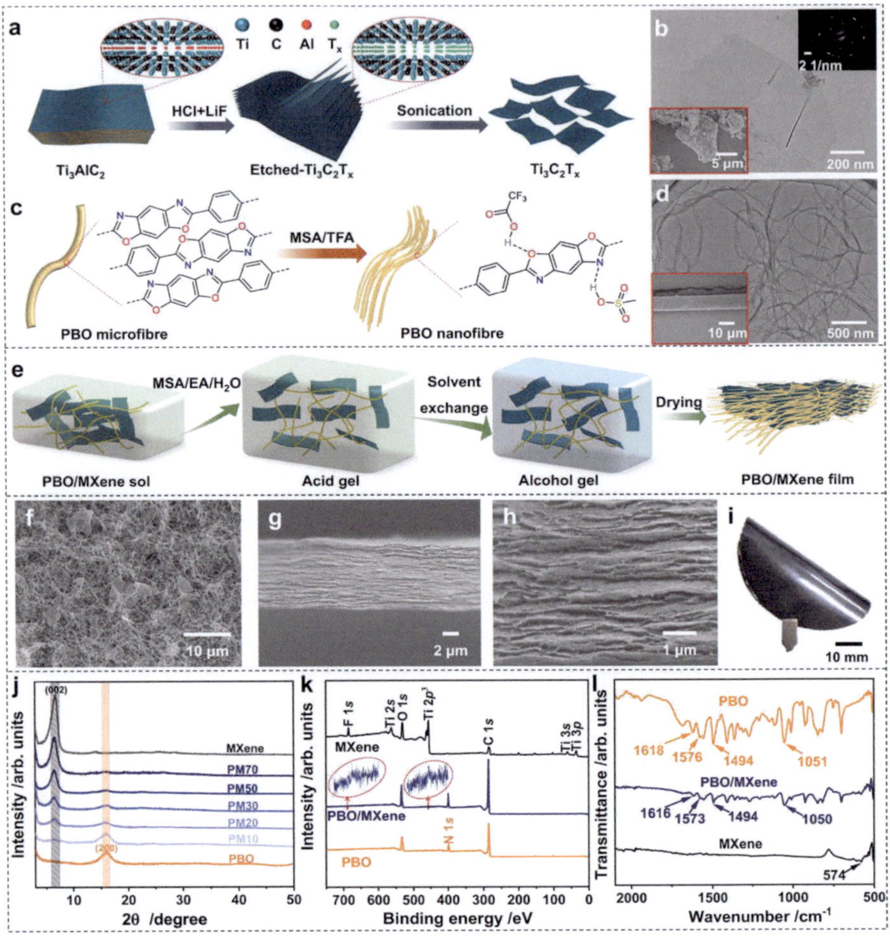

Fig. 11.2 Fabrication and characterization of nanocomposites [7]. Open Access

Structural parts of these materials are used, and their high strength-to-weight ratios and mechanical characteristics are used to add lightweight and durable structures. Nanomaterials offer improved corrosion protection, wear, and environmental degradation in coatings. These nanomaterials are suitable for thermal protection systems owing to their unique characteristics of thermal conductivity and heat resistance. Nanomaterials that have been integrated into smart materials have responsive behaviors toward diverted stimuli in which smart materials can exhibit active behavior, such as self-healing methods of materials, shape-memory alloys, and more sensitive and responsive sensors [8].

Figure 11.3 outlines the step-by-step process for preparing 3D printed filaments made from a thermoplastic polymer (TPI) and carbon nanotubes (CNT). It begins by breaking down the raw polymer pellet into a gritty powder, followed by blending

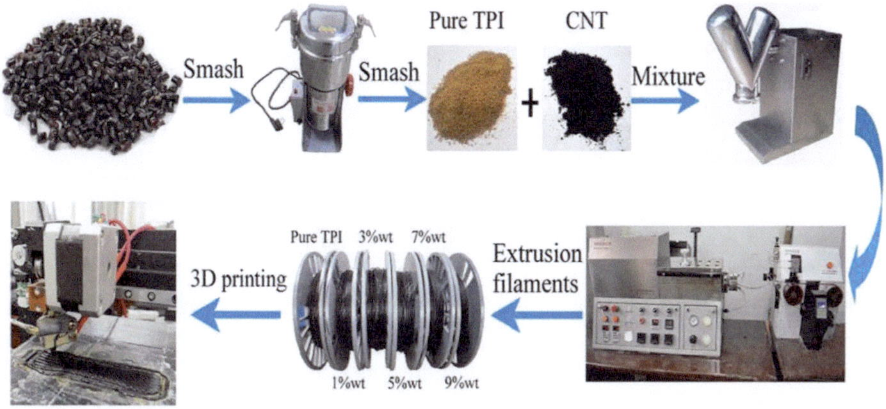

Fig. 11.3 Graphene–CNTs/alumina nanocomposites preparation [9]. Open access

with the CNT powder. Specialized equipment was used to mix the mixture to a uniform dispersion. Next, the blend was processed using an extrusion machine to form filaments containing different CNT weight percentages (1%, 3%, 5%, 7%, and 9%), as well as pure TPI. These filaments were then fed into a 3D printer, enabling the fabrication of custom nanocomposite parts with tailored mechanical and functional properties. [9]

The desire to have nanoparticles distributed in solutions or immobilized in an ideally uniform manner; the desire to have stable interfacial bonding between the nanoparticles and the other materials involved; the scale-up of manufacturing, health, and environmental issues; and the incorporation of new materials into existing corporate routines [10].

In this review, the present case of nanocomposites in the following high-performance areas is synthesized with special advantages, technical and performance issues, and research perspectives.

11.2 Nanocomposite Fundamentals

Nanocomposites are new-fangled materials consisting of a matrix, which may be polymeric, metallic, or ceramic, reinforced by nanoparticles, nanotubes, or nanosheets, and can significantly modify their characteristics. These fillers with small sizes of less than 100 nm are highly surface-to-volume ratio type of fillers, and because of this property, strong interfacial effects are achieved with increased mechanical strength, stiffness, and fracture toughness [11]. It also enhances the thermal stability, barrier activity, and electrical or magnetic conductivity, providing tortuous pathways or percolating networks. The ability to tune the type, size, shape, and dispersion of fillers, combined with nanocomposite multifunctional properties, offers a unique opportunity for the precise control of the properties and makes nanocomposites most suitable for the creation of lightweight, high-performance structures for aerospace, automotive, electronics, and biomedical applications [12–14].

11.2.1 Carbon-Based Nanomaterials

Nanocomposites are characterized by a wide range of matrices, with their structural classification split into polymer, metal, and ceramic matrix nanocomposites. The most popular and generally used are polymer matrix nanocomposites (PMNC) with a polymer component, such as a thermoplastic or thermoset, and nanoscale fillers, such as clay, carbon nanotubes, or metal oxides; they are highly desirable owing to their lightweight, flexibility, and improved mechanical strength, thermal resistance, and barrier properties [15].

Non-exhaustive list of carbon-based nanomaterials: carbon nanotubes and graphene/derivatives, nanodiamonds, fullerenes, and other nanocarbon allotropes. Studies of this kind have experienced a multifold increase over the past few years, with their distinct properties and the variety of applications in most markets, such as electronics, materials science, and medicine. These materials are also strong, conduct electricity, and are versatile; thus, they have been at the center of innovation and development in nanotechnology [16].

Carbon-based nanomaterials (CBNMs) constitute an extremely heterogeneous set of materials that have received considerable interest owing to their unusual characteristics and potential usefulness in many activities. Such materials include carbon nanotubes, graphene, graphene-based materials, nanodiamonds, fullerenes, and other carbon allotropic substances in nanosized form. A large variety of applications may exhibit these properties owing to their unique optical, electrical, structural, and chemical properties [17].

Figure 11.4 presents a comprehensive classification diagram of the carbon nanomaterials, showing six distinct structural forms arranged around a central "Carbon Nanomaterial" hub. All types have characteristic atomic structures: multiwalled carbon nanotubes are stacked concentrically like a pile of pipes, nanographite forms a planar hexagonal structure, single-walled carbon nanotubes have a smooth cylindrical cage, carbon black nanoparticles appear in irregular clumpy patterns, fullerene forms a soccer ball-like cage, and single-walled carbon nanohorns have a cone-shaped hollow structure. This diagram demonstrates clearly the structural variety of carbon-based nanomaterials when it shows how the organizations of the same carbon atoms may vary radically to provide a variety of different geometries, all with distinct properties that can find application in electronics, composites, energy storage, and catalysis [18].

Metal matrix nanocomposites (MMNC) are combinations of metallic matrices (e.g., aluminum and magnesium) with ceramic nanoparticles or carbon nanotube reinforcements, offering ductile and tough strengths of metals with high strength

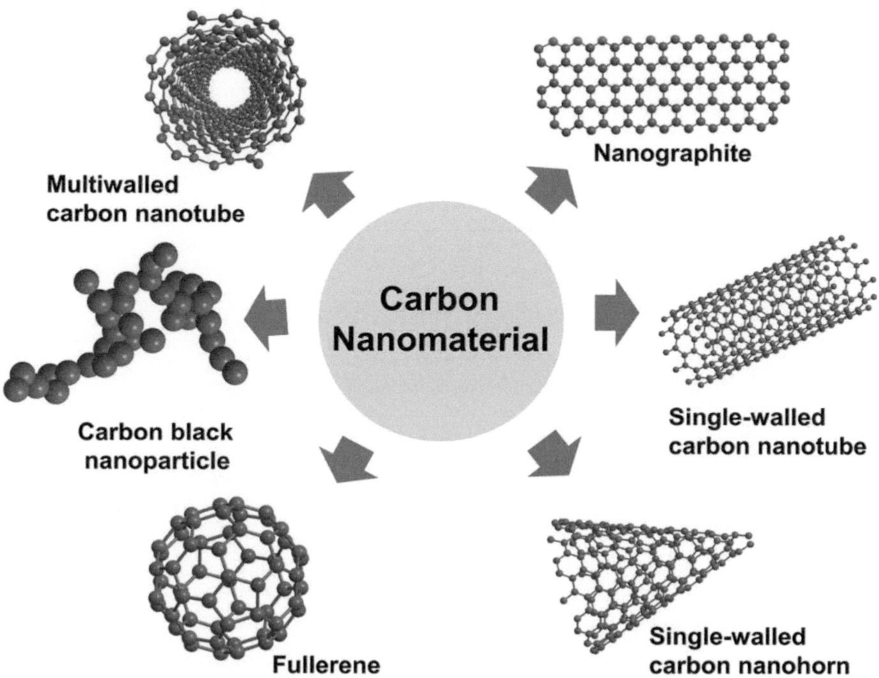

Fig. 11.4 Effects of carbon-based nanomaterials [18]. Open Access

and thermal resistance, and are useful in aircraft and automotive structural and automotive parts. The unique properties of MMNCs make them particularly attractive for applications requiring high strength-to-weight ratios and improved wear resistance. These materials have shown promise in reducing overall vehicle weight, leading to improved fuel efficiency and reduced emissions in the transportation sector. Additionally, ongoing research is focused on optimizing the manufacturing processes for MMNCs to enhance their scalability and cost-effectiveness for wider industrial adoption [19].

Ceramic matrix nanocomposites (CMNC) employ a ceramic matrix containing oxides or carbides and dispersed nanosized ceramic or metal particles, thereby exhibiting excellent hardness, corrosion resistance, and stability at high temperatures, which are vital for severe applications and novel engineering requirements. All types take advantage of the unique qualities of their matrix and the synergistic responses of nanoscale reinforcements to deliver customized solutions in many areas of endeavor, including electronics and biomedical devices, depending on the specific requirements of mechanical, thermal, and functional properties [20].

Table 11.1 shows that material selection and advanced coatings are critical for ensuring reliability, structural integrity, and mission longevity in space.

Nanocomposites are characterized by the form and dimensions of the reinforcing phase. This significantly contributes to their mechanical, thermal, and functional

Table 11.1 Material classes and their roles

Material type	Key properties and benefits	Typical applications	Citations
Titanium alloys	Corrosion resistant, strength to weight, thermal stability, AO resistance	Airframes, engines, spacecraft structures	[21]
Aluminum alloys	Lighter in weight, strong, easily made, and can be customized to fit certain requirements	Spacecraft surfaces, satellite subsystems	[22]
Ceramic composites	High resistances to oxidation/corrosion and extreme thermal stability, as well as low thermal conductivity	Thermal barrier coatings, heat shields	[23]
Polymer-based nanocomposites	Weight/flexibility, design, insufferable radiation/AO any resistance using nanofillers	Coatings, structural panels, electronics	[24]
Smart materials	Self-repair, form memory, environment adaptive	Sensors, actuators, vibration control	
Advanced coatings	Powerful reduction of AO sputtering, thermal fatigue, and radiation	Exposed surfaces, sensitive components	[25]

properties. When using nanoparticles as a reinforcement, they are generally 0-dimensional (almost spherical) entities, such as metal oxides, ceramic oxides, or quantum dots, uniformly dispersed in the matrix. The particles increased the stiffness, hardness, and wear resistance through dislocation impedance and microstructure refinement of the composite, often at a loading as low as 5 wt. %, particularly to preserve processability. In contrast, nanofiber-reinforced composites also embed one-dimensional fibers or fibrous structures (e.g., electrospun polymeric nanofibers or carbon nanotubes) that can percolate into networks at low volume fractions [26].

11.2.2 Hybrid Nanocomposites and Their Unique Characteristics

Hybrid nanocomposites are fabricated through engineering with two or more different types of nanoscale fillers, such as nanoparticles, nanofibers, lattelets, or a combination of all, dispersed in a single matrix to produce synergistic effects that cannot be produced by a single filler alone. A combination of carbon nanotubes, ceramic oxide nanoparticles, graphene nanosheets, and metallic nanoparticles can provide the best mechanical strength and fracture toughness, thermal stability, electrical conductivity, and barrier properties [27].

Figure 11.5 illustrates the evolution of nanocomposite structures from unary to increasingly complex hybrid systems, demonstrating the correlation between material sophistication and enhanced properties. This development proceeds through phases of the unary (single-component spherical nanoparticles), binary phase-separated materials (separate core–shell materials), binary core/shell structures (combined dual-phase systems), doped phases (incorporation of multiple colors in particles), and finally onto quinary CSS phase materials (rainbow-colored assemblies of 5-component materials). Underlying every representation of the structure is the associated binding energy spectroscopy of the typical peaks and intensities, which provides the chemical signature of each system. The spectral peaks are yellow and blue, which are examples of various bonding surroundings, and the complexity of the peaks is related to the complexity of the structure. The hierarchy of this classification proves the efficiency of hybrid nanocomposites in high-performance multifunctionality, which can be achieved by carefully matching several components at the nanoscale to control their properties by increasing synergies at the interface [28].

Groups of mechanical, thermal, and morphological characterization of graphene/Al_2O_3 reinforced epoxy hybrid nanocomposites reveal immense improvements in the properties of the materials and hence can be applied at advanced structural levels. These hybrid nanocomposites were formed because of the dispersion of different weight percent levels of graphene nanoplatelets (GNPs) and aluminum oxide (Al_2O_3) nanoparticles into the epoxy matrix. A combination of these materials leads to better

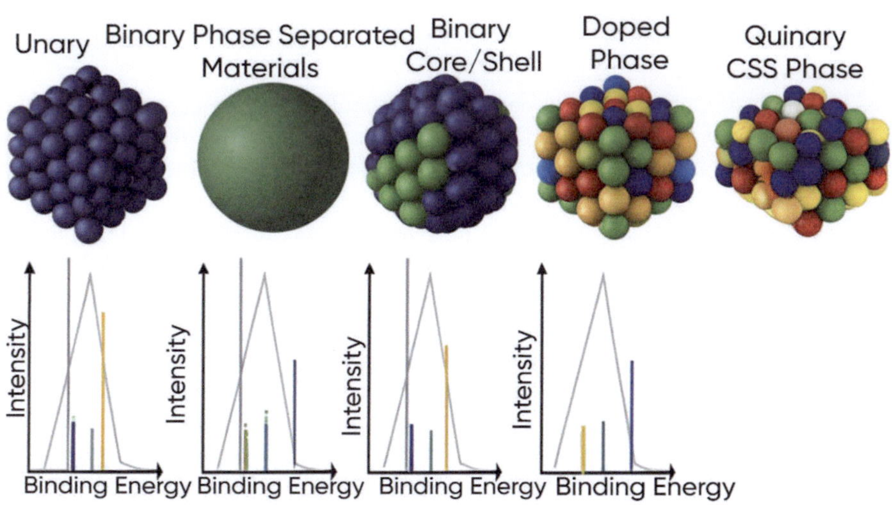

Fig. 11.5 Binding energies at active sites vary with increasing number of elements [28]. Open Access

mechanical strength, thermal stability, and characteristic morphological properties, which are important for applications in various industries, such as aerospace and automotive. The distribution of GNPs in epoxy plays a pivotal role in attaining the observed gains in the mechanical and thermal properties of the material [29].

Although the addition of graphene and Al_2O_3 imparts additional properties to epoxy nanocomposites, there are still issues that need to be addressed, such as nanoparticle agglomeration and interfacial debonding. These problems can hamper the uniformity and overall functioning of the composites and indicate that other studies on the methods of dispersion and interfacial bonding should be conducted to utilize the capabilities of these materials in more challenging products [30].

Figure 11.6 presents a scanning electron microscopy (SEM) analysis of the hybrid nanocomposite materials captured at four different magnification levels, revealing progressively finer structural details as the magnification increases. The surface topography at lower magnifications (10 μm scale) was noted to be rough with clear-cut void formation and coarse structural features on the material matrix. At greater magnification to intermediate scales (2 μm scale), the hybrid nature of the reinforcement system is evident, as carbon nanotube (CNT) networks and aluminum oxide nanofibers can be seen as fibrous and embedded within the composite. At higher magnifications (approaching the 1 μm scale), the individual voids are visible as dark circular holes, whereas the surface morphology exhibits details of the nanoparticle dispersion and interfacial interaction between various phases. The largest magnification presents the longest microstructural details, such as surface topography features, nanoparticle agglomeration, and intricate three-dimensional network architecture, which distinguishes hybrid nanocomposite materials, whereby synergistic effects are found in the interpenetrating network of several types of nanoscale reinforcements [31].

A unique feature of these materials is that they are multifunctional, and one hybrid formulation can strengthen the stiffness, create wear resistance, improve thermal conductivity, and confer electrical sensitivity, reactivity, or magnetism. In addition, there can be varied interfacial associations between contrasting nanofillers and the matrix, covering weak van der Waals forces to strong covalent binding, which allows superior dispersion to overcome undesirable agglomeration and the essence of the macroscopic castoff. Such a degree of adaptation allows hybrid nanocomposites to serve as irreplaceable materials in working assignments in industries that require high-performance levels, such as aerospace, automotive, electronics, and biomedical devices, where the materials are expected to work dependably under harsh performance conditions with multifunctional advantages [32].

Nanocomposites are also characterized by improved mechanical behavior derived from the high-level increase in strength and stiffness owing to the enabled high level of load transfer between the nanoscale fillers and the matrix, along with an increase in toughness caused by phenomena that prevent crack propagation through nanoparticle pull-out and cracklection. Thermal conductivity is supplied through percolating networks of conductive fillers (e.g., nanometer graphene or metal nanoparticles), thermal stability is reinforced by the high degradation temperatures of ceramic or inorganic reinforcements, and flame resistance is supported by the formation of char

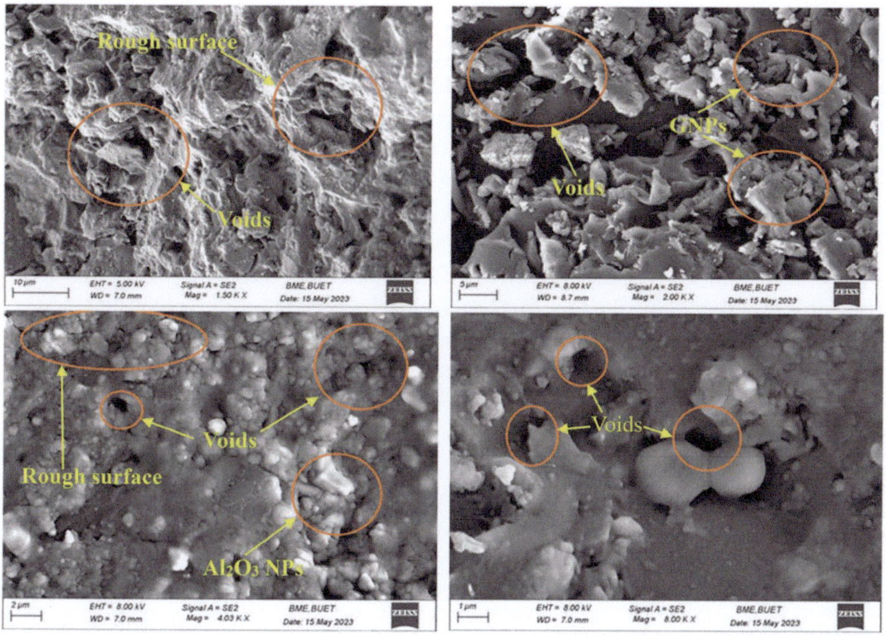

Fig. 11.6 SEM micrographs of hybrid nanocomposites at various magnifications [31]. Open Access

by both nanoparticles, which prevents both burning and mixing. Their electrical/magnetic characteristics could be readily adjusted. Conductive fillers can provide either semiconductoral-to-metallic conductivity or electromagnetic shielding, which can confer customized susceptibility to data storage and medical applications. The self-healing (via heat-activated polymerization) may manifest itself through multifunctional properties by adding stimuli-responsive fillers to enable a single nanocomposite to determine when it is damaged and to repair itself (heat-activated polymerization) [33].

11.3 Applications in Defined, Aeronautical, and Aerospace Sectors

Nanocomposites are essential in advanced technologies because their functionality is based on customized nanofiller networks (in heterogeneous matrices). In electromagnetic shielding and stealth and metal-matrix nanocomposites, owing to the multiple reflections, absorption, and polarization at the conductive nanofiller/polymer interface, a high shielding effectiveness (> 50 dB) can be achieved, which is lightweight and combines stealth in aerospace and defense applications against radar detection by stealth skins. Compared with sensors, actuators, and actuators in smart structures, piezoelectric polymer nanocomposites (especially PVDF or epoxy matrices laden with ceramic nanopowders, such as PZT or CCTO) or magnetostrictive composites support in situ transduction in the distributed self-sensing, vibration control, and morphing of smart structures, serving real-time structural health monitoring and adaptive control of civil or aerospace structures. Finally, carbon-polymer nanocomposites (i.e., composites of graphene, carbon black, or CNTs with dielectric or piezoelectric matrices) act as high-performance supercapacitor electrodes in energy storage and harvesting, as flexible dielectric generators in self-powered electronics and the IoT, and hybrid piezoelectric and triboelectric nanocomposite harvesters allow ambient mechanical or thermal energy to be converted into electricity [34].

Advanced metal matrix nanocomposites consist of nanoscale reinforcing materials, such as ceramic particles, carbon nanotubes, or graphene, which are integrated into a metallic host, aluminum or magnesium, to achieve a synergistic combination of important performance parameters, as presented in Fig. 11.7. These materials also display extremely high mechanical properties, such that lightweight structural segments within the aerospace industry are viable options without losing their durability. They are low-density in nature, thus contributing to fuel efficiency, and flexible processing routes facilitate easy shaping and joining. Metal continuity also leads to robust recyclability, and natural sustainability is complemented by lower material consumption and longer service life. Advanced metal matrix nanocomposites will prove to be a multipurpose solution to next-generation engineering challenges owing to the simple dispersion techniques used in conventional metallurgy, which allow cost-effective scale-up [35].

The sphere of propulsion and power systems is changing owing to innovations in nanocomposites, which improve the energy density, thermal stability, and mechanical strength. Propellants and nanosized metal powders (Al, Zr, Ti) or carbon nanoparticles were incorporated into conventional formulations, greatly improving the burning rates and specific impulse. The use of 5 wt% nanoparticles of coconut shell-derived carbon in ammonium perchlorate and HTPB solid propellants increased the specific impulse by as much as 37 s, due of the enhanced heat release and more uniform combustion of all propellant particles. Metallic and ceramic nanocomposites applied as coatings on engine components operating at high temperatures contain nanograin-pinned grain boundaries and are creep-resistant at temperatures above 1400 °C, developed through electrolytic co-deposition of nanoparticles in Ni or Co matrices on components [36].

Figure 11.8 illustrates the photocatalytic process in zinc oxide (ZnO) nanostructures used for pollutant degradation. When light of energy amounting to or above the bandgap of ZnO is introduced into the substance, the electrons in the valence band are excited to the conduction band, creating a pair of electrons and holes. These charge carriers diffuse to the surface and are recombined in either the surface or volume; however, with effective separation, they are engaged in redox reactions. Depending on the electron and hole concentrations, oxygen is reduced to form reactive oxygen species, whereas water or hydroxide ions are oxidized by holes to produce hydroxyl radicals. These reactive species destroy the pollutants and reduce them to intermediates and end products, which are harmless to CO_2 and H_2O [37].

Fig. 11.7 Advanced metal matrix nanocomposites [35]. MDPI

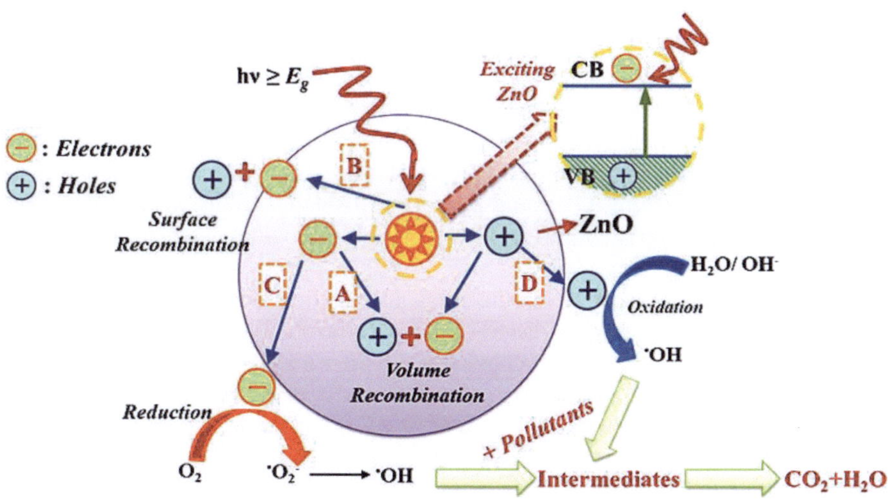

Fig. 11.8 Schematic illustrating the operational principle of a MOS (ZnO) photocatalyst [37]. Open Access

11.4 Technical and Operational Challenges

The scalability of nanocomposite synthesis poses a major challenge in scaling up the production of nanocomposites in the laboratory to an industrial scale, as many processes that are very fine in making the features at the nanoscale become nonuniform and inefficient when applied on the meso- or macroscale. The need to reconcile throughput with nanoscale control can limit continuous procedures such as roll-to-roll processing or large-area parallel deposition. Although it is now easy to produce materials with consistent particle size, morphology, and distributions over kilograms of volume, deposition must be performed in specialized reactors, and precision metrology is costly [38].

Another fundamental issue is the uniform dispersion and high interfacial adhesion of nanofillers, which are likely to be aggregated owing to their relatively high surface energy; as the mixing is incomplete or the filler is not sufficiently functionalized, the filler-matrix bonds can be too weak to enhance mechanical reinforcement and thermo-electrical diffusion, which reduces the effectiveness of the operation of all subsequent components. Finally, quality control and reproducibility require thorough multi-step characterization, particle sizing, surface chemistry, rheology, and final composite testing, but few standardized procedures and in-line quality control tools are available. The variation in raw material properties, mixing conditions, and instrument calibration can be nearly invisible but have real effects on performance; therefore, batch-to-batch variability is challenging to guarantee the consistency of high-volume production [39].

The properties of nanocomposites, performance, and durability rely on their capacity to sustain functionality with exceptionally long service lifetimes with regard to mechanical, chemical, and thermal loads. Long-term stability and aging depend on the gradual formation of polymer chains and oxidation of metal matrices or grain growth in ceramics, which may result in embrittlement, decreased interfacial bonding, and nanoreinforcement. Fatigue behavior and environmental degradation occur when bulk moisture, UV radiation, or corrosive reactions with a composite result in hydrolytic degradation, filler leaching, or microcrack nucleation, which is worsened by cyclic loading [40].

This leads to an enhancement in the crack propagation along failure-prone interfaces, resulting in strength and stiffness fading. To guarantee damage tolerance and impact resistance, it is necessary to optimize the filler dispersion, matrix toughness, and interfacial adhesion to allow the energy to be dispersed into mechanisms such as fiber pull-out, plastic deformation, and crack deflection, as opposed to catastrophic crack propagation. By designing surface treatment, hybrid reinforcements, and overcoating techniques, such issues can be overcome, and nanocomposites can be operated in a stiff environment without sacrificing their mechanical properties [41].

A systematic study of the features of the three nanocomposite structures (thin, thick, and emulsion-based) was performed using schematic diagrams and high-resolution microscopy (Fig. 11.9). SEM-top views show featureless surfaces on thin films, granular surfaces on thick coatings, and a porous honeycomb structure in emulsion-templated structures, which demonstrate a stepwise increase in the thickness of layers and pores present in the thick sections. Electrochemical cyclic voltammetry studies showed increased redox currents and a diffusion-limited nature with increasing porosity. The use of elemental mapping also outlines the consistent distribution of Au in the emulsion pores. Mass spectrometry also separates the protein-derived fragments of the emulsion system, and the topography quantifies the AFM nanoscale roughness. Lastly, the development of meso- and macropores in emulsion-based composites was verified using mercury porosimetry curves, which identify the significance of structural manipulation at multiple lengths in order to specify functionality [42].

An important factor is the safety and environmental implications of the development and deployment of nanocomposites, given that nanoscale materials can be

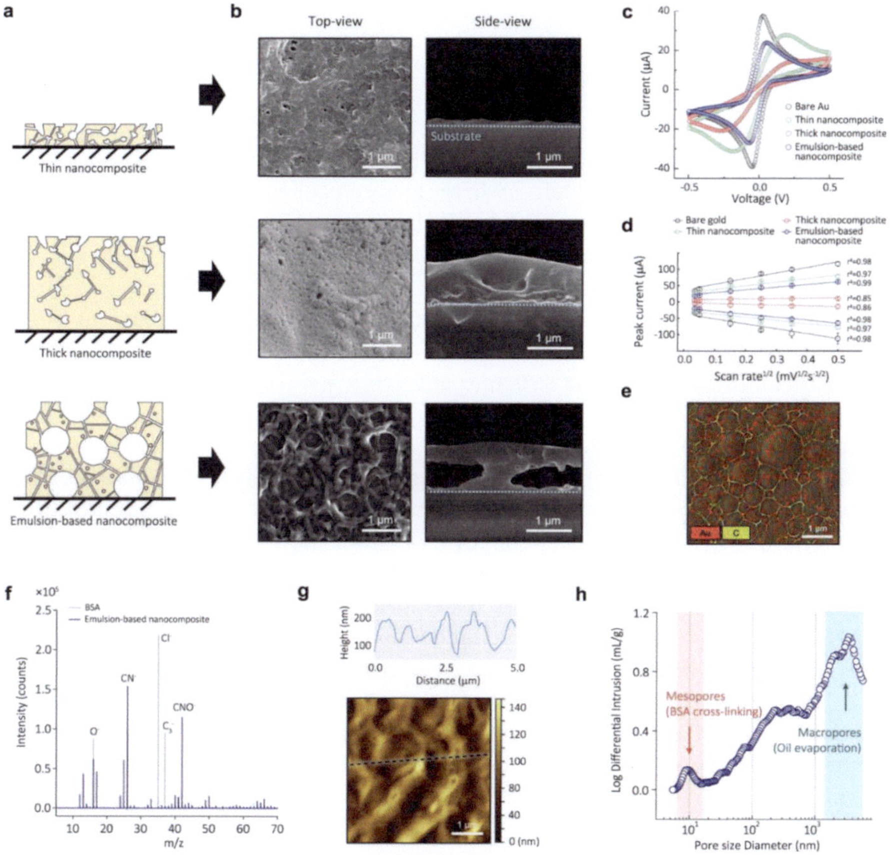

Fig. 11.9 Characterization of cross-linked nanocomposites [42]. Open Access

extremely different from those at the macro level. There is also the possibility of toxicity and health risks in the case of inhalation, ingestion, or contact of nanoparticles with the skin because of the large surface area and reactivity they could produce reactive oxygen species or pass biological membranes, resulting in inflammation, genotoxicity, or organ deposition. Environmental impact: As nanoparticles are released during manufacturing, use, and disposal, environmental impact and lifecycle assessment must consider the toxicity of the nanoparticles as well as the possibility of bioaccumulation in plants and animals or changes to microbial life once released into air, water, or soil, and the potential concentration of long-lived nanomaterials in landfills or wastewaters at end-of-life disposal. Regulatory issues and problems related to standardization exist between what constitutes a nanomaterial and the lack of harmonization in test methods used in exposure and hazard assessment, as well as in labeling and causing impediments in the comprehensive management of risks and instilling trust among the masses [43].

11.5 Future Research Directions

In the future, advancements in nanocomposite science will concentrate on systems inspired by biology and nature, where hierarchical structures and interfacial chemistry mimic elements such as bones or the dermis of sea cucumbers. This approach aims to achieve remarkable toughness, self-repair capability, and adaptive responses under normal conditions. Recent studies have shown that multilength-scale bouligand-type structures and adjustable interphases contribute to strong multifunctional performance [44].

Figure 11.10 illustrates the structure and fabrication of the multilayer ceramic nanocomposites used in oxide fuel cells. At the top, there is a diagram that shows the layered assembly into steps: starting by compressing the layered materials and subsequent lamination to form a final product, that is, a multilayer. Cross-sectional drawings indicate the existence of several different functional layers: a dense central electrolyte of yttria-stabilized zirconia (YSZ), transition layers mixing YSZ with lanthanum strontium manganite (LSM) or nickel oxide (NiO), and pure lanthanum strontium manganite (LSM) or nickel oxide/yttria-stabilized zirconia (Ni/YSZ) electrodes. The real multilayer microstructure is shown in the image of a scanning electron microscope (SEM) below, and each region is clearly labeled: the support electrode, graded transition layers, and dense electrolyte. The thickness, uniformity, and well-bonded interfaces necessary for efficient ion transport and mechanical device integrity are easily observed [45].

Machine learning for the classification of reinforcement mechanisms at the fiber–matrix interface using nanoindentation data is likely to improve our understanding and optimization of composite materials. This approach builds on the accuracy of

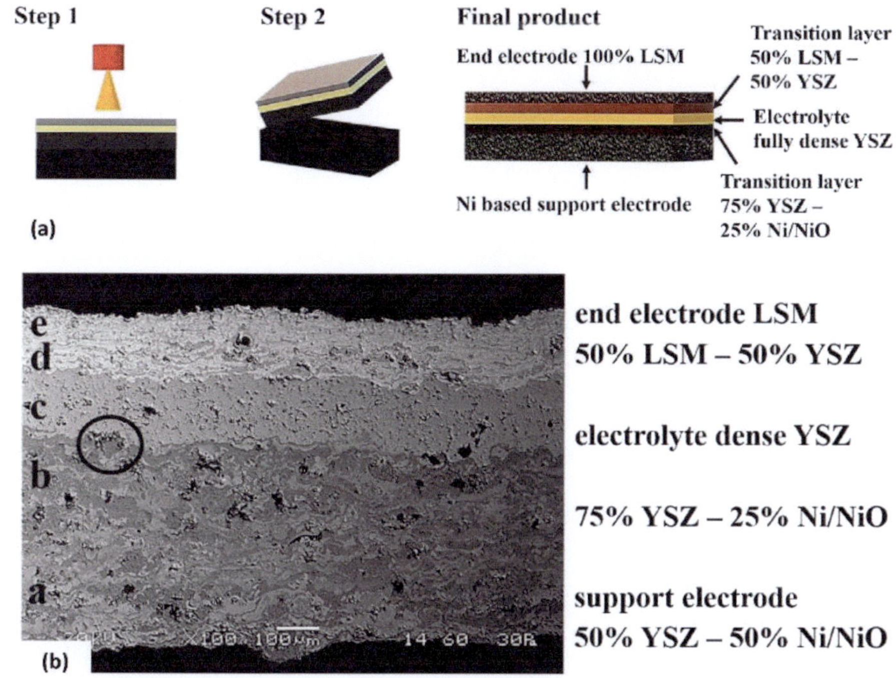

Fig. 11.10 Multilayer ceramic structures created by thermal spray [45]. MDPI

nanoindentation to generate quality data on mechanical properties, which are subsequently processed using machine learning algorithms to categorize and forecast reinforcement mechanisms. The incorporation of machine learning also offers a data-based method to build the relationship between structures and properties, with a focus on providing information about the interfacial properties of the composites [46, 47].

The use of these machine learning techniques has shown promising results in accurately identifying and categorizing different reinforcement mechanisms within CFRPs. This approach allows for a more detailed understanding of the material's microstructure and its impact on overall mechanical properties. Furthermore, the integration of machine learning with nanoindentation data analysis opens up new possibilities for rapid and automated characterization of composite materials, potentially accelerating the development of advanced CFRPs for various applications [48].

Figure 11.11 presents a comparative evaluation of three machine learning models—artificial neural network (ANN MLP), XGBoost, and radial SVM—in the classification of nanocomposite data. On the left side, schematic diagrams illustrate the basic structure of each algorithm: a neural network with layered nodes, boosted decision tree, and kernel-based SVM. Dotted lines direct the viewer to the two bar charts on the right. The first chart contrasts the overall accuracy, indicating that the radial SVM achieves the highest accuracy at 67%. The second, more detailed

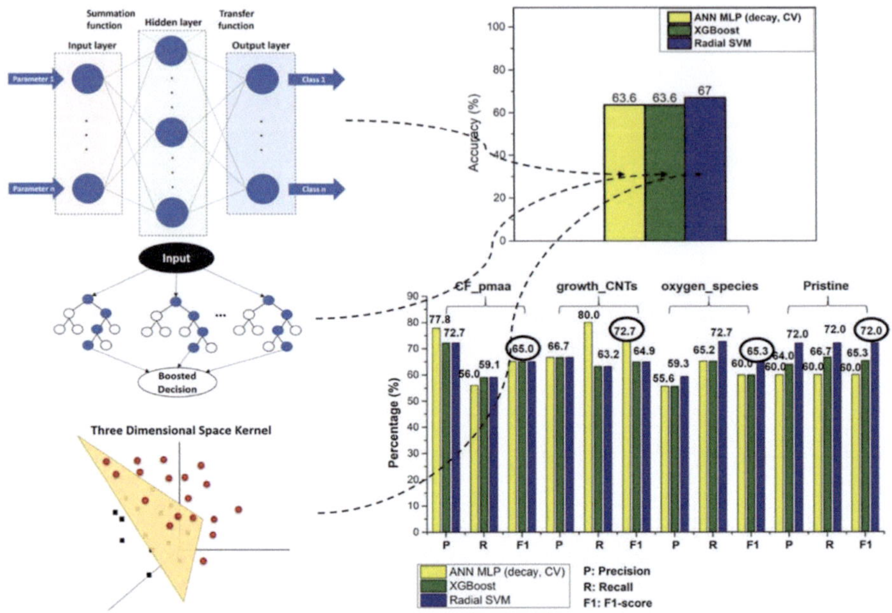

Fig. 11.11 Comparison of the average predictive ability of best-performing algorithms [49]. Open access

bar chart displays precision, recall, and F1-score across four nanocomposite categories, showing that model performance varies by property, but radial SVM generally delivers the most consistent results, particularly in the "growth_CNTs" and "Pristine" categories [49].

Although machine learning has great benefits when it comes to dealing with the reinforcement mechanism category, there are still problems when it comes to the proper interpretation of nanoindentation data because of local characteristics and difficulties with the composite character of the materials. The combined use of machine learning and other methods, such as acoustic emission analysis, might help us gain a better insight into the damage mechanism in composites. In situ monitoring combined with real-time control of the processing process converts production lines to closed-loop systems: high-speed optical and infrared cameras, digital imaging, and array sensors send signals to machine learning models, PID, and other controllers to identify defects, alter laser power or extrusion rates, keep nanoparticles dispersed and layers intact on-the-fly, and create predictable quality and minimized waste [50, 51].

11.6 Conclusion

Nanocomposites are revolutionary materials in defense, aerospace, and aeronautical applications because they are lightweight, high-strength, and multifunctional. They are a continuous matrix phase dispersed with fillers at the nanoscale. Particularly interesting properties are those of carbon-based nanomaterials, such as carbon nanotubes and graphene derivatives. Synergistic effects can be obtained using hybrid nanocomposites of two or more nanoscale fillers. Nanocomposites have been applied in electromagnetic shielding, stealth, sensors, actuators, energy storage, and harvesting. Nonetheless, synthesis scalability, consistent mixing, interfacial adhesion, quality control, long-term stability, and safety/environmental consequences are major factors of concern. The next research project will be in the area of bio-inspired hierarchies, application of machine learning to classify the reinforced mechanisms, and in situ monitoring with real-time process control.

The possibility of improving the performance and features of military and aerospace systems with the help of these advanced materials is huge. Future research should aim to solve the existing limitations and accessibility of new functions by adopting innovative design solutions. With ever-changing nanocomposite technology, the technology is set to transform future defense and space technologies. Structural integrity, thermal management, and electromagnetic shielding are some of the areas in which nanocomposite integration with military and aerospace systems shows great improvement. Innovations in nanocomposite fabrication, material combinations, and their counterpointing are under study to exploit the best sectorial properties of nanocomposites. Moreover, one of the major concerns of future advances in this field is the optimization of nanomaterials so that they can possess the ability to simultaneously improve multiple performance characteristics in the form of multifunctional nanocomposites.

References

1. Njuguna J, Pielichowski K (2004) Polymer nanocomposites for aerospace applications: fabrication. Adv Eng Mater 6:193–203
2. Pastore R, Delfini A, Santoni F, Marchetti M, Albano M, Piergentili F, Matassa R (2021) Space environment exposure effects on ceramic coating for thermal protection systems. J Spacecr Rocket 58:1387–1393
3. Buttler A, Spliethoff H (2018) Current status of water electrolysis for energy storage, grid balancing and sector coupling via power-to-gas and power-to-liquids: a review. Renew Sustain Energy Rev 82:2440–2454
4. Al Rashid A, Khan SA, Al-Ghamdi SG, Koç M (2021) Additive manufacturing of polymer nanocomposites: needs and challenges in materials, processes, and applications. J Mater Res Technol 14:910–941
5. Al-Asadi AS, Henley LA, Wasala M, Muchharla B, Perea-Lopez N, Carozo V, Lin Z, Terrones M, Mondal K, Kordas K (2017) Aligned carbon nanotube/zinc oxide nanowire hybrids as high performance electrodes for supercapacitor applications. J Appl Phys 121

6. Julkapli NM, Bagheri S, Sapuan SM (2015) Multifunctionalized carbon nanotubes polymer composites: properties and applications. In: Thakur VK, Thakur MK (eds) Eco-friendly polymer nanocomposites: processing and properties. Springer India, New Delhi, pp 155–214
7. Liu Y, Zou W, Zhao N, Xu J (2023) Electrically insulating PBO/MXene film with superior thermal conductivity, mechanical properties, thermal stability, and flame retardancy. Nat Commun 14:5342
8. Almeida V, Balzaretti N, Costa T, Machado G, Gallas M (2013) Surfactants for CNTs dispersion in zirconia-based ceramic matrix by sol–gel method. J Sol-Gel Sci Technol 65:143–149
9. Ramachandran K, Boopalan V, Bear JC, Subramani R (2022) Multi-walled carbon nanotubes (MWCNTs)-reinforced ceramic nanocomposites for aerospace applications: a review. J Mater Sci 57:3923–3953
10. Ben J, Song Z, Liu X, Lü W, Li X (2020) Fabrication and electrochemical performance of PVA/CNT/PANI flexible films as electrodes for supercapacitors. Nanoscale Res Lett 15:151
11. Bahamirian M (2022) A comparative study on the phase stability of ZrO_2–8 wt.% Y_2O_3: nano-and micro-particles. Adv Ceram Prog 8:53–60
12. Alvarez R, González-Elipe AR, Palmero A (2021) Deposition of porous nanocolumnar thin films by magnetron sputtering. Applications for Material Modification: Jenny Stanford Publishing Plasma, pp 55–105
13. Ataie A, Mostaghimi J, Pershin L, Xu P (2017) Fabrication of nanostructured cobalt ferrite coatings using suspension plasma spraying (SPS) technique. Surf Coat Technol 328:451–461
14. Bakhsheshi-Rad H, Hamzah E, Ismail A, Daroonparvar M, Kasiri-Asgarani M, Jabbarzare S, Medraj M (2015) Microstructural, mechanical properties and corrosion behavior of plasma sprayed NiCrAlY/nano-YSZ duplex coating on Mg–1.2 Ca–3Zn alloy. Ceram Int 41:15272–15277
15. Amani A, Montazer M, Mahmoudirad M (2019) Synthesis of applicable hydrogel corn silk/ ZnO nanocomposites on polyester fabric with antimicrobial properties and low cytotoxicity. Int J Biol Macromol 123:1079–1090
16. Anand PB, Nagaraja S, Jayaram N, Sreenivasa SP, Almakayeel N, Khan TMY, Kumar R, Kumar R, Ammarullah MI (2023) Kenaf fiber and hemp fiber multi-walled carbon nanotube filler-reinforced epoxy-based hybrid composites for biomedical applications: morphological and mechanical characterization. J Compos Sci 7:324
17. Chu H, Li Y, Wang C, Zhang H, Li D (2020) Recent investigations on nonlinear absorption properties of carbon nanotubes. Nanophotonics 9:761–781
18. Yuan X, Zhang X, Sun L, Wei Y, Wei X (2019) Cellular toxicity and immunological effects of carbon-based nanomaterials. Part Fibre Toxicol 16:18
19. Monish P, Hari KKL, Rajkumar K (2023) Manufacturing and characterisation of magnesium composites reinforced by nanoparticles: a review. Mater Sci Technol 39:1858–1876
20. Al-Mutairi NH, Mehdi AH, Kadhim BJ (2022) Nanocomposites materials definitions, types and some of their applications: a review. Euro J Res Dev Sustain 3:102–108
21. Fan Z, Liou F (2012) Numerical modeling of the additive manufacturing (AM) processes of titanium alloy. In: Titanium alloys-towards achieving enhanced properties for diversified applications, pp 3–28
22. Ajdelsztajn L, Schoenung JM, Jodoin B, Kim GE (2005) Cold spray deposition of nanocrystalline aluminum alloys. Metall Mater Trans A 36:657–666
23. Fang G, Gao X, Song Y (2023) A review on ceramic matrix composites and environmental barrier coatings for aero-engine: material development and failure analysis. Coatings 13:357
24. Bakhsheshi-Rad H, Najafinezhad A, Hadisi Z, Iqbal N, Daroonparvar M, Sharif S, Ismail AF, Akbari M, RamaKrishna S, Berto F (2021) Characterization and biological properties of nanostructured clinoenstatite scaffolds for bone tissue engineering applications. Mater Chem Phys 259:123969
25. Basha GMT, Srikanth A, Venkateshwarlu B (2020) A critical review on nano structured coatings for alumina-titania ($Al2O_3$-TiO_2) deposited by air plasma spraying process (APS). Mater Today: Proc 22:1554–1562

26. Chabi D, Querlioz D, Zhao W, Klein J-O (2014) Robust learning approach for neuro-inspired nanoscale crossbar architecture. ACM J Emerg Technol Comput Syst 10:1–20
27. Hemath M, Mavinkere Rangappa S, Kushvaha V, Dhakal HN, Siengchin S (2020) A comprehensive review on mechanical, electromagnetic radiation shielding, and thermal conductivity of fibers/inorganic fillers reinforced hybrid polymer composites. Polym Compos 41:3940–3965
28. Löffler T, Ludwig A, Rossmeisl J, Schuhmann W (2021) What makes high-entropy alloys exceptional electrocatalysts? Angew Chem Int Ed 60:26894–26903
29. Thorat YV, Chavan SS, Mohite DD, Pawar US (2025) Development of banana fabric biocomposites reinforced with graphene nanoplatelets and polyacrylonitrile nanofibers for suspended-load backpack applications. Discover Mater 5:1–14
30. Mondal K, Nuñez Iii L, Downey CM, Van Rooyen IJ (2021) Recent advances in the thermal barrier coatings for extreme environments. Mater Sci Energy Technol 4:208–210
31. Nipu SMA, Rahman MZ, Alam SS, Dev B (2024) Mechanical, thermal and morphological characterization of graphene/Al_2O_3-reinforced epoxy hybrid nanocomposites. Macromol Mater Eng 309:2400180
32. Fan X, Ding X, Wang P, Li Z, Cheng Y, Liu J, Yu J, Zhai J, Pan Z, Li W (2024) Ultra-low loading fillers induced excellent capacitive performance in polymer-based multilayer nanocomposites under harsh environments. Small 20:2405786
33. Tarani E, Terzopoulou Z, Bikiaris DN, Kyratsi T, Chrissafis K, Vourlias G (2017) Thermal conductivity and degradation behavior of HDPE/graphene nanocomposites: pyrolysis, kinetics and mechanism. J Therm Anal Calorim 129:1715–1726
34. Gnanasekaran K, Heijmans T, Van Bennekom S, Woldhuis H, Wijnia S, De With G, Friedrich H (2017) 3D printing of CNT-and graphene-based conductive polymer nanocomposites by fused deposition modeling. Appl Mater Today 9:21–28
35. Malaki M, Xu W, Kasar AK, Menezes PL, Dieringa H, Varma RS, Gupta M (2019) Advanced metal matrix nanocomposites. Metals 9:330
36. Aati S, Shrestha B, Fawzy A (2022) Cytotoxicity and antimicrobial efficiency of ZrO_2 nanoparticles reinforced 3D printed resins. Dent Mater J 38:1432–1442
37. Yu Z, Wang B, Gui L, Liu M, Liu S, Moliere M, Liao H (2024) Development of photocatalytic and electrocatalytic coatings via thermal spraying for environmental and energy storage applications: a short review. Surf Sci Technol 2:7
38. Cauwenberghs G, Cong J, Hu XS, Joshi S, Mitra S, Porod W, Wong H-SP (2023) Micro/nano circuits and systems design and design automation: challenges and opportunities. Proc IEEE 111:561–574
39. Chanda S, Bajwa DS (2021) A review of current physical techniques for dispersion of cellulose nanomaterials in polymer matrices. Rev Adv Mater Sci 60:325–341
40. Chandrasekaran S, Jayakumar A, Velu R (2022) A comprehensive review on printed electronics: a technology drift towards a sustainable future. Nanomaterials 12:4251
41. Chen J, Bull SJN, Letters N (2013) Finite element modelling of delamination in multilayer coatings. Nanosci Nanotechnol Lett 5:795–800
42. Lee J-C, Kim SY, Song J, Jang H, Kim M, Kim H, Choi SQ, Kim S, Jolly P, Kang T (2024) Micrometer-thick and porous nanocomposite coating for electrochemical sensors with exceptional antifouling and electroconducting properties. Nat Commun 15:711
43. Chen J, Yin H, Sun Q (2020) Effects of grain size on fatigue crack growth behaviors of nanocrystalline superelastic NiTi shape memory alloys. Acta Mater 195:141–150
44. Chen S, Tao R, Tu J, Guo P, Yang G, Wang W, Liang J, Lu SY (2021) High performance flexible lithium-ion battery electrodes: ion exchange assisted fabrication of carbon coated nickel oxide nanosheet arrays on carbon cloth. Adv Func Mater 31:2101199
45. Vardavoulias M, Gkomoza P, Arkas M, Niakolas DK, Neophytides SG (2021) Thermal spray multilayer ceramic structures with potential for solid oxide cell applications. Coatings 11:682
46. Alkurdi AAH, Al-Mohair HK, Rodrigues P, Alazzawi M, Sharma MK, Oudah AY (2024) Enhancing mechanical behavior assessment in porous thermal barrier coatings using a machine learning fine-tuned with genetic algorithm. J Therm Spray Technol 33:824–838

47. Babu SS, Mourad A-HI, Harib KH, Vijayavenkataraman S (2023) Recent developments in the application of machine-learning towards accelerated predictive multiscale design and additive manufacturing. Virtual Phys Prototyping 18:e2141653
48. Koumoulos E, Konstantopoulos G, Charitidis C (2019) Applying machine learning to nanoindentation data of (nano-) enhanced composites. Fibers 8:3
49. Konstantopoulos G, Koumoulos EP, Charitidis CA (2020) Classification of mechanism of reinforcement in the fiber-matrix interface: application of machine learning on nanoindentation data. Mater Des 192:108705
50. Cao Y-b, Ren H-t, Hu C-s, Meng Q-x, Liu Q (2015) In-situ formation behavior of NbC-reinforced Fe-based laser cladding coatings. Mater Lett 147:61–63
51. Chen T, Li W, Liu D, Xiong Y, Zhu X (2021) Effects of heat treatment on microstructure and mechanical properties of TiC/TiB composite bioinert ceramic coatings in-situ synthesized by laser cladding on Ti6Al4V. Ceram Int 47:755–768

MIX
Papier aus verantwortungsvollen Quellen
Paper from responsible sources
FSC® C105338

If you have any concerns about our products,
you can contact us on
ProductSafety@springernature.com

In case Publisher is established outside the EU,
the EU authorized representative is:
Springer Nature Customer Service Center GmbH
Europaplatz 3, 69115 Heidelberg, Germany

Printed by Libri Plureos GmbH
in Hamburg, Germany